The Digital Dilemma

Intellectual Property
IN THE INFORMATION AGE

Committee on Intellectual Property Rights and the
Emerging Information Infrastructure

Computer Science and Telecommunications Board
Commission on Physical Sciences, Mathematics, and Applications
National Research Council

NATIONAL ACADEMY PRESS
Washington, D.C.

NATIONAL ACADEMY PRESS • 2101 Constitution Avenue, N.W. • Washington, D.C. 20418

NOTICE: The project that is the subject of this report was approved by the Governing Board of the National Research Council, whose members are drawn from the councils of the National Academy of Sciences, the National Academy of Engineering, and the Institute of Medicine. The members of the committee responsible for the report were chosen for their special competences and with regard for appropriate balance.

Support for this project was provided by the National Science Foundation. Any opinions, findings, conclusions, or recommendations expressed in this material are those of the authors and do not necessarily reflect the views of the sponsors.

Library of Congress Catalog Card Number 99-69855

International Standard Book Number 0-309-06499-6

Additional copies of this report are available from:

National Academy Press
2101 Constitution Ave., NW
Box 285
Washington, DC 20055
800-624-6242
202-334-3313 (in the Washington Metropolitan Area)
http://www.nap.edu

THE NATIONAL ACADEMIES

National Academy of Sciences
National Academy of Engineering
Institute of Medicine
National Research Council

The **National Academy of Sciences** is a private, nonprofit, self-perpetuating society of distinguished scholars engaged in scientific and engineering research, dedicated to the furtherance of science and technology and to their use for the general welfare. Upon the authority of the charter granted to it by the Congress in 1863, the Academy has a mandate that requires it to advise the federal government on scientific and technical matters. Dr. Bruce M. Alberts is president of the National Academy of Sciences.

The **National Academy of Engineering** was established in 1964, under the charter of the National Academy of Sciences, as a parallel organization of outstanding engineers. It is autonomous in its administration and in the selection of its members, sharing with the National Academy of Sciences the responsibility for advising the federal government. The National Academy of Engineering also sponsors engineering programs aimed at meeting national needs, encourages education and research, and recognizes the superior achievements of engineers. Dr. William A. Wulf is president of the National Academy of Engineering.

The **Institute of Medicine** was established in 1970 by the National Academy of Sciences to secure the services of eminent members of appropriate professions in the examination of policy matters pertaining to the health of the public. The Institute acts under the responsibility given to the National Academy of Sciences by its congressional charter to be an adviser to the federal government and, upon its own initiative, to identify issues of medical care, research, and education. Dr. Kenneth I. Shine is president of the Institute of Medicine.

The **National Research Council** was organized by the National Academy of Sciences in 1916 to associate the broad community of science and technology with the Academy's purposes of furthering knowledge and advising the federal government. Functioning in accordance with general policies determined by the Academy, the Council has become the principal operating agency of both the National Academy of Sciences and the National Academy of Engineering in providing services to the government, the public, and the scientific and engineering communities. The Council is administered jointly by both Academies and the Institute of Medicine. Dr. Bruce M. Alberts and Dr. William A. Wulf are chairman and vice chairman, respectively, of the National Research Council.

Preface

The revolution in information technology is changing access to information in fundamental ways. Increasing amounts of information are available in digital form; networks interconnect computers around the globe; and the World Wide Web provides a framework for access to a vast array of information, from favorite family recipes and newspaper articles to scholarly treatises and music, all available at the click of a mouse. Yet the same technologies that provide vastly enhanced access also raise difficult fundamental issues concerning intellectual property, because the technology that makes access so easy also greatly aids copying—both legal and illegal. As a result, many of the intellectual property rules and practices that evolved in the world of physical artifacts do not work well in the digital environment. The issues associated with computerization are also amplified by the rise of the Internet and broader and more pervasive networking. These are the issues that inspired *The Digital Dilemma*.

This project grew out of a long history of Computer Science and Telecommunications Board (CSTB) interest in the legal issues related to computer technology in general and to intellectual property in particular. In 1991, CSTB published *Intellectual Property Issues in Software*, the report of a strategic forum in which I participated, and in 1994, it published the report of its second strategic forum, addressing intellectual property and other issues, entitled *Rights and Responsibilities of Participants in Networked Communities*. Recognizing the growing questions about intellectual property in the networked environment, CSTB hosted a project-planning meeting in December 1994 chaired by Pamela Samuelson (now at the

University of California, Berkeley) and involving experts from the areas of law, computer science, technology, library science, and publishing. In spring 1996, the former Federal Networking Council Advisory Committee (FNCAC) recommended that CSTB be asked to undertake a project in this area. After clarifying a division of labor with another part of the National Research Council (NRC) regarding the issues related to scientific data-bases as intellectual property,[1] CSTB transmitted a proposal in late 1996 to the National Science Foundation (NSF), which then administered the FNCAC; the project was funded in the fall of 1997, and CSTB empaneled the Committee on Intellectual Property Rights and the Emerging Infor-mation Infrastructure at the end of 1997. The course of this project reflected the circumstances of the time in which it was undertaken: the climate in the late 1990s for thinking about intellectual property policy reflected the early and mid-1990s history of public debates associated with attention to national and global information infrastructure, a period in which information policy (which includes intellectual property, pri-vacy, and free speech issues) began to inspire unusually vigorous public-interest-group and commercial advocacy activity.

CSTB's project was designed to assess issues and derive research top-ics and policy recommendations related to the nature, evolution, and use of the Internet and other networks, and to the generation, distribution, and protection of content accessed through networks. Box P.1 outlines the statement of task.

COMMITTEE COMPOSITION AND PROCESS

The study committee convened by CSTB included experts from in-dustry, academia, and the library and information science community, with expertise that spanned networks, computer security, digital libraries, economics and public policy, public and academic libraries, intellectual property law, publishing, and the entertainment, software, and tele-communications industries (see Appendix A for the biographies of study committee members). It did its work through its own expert deliberations and by soliciting input and discussion from key officials from the spon-soring agencies, other government officials, technologists, legal experts, economists, social scientists, librarians, industry experts, and advocacy

---

[1]A concurrent NRC study produced *A Question of Balance: Private Rights and the Public Interest in Scientific and Technical Databases* (National Academy Press, Washington, D.C., 1999), which identifies and evaluates the various existing and proposed policy approaches (including related legal, economic, and technical considerations) for protecting the propri-etary rights of private-sector database rights holders while promoting and enhancing ac-cess to scientific and technical data for public-interest uses.

BOX P.1
Synopsis of the Statement of Task

1. Assess the state of the art and trends in network and document or content technologies relevant to intellectual property rights management. The challenge is to sort out which trends are relevant, enduring, and promising, and how new communications and information technology may vitiate existing protections for intellectual property that the law offers to creators, users, and distributors.

2. Identify emerging opportunities and forms of publishing that have no precedents in existing media or current copyright law that may present new needs and opportunities for managing intellectual property rights.

3. Describe how electronic distribution is changing the markets (scale, distribution, cost incidence) for information products, whether they are available in alternative media or only electronically. This includes the rapidly changing structure of information and communications industries that operate and provide content for networks.

4. Assess the kinds, quality, and sufficiency of available data for measuring and analyzing relevant trends in the supply and demand for networked information services and associated electronic publishing of various kinds.

5. Review the characteristics of existing and proposed intellectual property law for both copyrightable works and noncopyrightable databases, in the United States and internationally, and the potential impacts of the proposed legal changes on the nation's research, education, and federal networking communities as information providers, distributors, and users of content.

6. Consider the mapping of technology and content elements, their owners, and their rights and responsibilities (e.g., the changing nature of liability and responsibility for service providers). Given that understanding, develop recommendations on how new technology might provide new mechanisms and tools to protect both property rights and public interests. Also, recommend what legal changes are necessary to respond adequately to the changing networked environment, while maintaining a reasonable balance between the protection of property rights and public interests.

---

group spokespersons (see Appendix B for a list of briefers to the committee). The committee met first in February 1998 and five times subsequently; it revised and strengthened its report during mid-1999.

Central to the content and flavor of *The Digital Dilemma* is the fact that the authoring committee is, by design, a microcosm of the diverse community of interest. Because of the contentious nature of intellectual property issues, every effort was made to ensure that a broad range of perspectives was represented—on the membership of the study committee, in the solicitation of briefings and other inputs to committee meeting agendas, and in the materials distributed to the study committee. The contention was evident throughout the course of the study, beginning with adjustments to committee composition to assure balance and continuing through committee debates on the numerous issues it addressed. It is an accomplishment that the committee agreed on its characterization of key issues and on a number of recommendations. It is not surprising, however, that the committee could not agree on all of the recommendations that it contemplated. In Chapter 6, uncharacteristically for a CSTB report, a number of issues are presented by articulating the different schools of thought. In these areas, the committee sought to inform debates that must continue because coming to a national consensus now—and deciding on policy that will have far-reaching impacts—is premature. Among the contributions of the report, therefore, is an articulation of the nature and concerns of multiple stakeholders—whose involvement is important for sound policy making—and a description of the issues where progress may be difficult in the near term.

ACKNOWLEDGMENTS

The committee appreciates the financial support and guidance provided by the National Science Foundation. Within the Directorate for Computer and Information Sciences and Engineering, the Division of Information and Intelligent Systems (Programs on Computation and Social Systems and Information and Data Management), the Division of Experimental and Integrative Activities, and the Division of Advanced Networking Infrastructure and Research provided support for this study, coordinated through Suzanne Iacono and Les Gasser (formerly at NSF; now at the University of Illinois, Urbana-Champaign). In addition, the Directorate for Social, Behavioral, and Economic Sciences provided support for this study through the Division of Science Resources Studies and the Division of Social and Economic Sciences (Programs on the Law and Social Sciences and Societal Dimensions of Engineering, Science, and Technology), coordinated through Eileen Collins.

We would also like to acknowledge the role of the Large Scale Networking Group of the Subcommittee on Computing, Information, and Communications (formerly the Federal Networking Council) and the instrumental efforts of Carol C. Henderson (formerly of the American Library Association) and Frederick Weingarten (American Library Association) in helping to launch this study.

Throughout the course of this study, a number of individuals contributed their expertise to the committee's deliberations. The committee is grateful to those who agreed to provide testimony at its three open meetings (see Appendix B). In addition, the committee would like to acknowledge Rick Barker (Digital Stock), Steven M. Bellovin (AT&T Labs-Research), Bruce Bond (The Learning Company), Scott Carr (Digimarc),

Stephen Crocker (Steve Crocker Associates), William Densmore (Clickshare), Laurel Jamtgaard (Fenwick & West), Robert P. Merges (University of California, Berkeley), Steve Metalitz (International Intellectual Property Alliance), Diane Pearlman (Online Monitoring Services), Shira Perlmutter (U.S. Copyright Office), Burt Perry (Digimarc), Marybeth Peters (U.S. Copyright Office), Paul Schneck (MRJ Technology Solutions), John Schull (Softlock Services), Oz Shy (University of Haifa), Linda Stone (Mitretek), Robert Thibadeau (Television Computing Inc.), and David Van Wie (Intertrust).

The committee appreciates the thoughtful comments received from the reviewers of this report and the efforts of the review monitor and review coordinator (who represent the Report Review Committee and the Commission on Physical Sciences, Mathematics, and Applications, respectively). These comments were instrumental in helping the committee to sharpen and improve this report.

Finally, the committee would like to acknowledge the staff of the National Research Council for their hard work. As the primary professional staff member responsible for the study, Alan Inouye crafted meeting agendas; drafted, edited, and revised text; and completed numerous other tasks that were instrumental in moving the committee from its initial discussions to this final report of the committee. Alan's consistent and apparently bottomless energy, insight, dedication to the task, and willingness to nag when needed were instrumental in getting this project to completion; it would not have been done nearly so well without his involvement. Jerry Sheehan shared the primary staff responsibilities with Alan Inouye during the first half of the study and continued to provide comments on the report manuscript as it progressed. Marjory Blumenthal provided input and guidance that were valuable in improving the final drafts of this report. Margaret Marsh, Nicci Dowd, and Mickelle Rodgers provided the committee with excellent support for meetings during the course of the study. The contributions of editors Susan Maurizi and Kim Briggs are gratefully acknowledged. Angela Chuang and Tom Lee, doctoral candidates at the University of California, Berkeley and the Massachusetts Institute of Technology, respectively, Jim Igoe, and Margaret Marsh provided valuable research assistance. D.C. Drake and Suzanne Ossa of the CSTB and Theresa Fisher, Claudette Baylor-Fleming, and Sharon Seaward of the Space Studies Board assisted with the final preparation of this report.

Randall Davis, *Chair*
Committee on Intellectual Property Rights and the
Emerging Information Infrastructure

Acknowledgment of Reviewers

This report was reviewed by individuals chosen for their diverse perspectives and technical expertise, in accordance with procedures approved by the National Research Council's (NRC's) Report Review Committee. The purpose of this independent review is to provide candid and critical comments that will assist the authors and the NRC in making the published report as sound as possible and to ensure that the report meets institutional standards for objectivity, evidence, and responsiveness to the study charge. The contents of the review comments and draft manuscript remain confidential to protect the integrity of the deliberative process. We wish to thank the following individuals for their participation in the review of this report:

Stephen Berry, University of Chicago,
Lewis M. Branscomb, Harvard University,
Julie E. Cohen, Georgetown University Law Center,
Charles Ellis, John Wiley & Sons,
Edward W. Felten, Princeton University,
Laura Gasaway, University of North Carolina, Chapel Hill,
Jane C. Ginsburg, Columbia University School of Law,
Stuart Haber, InterTrust,
Trotter I. Hardy, College of William and Mary Law School,
Peter F. Harter, EMusic.com Inc.,
Michael Hawley, Massachusetts Institute of Technology,
James Horning, InterTrust,

Mitchell D. Kapor, Kapor Enterprises Inc.,
Kenneth H. Keller, University of Minnesota,
Eileen Kent, Consultant,
Andrew Lippman, Massachusetts Institute of Technology,
Deanna Marcum, Council on Library and Information Resources,
Michael Moradzadeh, Intel Corporation,
Andrew Odlyzko, AT&T Labs-Research,
Ann Okerson, Yale University,
Harlan Onsrud, University of Maine,
Bruce Owen, Economists Inc.,
Anthony Stonefield, Global Music One,
Morris Tanenbaum, AT&T (Ret.),
Hal Varian, University of California, Berkeley,
Frederick W. Weingarten, American Library Association,
Richard Weisgrau, American Society of Media Photographers,
Steven Wildman, Northwestern University, and
Kurt Wimmer, Covington & Burling.

Although the individuals listed above provided many constructive comments and suggestions, responsibility for the final content of this report rests solely with the study committee and the NRC.

Contents

xvii

The Digital Dilemma

Executive Summary

THE ORIGINS OF THE DIGITAL DILEMMA

Borrowing a book from a local public library would seem to be one of the most routine, familiar, and uncomplicated acts in modern civic life: A world of information is available with little effort and almost no out-of-pocket cost. Such access to information has played a central role in American education and civic life from the time of Thomas Jefferson, who believed in the crucial role that knowledge and an educated populace play in making democracy work. Yet the very possibility of borrowing a book, whether from a library or a friend, depends on a number of subtle, surprisingly complex, and at times conflicting elements of law, public policy, economics, and technology, elements that are in relative balance today but may well be thrown completely out of balance by the accelerating transformation of information into digital form.

The problem is illustrated simply enough: A printed book can be accessed by one or perhaps two people at once, people who must, of course, be in the same place as the book. But make that same text available in electronic form, and there is almost no technological limit to the number of people who can access it simultaneously, from literally anywhere on the planet where there is a telephone (and hence an Internet connection).

At first glance, this is wonderful news for the consumer and for society: The electronic holdings of libraries (and friends) around the world can become available from a home computer, 24 hours a day, year-round; they are never "checked out." These same advances in technology create new opportunities and markets for publishers.

But there is also a more troublesome side. For publishers and authors, the question is, How many copies of the work will be sold (or licensed) if networks make possible planet-wide access? Their nightmare is that the number is *one*. How many books (or movies, photographs, or musical pieces) will be created and published online if the entire market can be extinguished by the sale of the first electronic copy?

The nightmare of consumers is that the attempt to preserve the marketplaces leads to technical and legal protections that sharply reduce access to society's intellectual and cultural heritage, the resource that Jefferson saw as crucial to democracy.

This deceptively simple problem illustrates the combination of promise and peril that make up the digital dilemma. The information infrastructure—by which we mean information in digital form, computer networks, and the World Wide Web—has arrived accompanied by contradictory powers and promises. For intellectual property in particular it promises more—more quantity, quality, and access—while imperiling one means of rewarding those who create and publish. It is at once a remarkably powerful medium for publishing and distributing information, and the world's largest reproduction facility. It is a technology that can enormously improve access to information, yet can inhibit access in ways that were never before practical. It has the potential to be a vast leveler, bringing access to the world's information resources to millions who had little or no prior access, and the potential to be a stratifier, deepening the division between the information "haves" and "have-nots."

The information infrastructure has as well the potential to demolish a careful balancing of public good and private interest that has emerged from the evolution of U.S. intellectual property law over the past 200 years. The public good is the betterment of society that results from the constitutional mandate to promote the "progress of science and the useful arts"; the private interest is served by the time-limited monopoly (a copyright or patent) given to one who has made a contribution to that progress. The challenge is in striking and maintaining the balance, offering enough control to motivate authors, inventors, and publishers, but not so much control as to threaten important public policy goals (e.g., preservation of the cultural heritage of the nation, broad access to information, promotion of education and scholarship). As usual, the devil is in the details, and by and large the past 200 years of intellectual property history have seen a successful, albeit evolving, balancing of those details. But the evolving information infrastructure presents a leap in technology that may well upset the current balance, forcing a rethinking of many of the fundamental premises and practices associated with intellectual property.

The stakes involved in all this are high, both economically and in social terms. Decisions we make now will determine who will benefit from the technology and who will have access to what information on what terms—foundational elements of our future society.

The Committee on Intellectual Property Rights and the Emerging Information Infrastructure believes that fundamental change is afoot. As a society we need to ask whether the existing mechanisms still work, and if not, what should be done. What options exist for accomplishing the important goals of intellectual property law and policy in the digital age? Test cases are now the stuff of daily news, as for example the upheaval in music publishing and distribution caused by digital recording and the MP3 format. The committee believes that society needs to look further out than today's crisis, try to understand the nature of the changes taking place, and determine as best it can what their consequences might be, what it would wish them to be, and how it might steer toward fulfilling the promise and avoiding the perils. Stimulating that longer-range exploration is the purpose of this report.

Although the report builds on some past efforts, it takes a broader approach, analyzing the issues from the perspective of a multiplicity of relevant disciplines: law, technology, public policy, economics, sociology, and psychology. The committee strongly believes that attempts to consider digital intellectual property issues through a single lens will necessarily yield incomplete, and often incorrect, answers. The report is narrow in one sense, focusing primarily on copyright because it protects the intellectual property most frequently encountered by the general public.

Opinions run strong on almost every issue addressed in this report, in large part because the stakes are so high. If, as is often claimed, societies are seeing a shift in economies as significant as the industrial revolution, with the transition to knowledge and information as a major source of wealth, then intellectual property may well be the most important asset in the coming decades.

(WHY) IS THERE A PROBLEM?

Origins of the Issues

Two events motivate reexamining the concepts, policies, and practices associated with intellectual property:

• *Advances in technology have produced radical shifts in the ability to reproduce, distribute, control, and publish information.*

— Information in digital form has radically changed the economics and ease of reproduction. Reproduction costs are much lower for both rights holders (content owners) and infringers alike. Digital copies are also perfect replicas, each a seed for further perfect copies. One consequence is an erosion of what were once the natural barriers to infringement, such as the expense of reproduction and the decreasing quality of successive generations of copies in analog media. The average computer owner today can easily do the kind and the extent of copying that would have required a significant investment and perhaps criminal intent only a few years ago.

— Computer networks have radically changed the economics of distribution. With transmission speeds approaching a billion characters per second, networks enable sending information products worldwide, cheaply and almost instantaneously. As a consequence, it is easier and less expensive both for a rights holder to distribute a work and for individuals or pirates to make and distribute unauthorized copies.

— The World Wide Web has radically changed the economics of publication, allowing everyone to be a publisher with worldwide reach. The astonishing variety of documents, opinions, articles, and works of all sorts on the Web demonstrate that millions of people worldwide are making use of that capability.

• *With its commercialization and integration into everyday life, the information infrastructure has run headlong into intellectual property law.* Today, some actions that can be taken casually by the average citizen—downloading files, forwarding information found on the Web—can at times be blatant violations of intellectual property laws; others, such as making copies of information for private use, may require subtle and difficult interpretation of the law simply to determine their legality. Individuals in their daily lives have the capability and the opportunity to access and copy vast amounts of digital information, yet lack a clear picture of what is acceptable or legal. Nor is it easy to supply a clear, "bright-line" answer, because (among other things) current intellectual property law is complex.

Why the Issues Are Difficult

The issues associated with intellectual property (IP) in digital form addressed in this report are difficult for a number of reasons:

• *The stakeholders are many and varied.* A wide variety of stakeholders present a broad range of legitimate concerns about the impacts of information technology. It is important to understand what these different concerns are and how technology affects these stakeholders. For example, the ability to self-publish on the Web may change the interaction between authors and traditional publishers, leading to shifts in power (see, for example, the discussion in Chapter 2 on music).

• *Content creators have different agendas, handle IP according to varying strategies, and look for different kinds of return on their investment.* Authors have a variety of motivations, different notions of what constitutes a return on their investment, and as a consequence, different strategies for handling intellectual property. The traditional model—content produced and sold, either directly or with advertiser support—is the most familiar and encourages a view of IP law as the foundation that provides exclusive rights. But other models include giving intellectual property away in the expectation of obtaining indirect benefit in a positively correlated market (e.g., distributing free Web browser software in the expectation of building a market for Web server software), sharing IP to enhance the community (e.g., providing open source software such as Linux and the Apache Web server), or keeping it private (e.g., establishing trade secrets).

The multiplicity of actors, motivations, returns, and strategies matters because discussions concerning intellectual property (e.g., the effects of changes in levels of IP protection) are often set in the context of a single model, suggesting that all parties are affected equally by any change in IP law or policy. But the actors are not homogeneous, and the consequences of IP policy decisions will not be felt uniformly. Policy discussions must take into account the heterogeneity of strategies for IP (as Benkler, 1999, elaborates).

• *Fundamental legal concepts can be interpreted differently.* For example, significantly different (and emphatic) views exist on whether the notion of "fair use" is to be construed as a defense against a charge of infringement or an affirmative right that sanctions copying in specific circumstances.[1] The difference matters, for both theoretical and pragmatic reasons. If fair use is an affirmative right, for instance, then it ought to be acceptable to take positive actions, such as circumventing content protection mechanisms (e.g., decoding an encrypted file), in order to exercise fair use. But taking such positive actions may well be illegal under the regime of fair use as a defense. The basic point is very controversial; some legal scholars (and a reviewer of this report) have labeled as "absurd" the notion that fair use could be an affirmative right.

---

[1]When one author quotes another, some (presumably small) amount of literal copying has occurred. The "defense" view of fair use holds that the literal copying, while a violation of the original author's exclusive rights, is excused by fair use and its public policy goal (namely, that society benefits from authors building on and critiquing previous work, even if they have to copy a small part of it). The affirmative right view of fair use, by contrast, holds the public policy goal as key and sees the copying not as a violation to be excused, but as a right that later authors have with respect to work that preceded them (as long as the copying stays within fair use guidelines).

• *Laws and practices vary worldwide, yet networks have global reach.* The information infrastructure, like the communications networks on which it builds, is global, yet there is considerable variation in different countries' laws, enforcement policies, and even cultural attitudes toward IP. This report focuses on U.S. law and practices but acknowledges that larger global issues are important and in many ways unavoidable. For example, it is typically impossible to determine where a reader of electronic information happens to be physically (and hence whose laws apply), and at times quite easy to move information from a country where certain actions may be illegal to one where laws (or enforcement) are lax.

• *The economics of information products and IP can be subtle.* Although content-producing industries account for a sizable and growing portion of the nation's economy and international trade, the economic significance of protecting IP is not completely clear. Stronger IP protection could encourage increased levels of creative output, resulting in more rapid progress and additional information products. But protecting IP also entails costs, including costs for directly related activities such as enforcement, and other less obvious costs (such as decreased ability to build on the work of others and the increased expenditure of resources to reproduce a product without violating its IP protection). The net economic effects of changes in protection levels are difficult to assess.

ISSUES IN ACCESS TO INFORMATION

Public Access

Copying and Access

In the digital world, even the most routine access to information invariably involves making a copy: Computer programs are run by copying them from disk to memory, for example (an act that some courts have ruled to be "copying" for the purposes of copyright law), and Web pages are viewed by copying them from a remote computer to the local machine. But the exclusive right to copy is the first and perhaps most basic right of a copyright holder. How can the conflict be resolved between the desire to provide access to works and the desire to control copying, if, for digital information, access *is* copying?

This dilemma affects authors and publishers who wish to distribute digital works and need a way to accomplish this so that the work can be accessed, yet still be protected against unauthorized reproduction. The problem affects policy makers, because the traditional first-sale rule of copyright, an important element of public policy, is undermined by information in digital form. That rule works in the world of physical artifacts because they are not easily reproduced by individuals and are not accessible to multiple, distant viewers. But neither of these limitations holds for digital works. Consumers are affected as well, because access is accomplished by copying, and in the digital world copyright's traditional control of copying would mean control of access as well.

> *Conclusion:* **The tradition of providing for a limited degree of access to published materials that was established in the world of physical artifacts must be continued in the digital context. But the mechanisms for achieving this access and the definition of "limited degree" will need to evolve in response to the attributes of digital intellectual property and the information infrastructure.**

In the physical world, publication has three important characteristics: It is public, it is irrevocable, and it provides a fixed copy of the work. In the digital world, none of these may be true. In the physical world, publication is fundamentally public and irrevocable because, while the work does not become the property of the public, enough copies are usually purchased (e.g., by libraries and individuals) that it becomes part of the publicly available social and cultural record. Publication is irrevocable because once disseminated, the work is available. Works may go out of print, but they are never explicitly taken "out of publication" and made universally unavailable; copies of printed works persist. Publication also accomplishes a certain fixity of the work: Distributed copies represent an archival snapshot; subsequent editions may be published, but each of them adds to the public record.

Works published in electronic form are not necessarily irrevocable, fixed, or public. They can be withheld from scrutiny at the discretion of the rights holder. Nor are they inherently public: Software enables fine-grained control of access, making works as open or as restricted as the rights holder specifies, with considerable ability to fine-tune who has what kind of access. The information infrastructure also offers many options for distribution other than printing and selling copies, including distribution on electronic mailing lists, posting on a password-protected Web site, and posting on preprint servers, among others. Nor are works in electronic form fixed: Old versions are routinely overwritten with new ones, obliterating any historical record. (What is the value of citing a Web page, if the content there is easily changed?) In some ways, the properties of digital distribution are desirable; some material (e.g., privately produced reports, business data) may be distributed in digital form precisely because of these characteristics, where it would not have been published

at all in the traditional manner. But those properties can also cause difficulties.

Conclusion: **The information infrastructure blurs the distinction between publication and private distribution.**

Recommendation: **The concept of publication should be reevaluated and clarified (or reconceptualized) by the various
stakeholder groups in response to the fundamental changes
caused by the information infrastructure. The public policy
implications of a new concept of publication should also be
determined.**

Licensing and Technical Protection Services

Use of licensing is becoming more widespread, especially for information previously embedded in physical artifacts and sold under the
first-sale doctrine. Increasingly, digital information acquired by libraries,
for example, is available only by license. While some licenses may have
advantages (e.g., providing more rights than are normally available under
copyright), their use as a model for distribution of information raises a
number of concerns, particularly the potential for an adverse impact on
public access.

The trend toward licensing also means that digital information is in
some ways becoming a service rather than a product. Buy a book and you
own it forever; pay for access to a digital book and when the period of
service is over, you often retain nothing. This is acceptable in a variety of
circumstances but can be problematic for archival purposes.

Licensors are also under no obligation to incorporate the public policy
considerations (e.g., fair use) that have been carefully crafted into copyright law. Mass-marketed information products raise a more general
concern, as they may substitute a contract (over which consumers have
little control) for copyright law. Mass-market license terms also raise
concerns about the legal uses of works in schools, libraries, and archives.

Technical protection mechanisms currently being explored are similarly a two-edged sword. They make possible the distribution of some
digital information that rights holders would otherwise be reluctant to
release, but also have the potential for significant adverse impact on public
access. Encryption technologies under development could enable the
distribution of content in such a way that consumers would find it difficult to do anything but view it: The technology can make it very difficult
to save a decrypted digital version of information or even print it, should
a publisher choose to package the information under those conditions.

(The publisher need not set such terms, of course, or may choose simply to charge one price for viewing, an additional fee for saving, and yet another fee for printing.)

Distribution without the right to save and/or print would create a world in which information may be distributed but never easily shared. Some committee members believe that if copyright is truly to be a pact between society and authors to encourage the creation and dissemination of information for society's ultimate benefit, highly constrained models of distribution call this pact into question. Market forces may ultimately discourage this approach, but the committee believes that it is important for this issue to be tracked so that, should this more restrictive approach become widespread, consideration can be given to public policy responses.

Conclusion: **The confluence of three developments—the changing nature of publication in the digital world, the increasing use of licensing rather than sale, and the use of technical protection services—creates unprecedented opportunities for individuals to access information in improved and novel ways, but also could have a negative impact on public access to information. Developments over time should be monitored closely.**

Recommendation: **Representatives from government, rights holders, publishers, libraries and other cultural heritage institutions, the public, and technology providers should convene to begin a discussion of models for public access to information that are mutually workable in the context of the widespread use of licensing and technical protection services.**

Archiving and Preservation

The information infrastructure raises important concerns with respect to archiving and preservation. The maintenance of our history, record of social and cultural discourse, scholarship, and scientific debate and discovery are of fundamental importance to our society. In the print world, the act of publication automatically makes archiving possible, both legally and logistically. In the digital world, where licensing is increasingly prevalent, archiving is allowed by the licensee only if it is explicitly authorized in the terms of the contract. Although some publishers facilitate such provisions (e.g., a few scholarly publishers), many others have not. In addition to the issues raised by licensing, other challenges with respect to digital archiving include an inadequate base of technological knowledge, insufficient funding, concerns about copyright liability, and a lack of large-scale collective endeavors by the relevant institutions.

Recommendation: **A task force on electronic deposit should be chartered to determine the desirability, feasibility, shape, and funding requirements of a system for the deposit of digital files in multiple depositories.**

The task force membership should broadly represent the relevant stakeholders and should be organized by an unbiased entity with a national reputation, such as the Library of Congress or some other governmental organization that has a pertinent charter and relevant expertise. The task force should be assigned for a limited term (2 years maximum).

Recommendation: **Congress should enact legislation to permit copying of digital information for archival purposes, whether the copy is in the same format or migrated to a new format.**

Access to Federal Government Information

Widespread use of the Web has in general provided greatly expanded access to federal government information. However, in some parts of the government, the evolution of the information infrastructure has instead been associated with a trend toward the commercialization of government information, increasingly limiting the amounts of information that can be accessed inexpensively by the public.

Conclusion: **When commercial enterprises add value to basic data, the resulting products deserve copyright protection insofar as these products otherwise satisfy the legal requirements for copyright.**

Recommendation: **As a general principle, the basic data created or collected by the federal government should be available at a modest cost, usually not to exceed the direct costs associated with distribution of the data. When agencies contract with a commercial enterprise to make federally supported primary data available, and provide no other mechanism for access to the data, such agreements should provide for public access at a cost that does not exceed the direct costs associated with distribution.[2]**

---

[2]The committee did not address the status of the data and research created by federally supported researchers based at academic or other institutions outside the federal government.

Individual Access and Use

Private Use and Fair Use

The information infrastructure raises the stakes around questions of private use and fair use and has increased copyright law's concern with private behavior. One of the most contentious copyright issues concerns the legality of private, noncommercial copying. While the issue is applicable beyond the sphere of digital information, the risks to rights holders are especially acute when the information is in digital form. Some rights holders believe that nearly all unauthorized reproductions are infringements, while many members of the general public believe that virtually all private, noncommercial copying of copyrighted works is lawful. The true legal status of private copying is somewhere in between these extremes.

Copyright has traditionally been concerned with public acts, such as public display and public performances. But with the evolving information infrastructure, private behavior (e.g., private use copying) is having a larger impact on the market, and the distinction between public and private is (as noted above) blurred in the digital world.

Conclusion: **A widespread (and incorrect) belief prevails in society that private use copying is always or almost always lawful. This viewpoint is difficult to support on either legal or ethical grounds. It is important to find ways to convince the public to consider thoughtfully the legality, ethics, and economic implications of their acts of private copying.**

Conclusion: **Fair use and other exceptions to copyright law derive from the fundamental purpose of copyright law and the concomitant balancing of competing interests among stakeholder groups. Although the evolution of the information infrastructure changes the processes by which fair use and other exceptions to copyright are achieved, it does not challenge the underlying public policy motivations. Thus, fair use and other exceptions to copyright law should continue to play a role in the digital environment.**

The appropriate scope of fair use may be reduced by the development of new licensing regimes enabled by the digital environment that reduce transaction costs, thereby reducing market failures and some of the rationale for fair use. Even so, there are other public policy rationales for fair use that should not be overlooked.

Conclusion: **Providing additional statutory limitations on copyright and/or additional statutory protection may be necessary
over time to adapt copyright appropriately to the digital environment. The fair use doctrine may also prove useful as a
flexible mechanism for adapting copyright to the digital environment.**

Opportunities and Challenges for Authors and Publishers

Authors and publishers alike find both promise and peril in the information infrastructure. For authors, it expands the class of "published"
authors and makes available inexpensive distribution methods over which
authors can exercise direct control, which may well produce a realignment of interests. It has also led to changes in terms for the ownership of
digital rights, often to the disadvantage of the author. The greater malleability of works in digital as opposed to hard-copy form also raises new
concerns about the authenticity and integrity of the information.

Point of Discussion: **Many members of the committee believe
that a task force on the status of the author should be established, with the goal of preserving the spirit of the constitutional protection and incentives for authors and inventors. Such
a task force would evaluate the viability of mechanisms that
facilitate both distribution and control of work (e.g., rights
clearance mechanisms) and examine whether issues should be
addressed with government action or kept within the framework of private-sector bargaining.**

For publishers, the information infrastructure promises the possibility of reduced costs for new information products and distribution but
also brings uncertainty. Publishers are unsure, for example, what their
revenue models will be in the online environment and face the most
fundamental of issues in the digital dilemma: How can they distribute
digital information without losing control of it?

Mechanisms for Protecting Intellectual Property

Technical Protection Tools

Technical protection tools include a wide variety of software- and
hardware-based mechanisms that limit access to or use of information.
Although these technologies are not widely used for IP protection in 1999,
a few tools have been deployed to protect IP in certain niches with some

success, for example, the digital watermarking of images, and selected use of encryption, especially in the entertainment industry (e.g., the encryption used in cable TV delivery).[3] Software-based tools have the advantage of ease of distribution, installation, and use. They also have a major drawback because the protected content must eventually be displayed to the user (or somehow "consumed") for its value to be realized. If the content is delivered to an ordinary PC, the information displayed can today be captured and copied by anyone with sufficient technical knowledge.

A higher level of protection for valuable content in the face of determined adversaries requires special-purpose hardware. This is (in part) the inspiration behind some "information appliances" (e.g., portable players for digital music, portable electronic books) and behind so-called "trusted systems," a combination of software and special hardware. Information appliances are beginning to have an impact in the market and may provide an effective delivery vehicle because they are not general-purpose (i.e., programmable) computers, from which displayed content can be captured fairly easily. The trusted system approach, to the extent that it relies on special-purpose hardware incorporated into ordinary PCs, faces the problem of convincing the many users of existing PCs to set aside their investment in existing hardware and buy new devices that will, in some ways, be less capable.[4]

Conclusion: **Technical protection mechanisms are useful but are not a panacea.**

Whatever the mechanism used, it is important to keep in mind that no protection mechanism is perfect. As with any security system, defeating it is a matter of time, effort, and ingenuity. Yet, as with any security system, perfection is not required for real-world utility: Existing technical protection mechanisms can protect digital information to a degree that keeps fundamentally honest people honest; this appears to be sufficient for a wide range of uses. The deployment of mechanisms also involves trade-offs that must be judged carefully: Adding a protection mechanism involves costs to the vendor (software development and maintenance)

---

[3]Encryption is used widely for other purposes, such as the use of the secure socket layer in communication over the Web to protect the confidentiality of transactions, but to date has been used sparingly for IP protection.

[4]The Trusted Computing Platform Alliance, a collaborative effort founded by Compaq, HP, IBM, Intel, and Microsoft, is apparently aimed at just such a goal, trying to provide security at the level of the hardware, BIOS, and operating system. See <http://www.trustedpc.org>.

and to the consumer (e.g., time and inconvenience). Hence, as with any security mechanism, technical protection must be carefully matched to the need.

> *Recommendation:* **Rights holders might consider using techni-cal protection services to help manage digital intellectual prop-erty but should also bear in mind the potential for diminished public access and the costs involved, some of which are imposed on customers and society.**

The experimental circumvention of technologies used to protect intel-lectual property is a common practice in the cryptology and security R&D community, one that enables the development of more efficient and effec-tive protection technologies. This useful practice is threatened by recent developments, notably the Digital Millennium Copyright Act (DMCA), which makes circumvention illegal except under certain conditions. The overall approach favored by the cryptology and security community is to make circumvention legal, while making certain exploitations of success-ful circumventions illegal (including, of course, the theft of IP). Some members of the committee believe that a number of specific changes are needed to the DMCA (detailed in Chapter 6).

> *Conclusion:* **As cryptography is frequently a crucial enabling tech-nology for technical protection services, continued advances in technical protection services require a productive and leading-edge community of cryptography and security researchers and developers.**

Business Models

Intellectual property protection is often viewed in terms of just law and technology: The law indicates what may be done legally, while tech-nology provides some degree of on-the-spot enforcement. But law and technology are not the only tools available. A third, powerful factor in the mix is the business model. By selecting an appropriate business model, a rights holder can at times significantly influence the pressure for and degree of illegal commercial copying and unauthorized reproduction by individuals.

Business models that can contribute to the protection of IP include traditional sales models (low-priced mass-market distribution with con-venient purchasing, where the low price and ease of purchase make it more attractive to buy than to copy) and advertiser-supported models (selling readers' attention to keep the product price low), as well as the more radical step of giving away IP and selling a complementary product or service (e.g., open source software given away, with consulting and maintenance as the service). Simply put, because digital content is difficult to protect, it can be very profitable to find a business model that does not rely primarily on technical protection, or even one that exploits tendencies to share and redistribute content.

> *Recommendation:* **Rights holders should give careful consideration to the power that business models offer for dealing with distribution of digital information. The judicious selection of a business model may significantly reduce the need for technical protection or legal protection, thereby lowering development and enforcement costs. But the model must be carefully matched to the product: While the appropriate business model can for some products obviate the need for technical protection, for others (e.g., first-run movies) substantial protection may be necessary (and even the strongest protection mechanisms likely to be available soon may be inadequate).**

Alternatives to Networks for Distribution of Content

Not every information product need be distributed by digital networks, given the availability of alternative mechanisms offering most of the advantages and far fewer risks. High-value, long-lived products (e.g., classic movies like *The Wizard of Oz*) might never be made legally available on the Internet while protected by copyright, because the consequences of someone capturing the bits are simply too great, and the technical, legal, and social enforcement costs of ensuring that this does not happen are prohibitive. Simply put, the information infrastructure need not be made completely safe for the mass marketing of every form of content.

The pressure to do so is reduced by the possibility of developing special-purpose delivery devices (such as digital video disks (DVDs)) that combine both software encryption and specialized hardware in a manner that makes the decrypted digital content very difficult to capture. While the specific encryption system used in DVDs was cracked late in 1999, it is still the case that making the content accessible only with specialized hardware can offer substantially more security than is possible with the software-only solutions used when content is delivered to general-purpose PCs. Delivering digital content in a physical medium (like a DVD) offers a combination of the advantages of digital content (e.g., compactness, low manufacturing cost) and the advantages of previous distribution media, like books, in which the content is "bound to" the physical object and hence less easily reproduced. With media like this available, there may be no need to risk the consequences of networked distribution for every work.

> *Conclusion:* **Some digital information may be distributed more securely using physical substrates rather than by computer networks.**

Summary—Protecting Intellectual Property

Given the diversity of digital information products, from scholarly articles and single songs to encyclopedias and full-length movies, no single solution is likely to be a good match to the entire range, nor would it be useful to attempt to select just one: It would be as unreasonable to treat all IP as if it were an inexpensive, low-end product as it would be to treat it all as an expensive, high-value product.

> *Conclusion:* **There is great diversity in the kinds of digital intellectual property, business models, legal mechanisms, and technical protection services possible, making a one-size-fits-all solution too rigid. Currently, a wide variety of new models and mechanisms are being created, tried out, and in some cases discarded, at a furious pace. This process should be supported and encouraged, to allow all parties to find models and mechanisms well suited to their needs.**

> *Recommendation:* **Legislators should not contemplate an overhaul of intellectual property laws and public policy at this time, to permit the evolutionary process described above the time to play out.**

Copyright Education

The committee believes that the public welfare would be well served by a program of education explaining why respect for copyright is beneficial for society as a whole (i.e., that copyright has benefits for all stakeholders, not only for rights holders), and detailing both the privileges and limitations of copyright protection. Copyright is the focus here because it is the form of intellectual property law most routinely encountered by the general public.

> *Conclusion:* **A better understanding of the basic principles of copyright law would lead to greater respect for this law and greater willingness to abide by it, as well as produce a more**

informed public better able to engage in discussions about intellectual property and public policy.

Recommendation: An educational program should be undertaken that emphasizes the benefits that copyright law provides to all parties. Such a copyright education program needs to be planned and executed with care. Appendix F discusses the rationale for and the desirable characteristics of copyright education.

The committee could not decide how extensive copyright education should be, who should conduct this education, or who should pay for it. However, the committee agreed that copyright education should focus on the basic fairness of the copyright law, should not be oversimplified, and should not be mandated by the federal government.

RECOMMENDATIONS FOR RESEARCH AND DATA COLLECTION

There are substantial gaps in the knowledge base available to policy makers who must grapple with the problems raised by digital intellectual property. In some cases, there has been little or no inquiry, while in others there are questions about the reliability of the information available. The committee urges the funders and managers of research programs to give priority to the areas of inquiry described below.

Economics of Copyright

Recommendation: Research should be conducted to characterize the economic impacts of copyright. Such research might consider, among other things, the impact of network effects in information industries and how digital networks are changing transaction costs.

To date, the methodology employed in some of the studies of illegal commercial copying has produced high-end estimates of losses in gross revenue. Trade associations would make a more useful contribution to the debate if they revised their methodology so that their estimates better reflect the losses attributable to illegal commercial copying. Notwithstanding the methodological deficiencies of the reported information, the committee concluded that the volume and the cost of illegal commercial copying are substantial.

Recommendation: Research should be initiated to better assess the social and economic impacts of illegal commercial copying and how they interact with private noncommercial copying for personal use.

The Possibility of an Alternative Foundation for Copyright

Given the challenges to the copyright regime posed by digital information, the committee concluded that alternatives to a copy-based model for protection of digital information deserve consideration, even if the implementation of any new model is not likely to occur anytime soon.

Recommendation: The committee suggests exploring whether or not the notion of copy is an appropriate foundation for copyright law, and whether a new foundation can be constructed for copyright, based on the goal set forth in the Constitution ("promote the progress of science and the useful arts") and a tactic by which it is achieved, namely, providing incentive to authors and publishers. In this framework, the question would not be whether a copy had been made, but whether a use of a work was consistent with the goal and tactic (i.e., did it contribute to the desired "progress" and was it destructive, when taken alone or aggregated with other similar copies, of an author's incentive?). This concept is similar to fair use but broader in scope, as it requires considering the range of factors by which to measure the impact of the activity on authors, publishers, and others.

Operation of Copyright Law in the New Digital Environment

Digital technology enables the creation of new kinds of information products and services, which raises a multitude of legal issues. Digital repositories pose difficult questions about authorship, ownership, and the boundaries among protected works. Additional issues arise concerning the meaning of digital publication and the distinctions between fair use and private use.

Recommendation: Legal research should be undertaken on the status of temporary reproductions and derivative work rights to inform the process of adapting copyright law to the digital environment, and to assist policy makers and judges in their deliberations.

As one example of the utility of such research, only a few years ago, proxy caching by online service providers and linking from one Web site to another on the Web were the subjects of considerable debate. Both are now generally thought to be lawful as a matter of U.S. copyright law, a position enabled in part by legal research that has explored the implications of alternative resolutions.

Recommendation: **Legal, economic, and public policy research should be undertaken to help determine the extent to which fair use and other exceptions and limitations to copyright should apply in the digital environment. As public policy research, legal developments, and the marketplace shape the scope of fair use and other limitations on copyright, and/or demonstrate a need for additional protections, any additional actions that may be needed to adapt the law, educate the public about it, or enforce the law may become clearer.**

Recommendation: **Research should be undertaken in the areas that are most likely to intersect with intellectual property law, namely, contract law, communications policy, privacy policy, and First Amendment policy. The interaction of intellectual property law and contract law is likely to be of particular significance in the relatively near future, as licensing becomes a more common means of information distribution, leading to potential conflicts with the goals of IP law.**

Impacts of the Broadening Use of Patents for Information Inventions

The long-term effects of the substantial de facto broadening of patent subject matter to cover information inventions such as computer programs, information design, and business methods (e.g., Internet business models) are as yet unclear, although the committee is concerned about the effects to date. Because this expansion has occurred without any oversight from the legislative branch and takes patent law into uncharted territory, this phenomenon needs to be studied on a systematic basis, empirically and theoretically, to ensure that expansion of patent protection is fulfilling its fundamental goal of promoting progress.

Recommendation: **Research should be conducted to ensure that expansion of patent protection for information inventions is aligned with the constitutional intent of promoting the progress of science and the useful arts.**

Improved Information on Perceptions and Behavior of the General Public Toward Intellectual Property

Little is known about how frequently individuals duplicate copyrighted materials and whether they even pause to question whether this activity may be illegal.

Recommendation: **Research and data collection should be pursued to develop a better understanding of what types of digital copying people think are permissible, what they regard as infringements, and what falls into murky ill-defined areas. Such research should address how these views differ from one community to another, how they differ according to type of material, how user behavior follows user beliefs, and to what extent further knowledge about copyright law is likely to change user behavior.**

A series of careful studies would help in assessing how various groups of individuals perceive copyright, and aid in determining when the law is violated through lack of knowledge versus when it is violated knowingly. Such studies are critical to shaping workable laws and designing educational campaigns to promote compliance.

GUIDELINES FOR USE IN FORMULATING LAW AND PUBLIC POLICY

The committee tried to develop several recommendations for specific changes to laws and public policy. This proved to be a formidable and often frustrating process and perhaps, in retrospect, an imprudent effort, because of the uncertainty created by the evolving information infrastructure, business models, and social responses to the uncertainty. A significant portion of the committee's deliberations can be characterized as spirited and energetic discussions. That this committee, a diverse and balanced group of experts, had difficulty in achieving consensus in many areas, despite extensive briefings, background reading, and deliberations, should serve as a caution to policy makers to contemplate changes to law or policy with the utmost care. Nevertheless, the committee offers some general principles that should prove useful in the formulation of law and public policy in the future.

Law constrains (or at least affects) behavior, but so do markets, social norms, and the constraints embedded in software, as for example, the computer program that controls access to a Web site. Being aware of the multiplicity of forces aids in understanding and analyzing issues and may open up additional routes for dealing with issues; not every problem need be legislated (or priced) into submission. In addition, content itself should not be viewed as a monolith: Some content (e.g., classic movies) has a high and persistent value; other content (e.g., today's traffic report) may have a modest value for a limited time period before becoming economically worthless (though perhaps of historical value later on). Thus, not all content needs the same kind of IP protection.

Conclusion: **Law and public policy must be crafted to consider all the relevant forces in the digital environment. Initiatives that consider or rely on only one or a subset of the relevant forces are not likely to serve the nation well.**

The rapid evolution in technology will be an ongoing source of uncertainty and, likely, frustration for policy makers who conceive of and attempt to deal with issues narrowly, in terms of the extant technology.

Conclusion: **Policy makers must conceive of and analyze issues in a manner that is as technology-independent as possible, drafting policies and legislation in a similar fashion. The question to focus on is not so much exactly what device is causing a problem today, as what the underlying issue is. Nor should policy makers base their decisions on the specifics of any particular business model.**

The information infrastructure makes private infringement of IP rights vastly easier to carry out and correspondingly more difficult to detect and prevent. As a result, individual standards of moral and ethical conduct, and individual perceptions of right and wrong, become more important. Laws that are simple, clear, and comprehensible are needed, particularly those parts of the IP law that are most directly relevant to consumer behavior in daily life. Support and adherence are far more likely if IP law is clearly understood and viewed by the general public as embodying reasonable standards of normative behavior. If intellectual property law is perceived as being so absolute in its prohibitions as to sweep within those prohibitions behavior that most individuals feel is not morally or ethically culpable, then even the more reasonable restrictions contained in the same body of law may be painted with the same brush and viewed as illegitimate.

Conclusion: **Public compliance with intellectual property law requires a high degree of simplicity, clarity, straightforwardness, and comprehensibility for all aspects of copyright law** that deal with individual behavior. New or revised intellectual property laws should be drafted accordingly.

Recommendation: Policy makers should use the principles outlined in Box 6.2 [Chapter 6] in the formulation of intellectual property law and public policy.

A FINAL WORD

Intellectual property will surely survive the digital age, although substantial time and effort may be required to achieve a workable balance between private rights and the public interest in information. Major adaptations may need to take place to ensure that content creators and rights holders have sufficient incentives to produce an extensive and diverse supply of intellectual property. Policy makers and stakeholders will have to work together to ensure that the important public purposes embodied in copyright law continue to be fulfilled in the digital context. The information infrastructure promises the possibility of greatly improved access to information for all of society. We as a society share the responsibility for developing reasonable compromises to allow the nation to benefit from the opportunities it can bring.

1

The Emergence of the Digital Dilemma

The role of information products and services in the U.S. economy is vast and still growing rapidly. The addition of an "Information Sector" category to the federal government's new industry classification system is recognition of both the sector's economic importance and the fundamental kinship of publishing (print and software), motion picture and sound recording, radio and television broadcasting, libraries, and information and data processing services.[1]

The widespread use of computer networks and the global reach of the World Wide Web have added substantially to the information sector's production of an astonishing abundance of information in digital form, as well as offering unprecedented ease of access to it. Creating, publishing, distributing, using, and reusing information have become many times easier and faster in the past decade. The good news is the enrichment that this explosive growth in information brings to society as a whole. The bad news is the enrichment that it can also bring to those who take advantage of the properties of digital information and the Web to copy, distribute, and use information illegally. The Web is an information resource of extraordinary size and depth, yet it is also an information reproduction and dissemination facility of great reach and capability; it is at once one of the world's largest libraries and surely the world's largest copying machine.

---

[1]See Murphy (1998).

The traditional tool for dealing with use and misuse of information is intellectual property law, the constellation of statutes and case law that govern copyrights, patents, and trade secrets. Part of the case for granting rights in intellectual property (IP) is the belief that protecting IP promotes the development of new products and services, and that erosion of those rights could threaten the economic performance of the information sector and curtail the major benefits it has brought.[2] But as this report argues, with this new abundance of information and the ease with which it can be accessed, reproduced, and distributed have come problems that must be seen in all of their complexity, including related economic, social, technical, and philosophical concerns, as well as the accompanying legal and policy challenges. Debates over these issues matter because the outcome will have a significant impact on today's information sector companies and will help determine the character of the digital economy of the future.[3]

AN ENDURING BALANCE UPSET?

The task of intellectual property protection has always been difficult, attempting as it does to achieve a finely tuned balance: providing authors and publishers enough control over their work that they are motivated to create and disseminate, while seeking to limit that control so that society as a whole benefits from access to the work. The challenge was elegantly stated some 200 years ago in a legal case in Great Britain:

> We must take care to guard against two extremes equally prejudicial; the one, that men of ability who have employed their time for the ser-

---

[2]A second argument in support of IP law is the principle that the creator of an information product ought to be entitled to control the dissemination and use of that information, an issue that is considered throughout this report. For the moment, note that the constitutional language granting Congress the authority to create copyright and patent protection mentions only an instrumental purpose: "To promote the progress of science and the useful arts" (U.S. Constitution, Art. 1, Sec. 8, Para. 8).

[3]Those debates include the Digital Millennium Copyright Act of 1998 (P.L. 105-304), which amends the Copyright Act, title 17 U.S.C., to legislate new rights in copyrighted works, and limitations on those rights, when copyrighted works are used on the Internet or in other digital, electronic environments. Efforts to enact legislation to provide protection for databases that do not qualify for copyright are taking place in the 106th Congress through H.R. 354, the Collections of Information Antipiracy Act, and H.R. 1858, the Consumer and Investor Access to Information Act. In its Treasury and General Government Appropriation Bill for FY2000 (S. 1282), the Senate Appropriations Committee endorsed the creation of an interagency federal office to fight against the infringement of IP rights of U.S. entertainment and computer companies. This action came in response to requests from industry executives such as Bill Gates, chairman and CEO of Microsoft Corporation, and Jack Valenti, chairman and CEO of the Motion Picture Association of America (Rogers, 1999).

vice of the community, may not be deprived of their just merits, and the reward of their ingenuity and labour; the other, that the world may not be deprived of improvements, nor the progress of the arts be retarded.[4]

In more recent times, a U.S. court reiterated the significance of balancing rights and access:

> We must remember that the purpose of the copyright law is to create the most efficient and productive balance between protection (incentive) and dissemination of information, to promote learning, culture and development.[5]

In the two centuries between those two statements, the United States has changed enormously, moving from an agrarian society to one heavily dependent on information and high technology. Yet many of the fundamental concepts of U.S. intellectual property have been in place for those 200 years and have, with some success, weathered substantial changes in technology and society. The first U.S. copyright statute was enacted in 1793 and protected only maps, charts, and books. Yet it has been adapted successfully over the past 200 years, in part by expanding both the set of exclusive rights conferred by copyright and the scope of the subject matter (embracing photographs, sound recordings, motion pictures, software, and more) and by qualifying those rights with exceptions such as the fair use rule. (Box 1.1 defines "fair use" and other key terms.)

During that time, copyright and patent law have had an instrumental role in the promotion and creation of a vast array of informational works, resulting in vibrant markets for IP. But copyright and patent laws have also defined limits on protection in order to facilitate the public interest in and benefit from shared information. Over time, compromises have evolved to balance the interests of the creators and consumers of intellectual work, fulfilling a number of important public policy objectives.

But the carefully crafted balance may be in danger of being upset. The emergence in the past 10 years of a new information infrastructure marked by the proliferation of personal computers, networks that connect them, and the World Wide Web has led to radical changes in how informational works are created and distributed, offering both enormous new opportunities and substantial challenges to the current model of intellectual property.

---

[4]Lord Mansfield in *Sayre v. Moore*, 1785, cited in Kaplan (1967), p. 17.

[5]*Whelan v. Jaslow*, 797 F.2d 1222; 21 Fed. R. Evid. Serv. (Callaghan) 571; U.S. Court of Appeals for the Third Circuit, March 3, 1986, Argued, August 4, 1986, Filed.

BOX 1.1
Key Terms Used in This Report

Author: Used generically to refer to a person or legal entity creating any variety of intellectual property, including books, music, films, and so forth.

Bootleg: An unauthorized recording of a broadcast or live performance.

Content: Used generically to indicate any work produced by an author, whatever the medium of expression (text, pictures, music or musical performances, computer programs, and so on).

Copy: A reproduction of a work. The definition of "copy" is complex and is discussed throughout the report.

Copyright: Copyright law protects artistic and expressive work; a copyright on a work provides the rights holder with exclusive rights to control certain uses of the work (e.g., reproduction, distribution to the public, public performances and public displays, and adaptation).

Counterfeit: An unauthorized reproduction of both the content and packaging of a work.

Fair use: The use of a copyrighted work for purposes such as criticism, comment, news reporting, teaching, scholarship, or research as permitted under 17 U.S.C. sec. 107.

Intellectual Property (IP): Intellectual property is the generic descriptor of the work product of authors and inventors. In the United States, intellectual property is protected by copyright, patent, trademark, and trade secret law.

License: Grant of permission from the rights holder of a work to engage in acts that in the absence of permission would be infringing.

Patent: Patent law protects useful inventions and discoveries, requiring them to be novel, useful, and nonobvious. A patent gives its owner sole right to control the gainful application of the specific ideas the patent discloses.

Piracy: Unauthorized duplication on a commercial scale of a copyrighted work with the intention to defraud the rights holder.

Rights holder: Used to indicate someone holding the IP rights to a work, whether the author, publisher, inventor, or some person or other legal entity to which the rights have been transferred.

Trademark: Trademark law covers the uses of trademarks, the compact patterns associated with an enterprise or a product line. A trademark is intended to unequivocally distinguish the marked objects from similar objects from different sources.

SCOPE OF THE REPORT

There is a healthy ferment of experimentation and debate going on in attempts to realize the promise of the digital age. This report seeks to explain and demystify the underlying technology trends, explore the range of technological and business tools that may be useful, and recommend a variety of actions that can be taken to help ensure that the benefits of the information infrastructure are realized for rights holders and society as a whole.

This report builds on recent previous studies in the area of intellectual property and digital technology. Perhaps the most visible effort was undertaken by the Information Infrastructure Task Force, which issued the report *Intellectual Property and the National Information Infrastructure* (IITF, 1995),[6] sometimes referred to as the IITF white paper. The IITF white paper presents the detailed legal issues concerning copyright and digital technology, but it does not address business models, protection technologies, or other issues in any particular depth. More recently, the U.S. Copyright Office commissioned a study on the future of copyright in the networked world (Hardy, 1998). That report, which provides good descriptive coverage of the relevant technologies and trends and some discussion of the pertinent economic and legal issues, identified trends but did not provide conclusions and recommendations.

This report of the Committee on Intellectual Property Rights in the Emerging Information Infrastructure does not duplicate the detailed legal analyses of the IITF white paper or the extensive review of technologies in the Hardy report. Instead, it offers a framework for the evaluation and construction of public policy, as well as a variety of specific conclusions and recommendations designed to help legislators, courts, administrators, and the public to understand what is at issue, to formulate questions clearly, and to assess alternatives. The focus on copyright derives from the observation that copyright protects a large variety of the IP frequently encountered by the public and has the highest visibility in the debates over IP and the information infrastructure. The members of the study committee were selected to provide the diverse expertise needed to ensure that stakeholders' wide-ranging perspectives were represented.[7]

For the most part, this report focuses on circumstances and actions possible in the United States. However, as discussed below, the study committee's conclusions and recommendations need to be considered in

---

[6]The Working Group on Intellectual Property Rights, chaired by Bruce A. Lehman, assistant secretary of commerce and commissioner of patents and trademarks, was established within the Information Policy Committee of the Information Infrastructure Task Force.

[7]See Appendix A for the biographies of study committee members.

a worldwide context. One of the consequences of global networks is the inevitable interaction between U.S. law and culture and those of other countries. This can be problematic because laws and IP practice differ widely across countries and are likely to remain different despite efforts at harmonization.

ORIGINS OF THE ISSUES

Given the successful growth and adaptation of intellectual property law over the years, any claim that the established balance is in danger of being upset must be clear and convincing about the origins of that danger. The committee identified problems arising from two primary sources: changes in technology and the availability of the digital information infrastructure as a routine part of everyday life. Three technological changes in particular—the increased use of information in digital form, the rapid growth of computer networks, and the creation of the World Wide Web— have fundamentally altered the landscape and lie at the heart of many of the issues presented by the evolving information infrastructure. These changes, coupled with the emergence of the information infrastructure as a part of daily life, present significant legal, social, economic, and policy challenges.

Technology Has Changed:
Digital Information, Networks, and the Web

Representing information in digital form, as opposed to the more traditional analog form, means using numbers to capture and convey the information. Music offers a clear example of the difference between the two. Capturing musical sounds requires describing the shape of the vibrations in air that are the sound. Records capture that information in the shape of the groove in the vinyl. CDs, by contrast, capture the same information as a large collection of numbers (see Box 1.2). Digital information has a remarkable breadth of descriptive ability, including text, audio (music, speech), video (still and moving pictures), software, and even shape (e.g., in computer-aided design).

Why Digital Information Matters

Access Is by Copying. When information is represented digitally, access inevitably means making a copy, even if only an ephemeral (temporary) copy. This copying action is deeply rooted in the way computers work: Even an action as simple as examining a document stored on your own disk means copying it, in this case twice—from the disk to the computer's

BOX 1.2
Capturing and Compressing Information in Digital Form

Representing information in digital form means capturing it as a collection of numbers. Music offers an easy example. Music (or any other sound) is a vibration that can be described by a sound wave (Figure 1.2.1). Traditional vinyl records capture the sound by putting a groove in the vinyl that has the same shape as the sound wave.

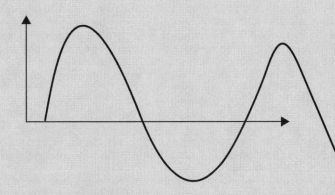

FIGURE 1.2.1 A sound wave.

Digitizing a sound wave is done by measuring its height thousands of times a second (Figure 1.2.2). Measuring the wave at such closely spaced intervals provides a reasonably accurate approximation to the shape of the wave and, hence, the sound.

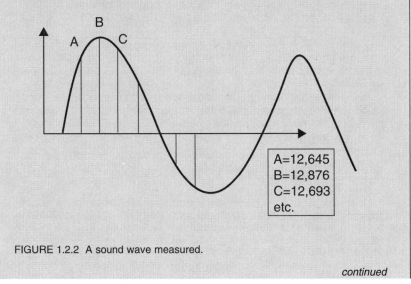

A=12,645
B=12,876
C=12,693
etc.

FIGURE 1.2.2 A sound wave measured.

continued

BOX 1.2 Continued

For a standard music CD, measurements are done 44,100 times/second, using numbers that range from 0 to 65,535 (i.e., 16-bit samples), with one measurement for each channel of music. As 16 bits is 2 bytes, a stereo recording thus requires 2 bytes/sample * 44,100 samples/second * 2 channels = 176,400 bytes/second of music, or roughly 10 megabytes/minute. A standard music CD is around 1 hour of music encoded this way and contains about 320,000,000 samples (i.e., 320,000,000 numbers).[1]

There are other ways of digitizing music, for example, using the Musical Instrument Digital Interface (MIDI) format typically employed with musical synthesizers. Where techniques like MP3 digitize the actual *sound* of the music, MIDI captures how the song was *played* on the synthesizer. Hence, while a MIDI file is also a collection of numbers, those numbers indicate which notes were played and when, how long they lasted, their volume, and so on. MIDI was created as a way of allowing music composed on one synthesizer to be played back on another, but it became for a time a popular way to digitize music.

Pictures can be digitized by measuring the color at closely spaced dots (often 600 or 1,200 dots per inch), then representing the color at each dot with a triple of numbers indicating what combination of red, green, and blue will produce the color found at that spot. The picture is reproduced by putting the appropriate dots of color at the right places on paper or on the screen. Video can be digitized as a sequence of digitized frames.

Text is the simplest thing to "digitize," as there is already a code—the ASCII code—that assigns code numbers to each typewriter character (an "A" for example is given the code 65, "B" is 66, while an "a" is 97); ASCII is used almost universally.

In all of these cases, the size of the digitized file of information can be made considerably smaller by compressing it. The simplest compression techniques rely on finding more compact ways to capture the same information. One technique, called "run length encoding," takes advantage of the fact that numbers can repeat. Consider a very simple example of a file containing a sequence of 1s and 0s and imagine it contains the following: 1 1 1 1 1 0 0 0 0 0 0 0 0 0 0 0 0 1 0 1 0 0 0 0 1 1 1 1 1 1 1.

Exactly the same information could be captured by indicating that the file contains five 1s, twelve 0s, a 1, a 0, a 1, four 0s, eight 1s. There is no information lost in describing the bits this way, yet this description can take up considerably less space.

A variety of more sophisticated compression algorithms are available, many of which rely on specific properties of the information being compressed. Video compression, for example, often relies on the fact that typically very few things in a scene change from one frame to the next (i.e., in 1/30th of a second). This makes it possible to encode one frame by indicating only what changed compared to the previous frame.

---

[1]As accurate as this is, the human ear can still hear some of the distortion produced by approximating the wave with 44,100 samples per second. This has led to calls within the industry to move to 96,000 samples per second. This faster sampling, along with the desire for more channels (e.g., to support surround-sound), will mean future CDs may contain more data per second of music.

memory and then again onto the video display.[8] Before you can view a page from the World Wide Web, the remote computer must first send your computer a copy of the page. That copy is kept on your hard disk, copied again into memory, and then displayed on the screen.[9] In addition, intermediate copies of the page may have been made by other computers as the page is transported over the network from the remote computer to yours.[10]

Such copying occurs with all digital information. Use your computer to read a book, look at a picture, watch a movie, or listen to a song, and you inevitably make one or more copies. Contrast this with the use of traditional media: Reading a book does not involve making a copy of it, nor does watching a movie or listening to a song.

This intimate connection between access and copying has considerable significance in the context of intellectual property protection. One of the essential elements of copyright—the right to control reproduction—works as expected in the world of traditional media, where there is an obvious distinction between access and reproduction and where the copyright owner's control of reproduction provides just that. But in the digital world, where no access is possible except by copying, complete control of copying would mean control of access as well.

This intimate connection has consequences for all parties in the digital world. Rights holders may seek to control access to digital information, because access involves reproduction. Readers may find their traditional access to information susceptible to control in unprecedented ways. Policymakers, meanwhile, must consider how to maintain the appropriate balance between control and dissemination.

Economics, Character, and Speed of Digital Reproduction. Digital representation changes both the economics and the character of reproduction. Copying digital information, even on a home computer, is easy and inexpensive: A standard (1.44 megabyte) floppy disk, which holds the equivalent of about 500 pages of text, takes no more than a minute to duplicate and is treated as if it were a piece of paper (e.g., routinely given away). A CD, which holds 650 megabytes (the equivalent of about 220,000 pages, or 44 cartons), can be copied in 15 minutes to a blank compact disk that costs

---

[8]The information could be displayed directly from the disk to the screen, but disks are much too slow to make this practical. Main memory is thousands of times faster; hence, several pages of the information are copied there first. This copy is typically ephemeral, disappearing as soon as you view a different page.

[9]The copy on your disk may disappear from your disk when you exit your Web browser, but it can often easily be saved permanently.

[10]A primer on the operation of the Internet is given in Appendix C.

about $1.00, using equipment now widely available for PCs and costing only a few hundred dollars.[11]

Copying information has always been possible, but the advent of digital information brings an extraordinary increase in the amount of information that can be easily and inexpensively reproduced. Given the widespread availability of computers, many people now have the ability to casually reproduce vast amounts of information. Consequently, the traditional physical and economic impediments to copyright infringement have been considerably undermined. Its size once meant that a 30-volume encyclopedia could be reproduced only by those with considerable means and motive; now an encyclopedia on a CD can be reproduced in a few minutes on what is fast becoming ordinary technology.

The character of reproduction has changed as well. Although a photocopy often isn't as sharp as the original,[12] a digital copy is indistinguishable from the original as are all successive digital copies. For every form of digital information, every copy is as good as the original and can therefore be the source of additional perfect copies, which greatly reduces what was once a natural impediment to copyright infringement. With the traditional form of information, the successively lower quality of each generation of copy offered a natural limitation to redistribution. With digital information there is no such limitation.

Content Liberated from Medium. Information in digital form is largely liberated from the medium that carries it. When information is sent across networks, there is no need to ship a physical substrate; the information alone flows to the recipient. The liberation of content is also evident when bits are copied across media (disk to tape to CD to floppy) with the greatest of ease. The choice of media may have consequences for the amount of storage or speed of access, but the content of the information and its properties (e.g., the ability to make exact copies) are preserved perfectly across a variety of media.

Information in traditional analog forms (movies, paintings, sculpture) is, by contrast, far more tightly bound to the underlying physical media.

---

[11]Text is a particularly compelling example because it puts, relatively speaking, little information on a page. Graphic images contain far more information: a single 5" x 7" color photograph may require 14 megabytes to store digitally (300 bits per inch resolution and 36 bits of color); hence a CD might hold about 46 such photographs. There are also a variety of ways to compress digital information so that it takes up even less space; sometimes this means being able to fit 30 times more information onto a disk than it could hold without compression.

[12]The difference between a copy and an original depends of course on the quality of the equipment being used. Audio or video tapes copied using standard (i.e., analog) technology, for example, are markedly inferior in quality to digital copies.

It is not easily transported without the underlying medium, nor is it so easily extracted for copying (consider copying a sculpture). The point, of course, is comparative: Bits still need to be stored someplace, and even a sculpture can be copied, but the difference is so large—several orders of magnitude and constantly increasing with advancing technology—that the experience from the individual's viewpoint is qualitatively different.

The liberation of content from the medium has unsettling consequences for the protection of IP in digital form. Until very recently, intellectual works have been produced and distributed largely as analog works embedded in a physical artifact (e.g., printed books, movies on video tape). IP law and practice have been worked out in the context of such artifacts, and much of our comfort with IP law is based on the familiar properties of information closely bound to a physical substrate. Digital information changes those properties in substantial ways.

New Kinds and Uses of Information. Digital information is plastic, easily searched and indexed, and easily cross-indexed. It is plastic in the sense that it is easily changed. Although a paper book is difficult to alter and hard to search even with a good index, online text can be changed easily, for instance, by adding and rearranging paragraphs. Coupled with digital transmission, plasticity of information confers, along with great advantages, the potential for fraudulent acts such as plagiarism or forgery.[13]

In addition, although traditional documents are static—a printed book contains the same words from one moment to the next—digital documents can be dynamic, changing from moment to moment or offering different views. For example, articles posted on the Web often undergo revision in response to comments from readers. Short of making a (static) local copy, how does one cite such a thing, if it may say something different tomorrow? Even with a static local copy, who is to say what the document once said at a particular point in time, if there are at least two different versions? The plasticity of digital information could have a significant impact on the nature and value of citations and on scholarly research.

The ease of searching and indexing digital information enormously facilitates the creation of derivative works of unusual forms.[14] Consider

---

[13]The use of a digital signature (see Appendix E) may be of some assistance by providing a way to "sign" a digital document. If the document is subsequently altered, there will be a detectable mismatch between the signature and the document. Large-scale use of digital signatures requires a substantial infrastructure that is only now emerging, with the growth of e-commerce.

[14]A "derivative work" is a work based on one or more preexisting works, such as a translation, musical arrangement, dramatization, fictionalization, motion picture version,

an online textbook. Someone knowledgeable in the field covered by the textbook may, on reading the text, decide that there is a better order of presentation of the material and might indicate that by establishing a set of hyperlinks that effectively reorganize the book.[15] Is the set of links a derivative work?

In a similar vein, a practice on the Web known as "framing" has raised a number of IP-related questions, particularly in the commercial context. Framing refers to one Web page presenting information from another. When both pages are the work of the same author, no issue arises. Questions arise when the framed page is the work of a different author and when the information on that page is presented in less than its entirety (e.g., without advertisements that originally appeared there, or stripped of information identifying the author). In that case have the first author's rights been infringed by the second author's adaptation?

In the music world, the ease of searching, indexing, and reproducing digital information has led to enormous growth in *sampling*—the reuse of segments of previous works—leading to questions of intellectual property infringement and fair use.

Increasing Use of Licensing. From the early days of the software market to the present, commercial distribution of digital information typically has been through the use of licenses rather than by sale. Packaged software traditionally has had a shrink-wrap license, an agreement that purportedly goes into effect upon opening the (shrink-wrapped) package. More recently, a wide variety of digital information is being marketed on the Web with what are sometimes whimsically called "click-wrap" licenses, an agreement presented on the screen and "agreed to" by the click of a mouse. Negotiated licenses are also used to clarify the terms governing access to large databases.

The difference between selling a work and licensing it is significant. The sale of a physical copy of a work has been the dominant model for transferring IP to the consumer for more than 200 years. Sales involve the complete transfer of ownership rights in the copy. Copyright law explicitly anticipates the sale of intellectual property products and, by the "first-sale rule," constrains a copyright holder's rights in copies of the work that have been sold. For example, the purchaser is free to lend, rent, or resell

---

sound recording, art reproduction, abridgment, condensation, or any other form in which a work may be recast, transformed, or adapted. A work consisting of editorial revisions, annotations, elaborations, or other modifications, which, as a whole, represent an original work of authorship, is a derivative work.

[15]This tactic is not new, having been used for some years in educational "course packs" like the electronic field trips offered by the Colonial Williamsburg Foundation.

the purchased copy.[16] In that sense, copyright law follows IP products into the marketplace and promotes the continued dissemination of information.

Licensing, however, constitutes a limited transfer of rights to use an item on stated terms and conditions. Licenses are governed by contract law and, as such, are essentially a private agreement between two parties. That agreement can involve a wide range of terms and conditions (Box 1.3) and need not incorporate any public policy considerations, beyond some basic limits on what constitutes an enforceable contract.

Contracting has benefits; for example, it may enable distribution of some information products that would otherwise not come to market. But there are also drawbacks, particularly the possibility that the terms of a license may be far more restrictive than the provisions for access normally granted under copyright's first-sale doctrine. To the extent that highly restrictive licensing replaces the sale of copyrighted works, society may be the loser, especially if the public policy goals embodied in copyright law are omitted from contracts.

This issue's significance is underscored by the proposed Uniform Computer Information Transactions Act (UCITA),[17] which would validate mass market licenses for information, making the license terms enforceable if the consumer has given some sort of token assent (e.g., by clicking a mouse or installing the software). Such proposals seek to establish more validity for shrink-wrap and click-wrap agreements currently regarded as questionable. There is promise in the potential to reduce the overhead for making things available through licensing (just a mouse click) and reduce uncertainty by establishing whether the agreement is enforceable—additional information products may appear in the marketplace as a result. The peril lies in the possibility noted above that licensing could become a replacement for sale of all manner of copyrighted works, without provision for the public policy goals embodied in copyright law.

Multiplicity of Access and Access at a Distance. Information in digital form is accessible to thousands of people virtually simultaneously, because multiple users of a server can read the same file at their own indi-

---

[16]There are certain exceptions for phonorecords and computer programs; see section 109 of the copyright law.

[17]UCITA is the new name for the proposed Article 2B amendment to the Uniform Commercial Code (UCC). UCITA is no longer intended to be part of the UCC; however, UCITA's drafters seek to have it adopted by all state legislatures as a uniform state law. On July 29, 1999, the National Conference of Commissioners on Uniform State Laws passed UCITA. State commissioners will now send UCITA to state legislatures for approval as a uniform state law (see Chapter 3).

BOX 1.3
When Does a Contract Cover Future Media and Technology That Do Not Yet Exist?

The history of 20th-century intellectual property law is replete with controversies arising from a recurring fact pattern—whether the scope of rights granted to use intellectual property enables the licensee to exploit that property by means of a new medium or technology that was not yet invented at the time the contract was entered into. The many cases dealing with this question read like a history of the development of communications media over the last century. There are cases dealing with the question of whether a grant of performance rights included, after Edison invented motion pictures, the right to make a motion picture based on the dramatization of the initial work; whether a grant of motion rights during the silent film era included the right to add a sound track when talking motion pictures were introduced; whether the right to make a motion picture granted at a time when television was not yet invented (or at least widely known) included the later right to exhibit the motion picture on television; and whether grants of the right to produce either a motion picture or a television program in the era before the introduction of the VCR included the right on the part of the motion picture studio to rent and sell copies of the motion picture or television program on videocassettes and video disks.[1]

The cases considering these "after invented media" or "new technology" scenarios tend to apply conventional principles of general contract law, including a search for the true "intent" of the parties as reflected in the specific contract at issue, a close analysis of the wording that was used, and the application of rules of strict construction (e.g., construing ambiguities against the interests of the party who drafted the contract).

Some legal commentators view these cases as inherently nonuniform and conflicting.[2] To the extent one can articulate a consistent thread that runs through the

---

[1]See, for example, *Manners v. Morosco* (252 U.S. 317 (1920)); *Kirke La Shelle Co. v. Paul Armstrong Co.* (263 N.Y. 79, 82-89 (1933)); *Bartsch v. Metro-Goldwyn-Mayer, Inc.* (391 F.2d 150 (2d Cir. 1968) cert. denied, 393 U.S. 826 (1968)); *Ettore v. Philco Television Broadcasting Corp.* (229 F.2d 481 (3d Cir. 1956), cert. denied, 351 U.S. 926, 109 U.S.P.Q. 517 (1956)); *Filmvideo Releasing Corp. v. Hastings* (446 F. Supp. 725 (S.D.N.Y. 1978)); *Cohen v. Paramount Pictures Corp.* (845 F.2d 851 (9th Cir. 1988)); and *Tele-Pac v. Grainger* (570 N.Y.S. 2d 521, appeal dismissed, 79 N.Y. 2d 822, 580 N.Y.S. 2d 201, 588 N.E. 2d 99 (1991)) and *Muller v. Walt Disney Prods.* (871 F. Supp. 678, 682-83 (S.D.N.Y. 1994)).

[2]Contrast, for example, *Cohen v. Paramount Pictures Corp.* (845 F.2d 851 (9th Cir. 1988)) and *Rey v. Lafferty* (990 F.2d 1379, 1382 (1st Cir. 1993)) (both holding that without a broad grant of rights or a "new technology" clause, the grantor will retain the benefit of exploitation in new after-invented media, and the licensee will obtain no inherent right to exploit the licensed property in new media) with *Rooney v. Columbia Pictures* (538 F. Supp. 211 (S.D.N.Y. 1982) aff'd, 714 F.2d 117 (2d Cir. 1982), cert. denied, 460 U.S. 1084 (1983)) and *Platinum Records v. Lucasfilms, Ltd.* (566 F. Supp. 226 (D. N.J. 1983)) (holding that an unambiguous broad grant of rights or a "new technology" clause are sufficient to accord to the licensee the benefit of new after-invented media exploitation rights, even if such new media were not, and could not have been, foreseen at the time of the original grant or license).

cases, it would be that in order to obtain distribution or other exploitation rights for yet to be invented media, a licensee must obtain either (1) a broad "blanket" grant of all rights, or (2), even better, a grant of rights containing a so-called "new technology" clause (i.e., a clause granting the right to exploit the intellectual property by any and all means and media, whether now known or hereafter invented or devised). The use of such new technology clauses is now common in contracts covering all kinds of intellectual property, across a wide range of industries. It is not uncommon for the new technology clause to specify that the grant of future media or technology includes, without limitation, and only by way of example, virtually every then imaginable new technology or medium, in perpetuity and throughout the universe.

However, even the inclusion of a broad new technology clause is not a guarantee that the grant of new media rights will be upheld, nor is the lack of a new technology clause necessarily fatal to a licensee's claim that such rights were granted. The presence of an ambiguous term or ambiguous provision in an agreement opens the door to parol or other extrinsic evidence concerning the actual intention of the parties.[3]

The continuing evolution of the digital information infrastructure promises to keep future generations of lawyers busy litigating controversies over whether contracts written in the predigital era included within the scope of the rights granted the right to exploit the licensed intellectual property by means of the Internet. As the personal computer and the conventional television receiver converge into a single information appliance, the question of whether a grant of "television rights" includes the right to transmit a motion picture or television program to the household over the Internet, presents the kind of issue that will engage lawyers and judges well into the next century.

---

[3]Compare, for example, *Subafilms, Ltd. v. MGM-Pathe Communications Co.* (988 F.2d 122 (9th Cir. 1993) 1993 WL 39269, reh'g granted, 5 F.3d 452 (9th Cir. 1993), vacated in part, 24 F.3d 1088, 30 U.S.P.Q. 2d 1746 (9th Cir. 1994)) (holding that despite the presence of a "new technology" clause in the grant of rights, new media rights were not included) with *Bourne Co. v. Walt Disney Co.* (68 F.3d 621, 630 (2d Cir. 1995), cert. denied, 116 S. Ct. 1890 (1996)) (holding that despite the lack of a "new technology" clause and a broad grant of rights, the term "motion picture" was sufficiently ambiguous to admit parol or other extrinsic evidence which the jury verdict found, as a question of fact, meant the term "motion picture," under an original grant drafted in the 1930s, and included the right to exhibit the motion picture on videocassettes).

---

vidual pace without interfering with each other. This attribute of course makes digital information much more flexible than traditional media; a single copy of a book, for example, is not accessible to more than one or two people simultaneously.[18]

---

[18]Information that is broadcast (e.g., a movie, television show, radio show) may be viewable by many people simultaneously, but, unlike digital information, no viewer has control

Digital information can also be accessed remotely by, for example, using a modem that allows one computer to call another over ordinary phone lines. The ability to access information in this manner removes the need for geographical proximity, eliminating another of the familiar limitations of information in traditional forms. As a consequence, digital information presents opportunities for access that are vastly greater than those presented by traditional media.

Why Computer Networks Matter: Economics and Speed of Distribution

Today, computers are routinely connected to networks that enable rapid, inexpensive distribution of information. With speeds that reach a billion characters per second on single links, computer networks are drastically changing the economics of information distribution, lowering another of the natural barriers to violation of intellectual property rights. To profit from a book or video, the publisher or (pirate) must incur the costs of reproducing it and distributing the copies. But copying digital information costs almost nothing, and networks make worldwide distribution very inexpensive and very fast. Consequently, it is easier and less expensive for a content owner to distribute a work, and significantly easier and less expensive for a pirate to engage in illegal commercial copying and distribution.

Computer networks amplify the consequences of copyright violations that were previously tolerable. It usually made little difference, in terms of lost revenue, if someone made a photocopy of a book. Photocopying an entire book is inconvenient and often more costly than buying the book, so not very many photocopies are made and distributed. Digital information has radically altered the economics involved, leading to upheavals not only in the relationships among authors, publishers, distributors, and others, but perhaps also in the disappearance of some roles and the emergence of others. The beginning of such massive change can be seen in online publication of books, in bookstores, in new forms of contracts between research libraries and publishers of scientific periodicals, and in new kinds of scholarly offerings.

Publishers are understandably concerned and cautious. They and many authors see possible loss of revenue when a single copy of a work can be widely accessed from a digital library at no cost to the user. How

---

over the timing or sequencing of the presentation (e.g., rewinding to review a scene). Digitized movies can be seen concurrently by many viewers, each of whom has individual control of the scheduling and sequencing of what is shown.

many paying subscribers to a technical journal will there be if the articles are easily available from even one online library? Although professional periodicals published on paper seem likely to persist indefinitely, it is not clear how many subscriptions will be purchased and at what price. If the number of subscriptions drops significantly, readers may enjoy the benefits of digital distribution in the short term, only to find fewer publications available in the longer term.[19]

The speed of digital network distribution has consequences for enforcement as well. When physical copies must be produced and distributed, the process is spread over time and can be interrupted. When information is disseminated by computer networks, it travels to sites around the world in moments. Temporary restraining orders are of little use in forestalling deeds done in minutes.

Why the Web Matters

A Worldwide Publishing Medium. The World Wide Web is a vast collection of electronic documents formatted using special languages (e.g., Hypertext Markup Language (HTML)). Documents formatted in these languages have a number of properties, the most important of which are that they contain multimedia (text, graphics, audio, video) and they link to other documents (or other digital information, including databases) in a way that makes it effortless for readers to access other information. This vast collection of interconnections is what gives the Web its name and much of its interesting character.

The Internet makes it possible for computers to exchange information, while the Web provides the superstructure in which that information can be organized and published. It is, among other things, a giant (and growing) worldwide bulletin board that can be scanned for information of interest, and on which additional information can be posted.[20]

This superstructure is the last in the triumvirate of the impacts of digitization. Digital information radically changes the economics and character of *reproduction,* computer networks radically change the economics and character of *distribution,* and the Web radically changes the

---

[19]Concern over wide-scale access to digital information is not new: The Commission on New Technological Uses of Information (CONTU) hearings raised the issue in 1976. The issue has become real and pressing now as a consequence of the World Wide Web and rapidly growing amounts of information in digital form.

[20]The Web is of course considerably more interesting and powerful than the bulletin board metaphor conveys. It is also a vehicle for active dissemination of information, a means of connecting people with each other, a "space" in which communities have arisen and grown, and more.

economics and character of *publication*. Reproduction and distribution put information in the hands of those who know they want it, but publication makes people aware of available information. The Web, as a publicly accessible resource, functions as a publication medium for all who have access to the Internet, allowing people to make known the existence of their work and perhaps link it to other relevant works. The Web also makes the mechanics of replication and distribution accessible to the user, who has only to click on a document to have the content delivered to his or her own computer.

One consequence is that anyone can be a publisher, and indeed the astonishing variety of documents, opinions, articles, and works of all sorts on the Web demonstrates that millions of people worldwide are making use of that capability.[21] The ability of everyone to publish in turn leads to the possibility of "disintermediation," the reduction of the role of the intermediary, as authors and consumers gain the ability to connect more directly, without traditional intermediaries such as publishers (see Chapter 2 on music and MP3 for further discussion of this issue). The ability of everyone to publish may also shift the publishing bottleneck: Previously the difficulty was in getting work published; in the future, the difficulty may be in getting noticed amid the profusion of works available.

The consequences for traditional publishers are still being worked out. There is both opportunity and upheaval inherent in the new technology. The opportunity comes as publishers use the Web as another medium to advertise and, in some cases, distribute their works. The upheaval comes from the difficulties of publishing information electronically without also losing control over reproduction and distribution (e.g., from copies made by those viewing the information) and from the possibility of fundamental changes in the role of publishers if authors find they can reach their audiences via the Web with considerably less assistance from publishers.

The effect of the digitization of information on the economics of publication appears to be substantial, but not necessarily obvious (Box 1.4). Part of the phenomenon has been described by the observation that "shipping atoms is different from shipping bits" (Negroponte, 1995). To some degree this is true. In the print world, publishers incur substantial costs in manufacturing multiple tangible copies of copyrighted works—whether CDs or books—and shipping them to various points in a distribution chain. Publishers have to pay for storage and display of tangible copies; there also may be costs in remaindering or destroying extra copies if too

---

[21]As of February 1999, the Web was estimated to contain 800 million pages, holding some 6 terabytes of text (Lawrence and Giles, 1999). Even with automated tools helping to generate Web pages, 800 million pages require quite a few authors.

BOX 1.4
The Economics of Information

The principles of the economics of information provide a useful analytical framework for understanding the consequences of proposals to change intellectual property law and its enforcement. There are three central economic problems:

- Adequately compensating those who create new information products and services,
- Maximally disseminating and using the new information in the economy, and
- Determining the most valuable information products and services for production.

There are no perfect solutions; developing intellectual property policy inevitably involves trade-offs among these problems.

In some cases, the data needed to inform these trade-offs are not available. For example, there is little hard evidence about the quantitative significance of a given level of copyright protection. Also, little is known about the relationship between the strength of copyright protection and the amount of works produced (e.g., does stronger copyright protection lead to an increase in the amount of works produced?).

By contrast, there is a substantial body of literature from the 1950s and 1960s on the economics of innovation, focused on the role of patents in fostering change.[1] The work of Griliches, Mansfield, Scherer, Schmookler, and Williamson was especially notable in establishing the role of patents in fostering innovation and economic development.[2] Their work also examined how various institutional arrangements and industrial structures affected the rates of innovation and patenting. This body of work laid down a solid base of knowledge that supported the development of patent policies between 1960 and 1980 era. No comparable body of work exists with respect to the importance of copyright in fostering information creation and use.

See Appendix D for a discussion of information economics.

---

[1]See, for example, Kamien and Schwartz (1975).
[2]See the references in Kamien and Schwartz (1975).

many were printed. Because replicating digital products and shipping them via digital networks is cheap, rapid, and easy, the economics of digital publishing is different.

It is tempting to focus on the reproduction and distribution aspects of digital publishing where it is clear that there are very substantial reductions in the cost of publishing attributable to digitization. But the analysis doesn't end there; there is more to publishing than manufacturing, distribution, and storage. For example, many of the costs of publishing are the same whether the work is published digitally or in print form, including the costs of getting reviewers for the work, editing it, preparing it for publication, and, importantly, promoting the work once it gets published. Digital publishing also involves a number of new costs that may not be readily evident: costly hardware and software, retraining of publishing staff, retooling of the publication process, and an increased investment in the selling and customer support processes. In addition, it is relatively common for publishers to engage in dual publication, with one version of a work in print and one in digital form. This is necessarily more expensive than print publication only.

One final consequence of the Web as a publication medium arises from the routine encounters with international variations in laws and enforcement practices that result from the Web's global reach, an issue explored in more detail below. Sites containing illegal copies of music, for example, are quite popular and are found around the world, raising issues of jurisdiction and presenting great difficulties in enforcement. Files containing unauthorized copies of information can easily be moved from one computer to another elsewhere in the world, where laws are less stringent or enforcement is lax, yet the information remains as accessible (through the Internet) as if the host were down the block. Government activities, such as censorship and taxation, also become more difficult in the global environment of the Web.

Digital Distribution and the Changing Nature of Publication. The liberation of content from medium represented by digital information challenges many things we have come to assume about publication and copyrighted works, both economically and socially. Among these are the nature and character of publication. In the world of physical artifacts, the act of publication—the offering of a work for general sale—has three important characteristics: It is public, it is irrevocable, and it provides a fixed copy of the work; in the digital world none of these may be true.

In the world of physical artifacts, publication is fundamentally public and irrevocable. Once published in the traditional sense, copies of most works may be accessed from a library or other cultural institutions. Publication is irrevocable in the sense that once a work is published, it is virtually impossible to withdraw it completely from the public. Works may go out of print (no new copies made), but there is no notion of their being explicitly and overtly taken "out of publication" and hence becoming universally unavailable through withdrawing of all or most of the copies in existence.

The act of publication also implies a certain fixity of the work. Multiple physical copies are not easily modified, so those distributed represent an archival snapshot of the work at a particular moment in time.

Subsequent editions may be published, but each of them adds to the public record, producing a history of the evolution of the work. There is no plausible notion of being able to revise or expunge earlier versions.

Works in electronic form on the Web may have none of these characteristics. Posting does not make them irrevocably available; they can be removed from scrutiny at the pleasure of the rights holder.[22] Nor are they inherently public: Network access can be controlled to permit as restricted a distribution as the rights holder cares to enforce.[23] Nor are they fixed: Old versions are routinely overwritten with new ones, obliterating any historical record.

In a variety of circumstances such properties may be desirable and may permit some material (e.g., privately produced reports or business data) to be made available that would not have been published at all in the traditional manner. Restricted distribution of information is an important option. But, as discussed below, the widespread use of restricted distribution may also have undesirable consequences for public access to our intellectual heritage.

The Programmable Computer Makes a Difference

The computer is unique among electronic devices in being a programmable, general-purpose information processor. Any information it receives can be modified in virtually any manner. Other popular consumer electronic devices typically perform one or a few functions that are built into their hardware. A digital audiotape (DAT) player, for example, will play back the information on its tape and can make a first-generation copy, but the player cannot edit the information, redistribute it, or transcribe the words the information contains.[24] By contrast, a computer that receives audio information (e.g., an audio file) can play, record (and duplicate), edit, and redistribute (e.g., over the Internet) that information, and, with the right program, transcribe any words found in the signal.

The distinction between special-purpose devices like the DAT player and general-purpose computers has substantial significance for enforce-

---

[22]Although viewers might, of course, have made their own copies in the interim, this still does not make publication irrevocable, as the viewers may not make their copies accessible.

[23]Restricted distribution of information is not new, of course; it has been accomplished previously through both legal and economic means. The issue here is the ease and the precision (specifying who can have access) with which restricted distribution can be accomplished, making it a far more usable option.

[24]DAT players have a serial copy management system, an electronic mechanism that prevents a DAT player from making a copy of a copy. Hence, although you can copy a CD onto a DAT tape, the DAT player will not allow you to make a copy of that tape.

ment of intellectual property rights in the digital world. Although the behavior of special-purpose devices is restricted by design (the hardware of a DAT player, for example, prevents making a second- or higher-generation copy of a tape), it is, by contrast, very difficult to *limit* the behavior of a computer. A computer's behavior is changed easily by loading in a new program or modifying an existing program; as a result, it is very difficult to enforce limitations on what a computer can do.

The generality and flexibility of a computer's capability are extremely valuable; these attributes are basic to the information revolution and its success and cannot be casually removed. But they are also key factors in many of the current concerns about IP, given that information received by a computer can be manipulated and redistributed without practical limitations.

This issue is likely to grow more pressing in the future as the computer, the Internet, and consumer electronics technologies coverge, and the computer-like capabilities of these devices increase: For example, VCRs are beginning to get computation and storage capabilities that allow them to access and store information from the Web. The increasing presence of computer-like devices may herald significant benefits for users and the U.S. economy as a whole. From an IP perspective, however, it means an increasing proliferation of devices capable of replicating and distributing valuable digital information. It also means that when some type of technical protection is required, implementing a generic system that protects information across multiple-device platforms and applications may become increasingly difficult.

For example, encryption, either in software or specialized hardware, offers only a partial solution to the problem of protecting information, because much of the information distributed digitally must be comprehensible to the viewer.[25] Encryption can protect information on the way to and from the consumer and during storage. But documents must at some point be read, data in databases displayed, songs listened to, movies viewed, and so on. Simply put, users must be able to get at information, and once they have, making an electronic copy of it is often not difficult. One proposed solution—restricting the user's ability to get at and record "in-the-clear" information—may meet with resistance from users, because capturing and saving the output of a program are frequent and appropriate actions across a wide range of computing.

For some uses of digital information, special-purpose hardware will be appropriate and economical and can be designed so that consumers

---

[25]Encryption, which allows owners of digital works to scramble them so that they can only be unscrambled by legitimate users, is discussed in some detail in Chapter 5 and Appendix E.

will prefer its convenience over general-purpose computers (such devices have been called "information appliances" (Norman, 1998)). Encrypted cable television is the best known current example: DVDs, portable digital music players (e.g., the Rio), and portable digital book readers (e.g., the Rocket eBook) offer other examples. Each of these devices uses special-purpose hardware that makes it impractical and uneconomical for most people to capture information in ways that permit unauthorized use. For such applications, technical means of protection may be practical. For uses in which unencrypted text must be manipulated in general-purpose computers, however, information owners will have to look to other than technical means to constrain what is done with the information.

Technology Has Emerged into Everyday Life, Running Headlong into Intellectual Property

As argued above, three recent technological trends are key to the possible upset of the delicate balance of interests in intellectual property—digital information, networks, and the Web. A second factor challenging the balance arises from the transformation of the digital information infrastructure into a routine part of everyday life. In the United States, computers and the Web have become commonplace in work settings and are fast becoming a routine presence in households; what was once the province of corporations and research laboratories has become a broadly available capability. One important consequence is that ordinary citizens are now faced with questions involving the subtleties of intellectual property law, questions they are ill-prepared to answer.

For example, citizens are routinely finding that they have the means and the opportunity to access and copy vast amounts of digital information, including software, text, and audio and video material, but no clear picture of what is legal or acceptable. As another example, people frequently use more than one computer (e.g., at home and at work), raising the question of whether the one copy of a program they purchased can be installed on both machines. Similarly, is it permissible to reproduce an audio CD to have a copy in the car as well as at home?[26] What if only one of them is ever in use at once? The issue here is not what the answers are; it is that everyday life has been intruded on by what can be subtle questions of intellectual property law.

---

[26]The 9th Circuit decision in *Recording Industry Association of America v. Diamond Multimedia Systems* (F.3d (9th Cir. 1999)) gave recognition in passing to the notion of "space-shifting" of music for personal use (i.e., an individual making a copy of a legally owned musical work in order to use the copy in a different place).

Additional examples abound in using the Web, where one of the most telling features of the new electronic landscape is the uncertainty that currently pervades it. For example, as pointed out above, merely viewing a Web page may involve making copies of that page. Users rely on a currently implicit understanding that the page wouldn't have been placed on the Web for public access unless the rights holder permitted it to be viewed, giving at least tacit permission for the viewing and the incidental copying that accompanied the viewing. But is this truly the case?

And suppose you want to share that page or ensure you can revisit it later? Given the volatility of the Web, the best way to do this may be to make a copy for your own use—but can you legally do this? Whom do you ask? Who is the author? Is the author living or dead? When did the author die? Who are the heirs, and how do you get in touch with them? Suddenly, an attempt to ensure continued access to a source of information becomes very complex, costly, and daunting. These uncertainties become more acute as Web users become more informed about intellectual property issues and as they become more willing to try to obey the sometimes intricate law.

A second consequence of the emergence of the information infrastructure into everyday life is that individuals find themselves capable of reproducing vast amounts of information, in private, using commonplace, privately owned equipment. A single individual can now do in private what once would have required substantial commercial equipment and perhaps criminal intent. One important consequence is that copyright law is becoming more concerned with regulating private behavior of individuals.

Traditionally, copyright has concerned public actions with public consequences, such as public performance, public display, and dissemination of copies (an inherently public act), and has focused on actions of organizations or individuals (like pirates) whose actions have large-scale public consequences. But with computer and communication equipment becoming commonplace in the home, the potential impact of the private behavior of individuals has grown, and so correspondingly has interest in regulating that behavior.[27] This represents an important consequence of information technology's emergence into everyday life and presents another social and policy challenge in managing the IP balance.

---

[27]Some of that interest in regulation has resulted in the acceptance of a variety of private-use copies, as, for example, the time-shifting use of VCRs articulated in the Sony Betamax case (*Sony Corp. v. Universal Studios Inc.* 464 U.S. 417 (1984)), the Audio Home Recording Act's protection of private-use copies of music (embodied in Chapter 10 of the copyright law), and the 9th Circuit decision in *RIAA v. Diamond* noted above.

Intellectual Property Law Is Complex

Complying with IP law presents difficulties in part because of its complexity. Intellectual property law is a compendium of general principles (e.g., the exclusive right to reproduction), subtle distinctions (e.g., "idea" versus "expression"), and numerous special-case exceptions (e.g., the right to play background music, royalty free, at agricultural fairs). Copyright is complex partly because it deals with intangible rights in intangible subject matters, partly because it regulates the activities of a wide variety of industries, and partly because it reflects the results of hard-fought negotiations and industry-specific compromises. Much of the complexity of this law is pertinent only to the specific industry-to-industry dealings it addresses and is irrelevant to the general public. When corporations and lawyers were the ones routinely grappling with copyright laws, the complexity was a burden handled by a relatively specialized audience with appropriate skills and training. Now that the issues have emerged into the mainstream of daily life, the same complexities are being faced by people unprepared for them.

Music offers one illustration of the complexity of copyright laws that consumers routinely encounter. Here, for example, is information posted by the Recording Industry Association of America (RIAA) in an attempt to let people know what they can and cannot do. Even the RIAA admits the complexity:

> First, for your personal use, you can make analog copies of music. For instance, you can make analog cassette tape recordings of music from another analog cassette, or from a CD, or from the radio, or basically from any source. Essentially, all copying onto analog media is generally allowed.
>
> Second, again for your personal use, you can make some digital copies of music, depending on the type of digital recorder used. For example, digitally copying music is generally allowed with minidisc recorders, DAT recorders, digital cassette tape recorders, and some (but not all) compact disc recorders (or CD-R recorders). As a general rule for CD-Rs, if the CD-R recorder is a stand-alone machine designed to copy primarily audio, rather than data or video, then the copying is allowed. If the CD-R recorder is a computer component, or a computer peripheral device designed to be a multipurpose recorder (in other words, if it will record data and video, as well as audio), then copying is not allowed.[28]

The commonsense view of intellectual property often conflicts with what the law actually says, leaving even those who wish to act appropri-

---

[28]This information was obtained from the RIAA Web site at <http://www.riaa.com/tech/techht.htm>.

ately at a loss to know what to do. The subtle and sometimes difficult concept of "fair use" provides one compelling example. Fair use permits reproduction of limited amounts of copyrighted material for restricted purposes, such as review, analysis, and commentary. An appropriate analysis of fair use requires consideration of four factors (the purpose and character of the use, including whether such use is of a commercial nature or is for nonprofit educational purposes; the nature of the copyrighted work; the amount and substantiality of the portion used in relation to the copyrighted work as a whole; and the effect of the use upon the potential market for or value of the copyrighted work). Even experts are challenged by the assessment of the four factors.

Yet the concept is central to the situations that a thoughtful consumer routinely encounters. What about installing software on two computers, or copying a CD? How does the consumer begin to answer the question of the "nature" of the copyrighted work, or evaluate the impact on the market? If the consumer would have objectively decided that buying a second copy of a CD for the car was not worth the cost, does it follow that making a personal copy is permissible because it would have no market impact?

The subtlety of these considerations sometimes contrasts sharply with consumers' reasoning. A column on software piracy in a trade newspaper elicited this response from a reader:

> If I think the publisher's terms are unfair, I treat myself to what I think is fair use. I don't care if it's legal. I care if it's fair. Let's say, for example, that I buy a PC that comes with NT and one that comes with Windows 98. . . . I later decide that I want to run NT on the second one and Win98 on the first. . . . No chance [that in this case more licenses would be bought]. . . . I don't care if this violates the terms of the license agreement or the law. It's fair.[29]

This consumer, having heard of "fair use," attempts to understand it in the ordinary way (i.e., by dissecting the meaning of the individual terms in the phrase, apparently unaware that the phrase is simply a name for a more complex concept), a common behavior for people faced with unfamiliar concepts. The response also demonstrates the frustration in the ordinary consumer's view of the distinction that can arise between law and perceived fairness. The consumer's attitude here may not be admirable, but it is not unusual and will continue to be a problem as long as the subtleties of intellectual property law intrude on everyday life and as long as those subtleties require difficult judgments.

---

[29]See Foster (1998), p. 79.

These problems are not easily remedied in part because of the pace and continuity of technological change. Answers may arise over time in piecemeal fashion. For example, some software vendors have modified their licenses to indicate whether they permit installing a program on multiple computers owned by a single individual. But there is a considerable lag time between the issue arising and an answer being formulated and accepted. Given that we can expect more technological change, additional issues will always arise. The core problem is not the individual issues, but the complexity of copyright law and the lack of a set of principles that are clear and comprehensible enough for the average person to apply them to the next new issue.

Cyberspace Is an Odd New World

This new world of digital technology presents difficulties for lawmakers as well, largely because this world of information differs in some important ways from the world we have grown used to inhabiting. Cyberspace is an odd new world, where our ordinary intuitions are not always reliable. It is also an area where legislation is not always easy to craft. For example, cyberspace currently hides virtually all physical cues about a person, making identity difficult to establish (Figure 1.1). The lack of cues matters for a variety of reasons, including enforcement. In the physical world, restricting certain material to adults is enforceable in part because adults are typically easily distinguished from children, for example at newsstands. This is less so in cyberspace.

Cyberspace also blurs cues about social contexts. Certain behaviors—such as performing a play or copying a work under fair use—are accepted in classrooms and libraries because the social good is served by allowing those exemptions to copyright protection. In return, the behavior is limited to those contexts. But if cyberspace blurs the issue of where you "are," it may be more difficult to keep in mind which behaviors are appropriate to their current context.

Cyberspace is also unfamiliar because it permits action at a distance. Computers can be broken into from half a world away, for instance. This leads both to problems of jurisdiction (whose laws apply?) and to problems of social norms—feelings of personal responsibility fade when the victims of the crime are never seen or met and reside thousands of miles away.

Cyberspace is a world in which every (digital) product carries with it the possibility of an almost magical speed, ease, and precision of replication. Imagine a world in which you had a home appliance capable of reproducing every physical good you bought, allowing you to make as many of the item as you wanted. That's the computer in today's world of information goods.

"On the Internet, nobody knows you're a dog."

FIGURE 1.1 "On the Internet, nobody knows you're a dog." Reprinted by permission. © The New Yorker Collection. 1993. Peter Steiner from cartoonbank.com. All rights reserved.

Cyberspace poses difficulties for those who understand something about copyright because, for the most part, the status quo of the print world doesn't carry over into the digital world. As noted, it is impossible to make any use of a copyrighted work in digital form without also making a number of temporary copies. Suddenly, the right to control reproduction of works would seem to encompass the right to control even reading and browsing protected works. The disconnect between the print and digital worlds shows up as well in the argument that the first-sale rule doesn't apply in the digital environment, because in that world lending a work necessarily involves making a copy of it, not just redistributing that copy (which is, strictly speaking, all that this rule allows). The disconnect is also evident in the more extensive use of licensing of digital works, rather than sales of copies, even to libraries. To the extent that these licenses restrict the degree to which the information can be shared with others, users can't look to first-sale rights to share digital copies, as first-sale rights accrue only to owners of copies, not to licensees.

In time, the territory may become familiar, and easier means may be found to craft social and legal procedures for it, but for the moment there is substantial challenge in learning how to cope with this odd new world.

WHAT MAKES PROGRESS DIFFICULT?

Stakeholders' Interests Are Diverse

The debate over intellectual property includes almost everyone, from authors and publishers, to consumers (e.g., the reading, listening, and viewing public), to libraries and educational institutions, to governmental and standards bodies. Each of the stakeholders has a variety of concerns (see the addendum to this chapter) that are at times aligned with those of other stakeholders, and at other times opposed. An individual stakeholder may also play multiple roles with various concerns. At different times, a single individual may be an author, reader, consumer, teacher, or shareholder in publishing or entertainment companies; a member of an editorial board; or an officer of a scholarly society that relies on publishing for revenue. The dominant concern will depend on the part played at the moment.

The questions raised by the growing use of the information infrastructure are also difficult because they raise the possibility of redistribution of economic power. For example, given the ease of reproduction and modification, authors are concerned about losing control over works that they make available in digital form. Publishers are similarly concerned about the ease of reproduction and are attempting to work out models of publication appropriate for the digital world. How, for example, does one charge for content when, once made available, the content can be easily reproduced and further distributed?

One response involves use of the kinds of technical protection mechanisms described in Chapter 5. But some legal and public policy scholars look at the trends toward licensing and the use of technical protection mechanisms and see the possibility of adverse impact on public access to information.

Information consumers are perplexed. They find themselves in a world with both enormous opportunity and considerable complexity, faced at times with difficult questions about how to do the right thing. The massive ambiguity and lack of clarity concerning intellectual property rights make it difficult at times for citizens to honor these rights. There is a strong public policy good in retaining and encouraging the rule of law and respect for law, yet the digital revolution has confused the rules for managing intellectual property to an extent that threatens to upset the long-standing but delicate balance of competing interests.

There Is a Variety of Forces at Work

Intellectual property is typically conceived of first and foremost as a legal construct. But as Lessig (1999a,b) has argued, although law may constrain what we may (legally) do, equally powerful constraints arise from forces such as markets, social norms, and the values embedded in hardware and software. Markets put things within or out of our economic reach, social norms urge conformity with group values, and hardware and software encourage some behaviors and make others impossible.

The various forces differ in the explicitness of the values underlying them. Although laws generally result from a process that, in principle, is public and encourages examination of values and motivations, the same cannot be said of technology. The software written to control access to a Web site is a form of private regulation, and the process that created it rarely involves explicit discussion of the values embedded in and enforced by the program. Technical protection services (i.e., hardware and software used to protect IP) may offer content producers and distributors the important ability to manage access to their intellectual property, but those mechanisms may also enforce restrictions on the use of content that do not align with the (limited) rights of authors specified in copyright law. That law explicitly embraces certain public policy goals, but technical protection mechanisms and the policies they enforce may reflect a choice to overlook or to ignore those goals, with little or no opportunity for the public discussion or evaluation that goes into the creation of statutes.

The more general phenomenon here is the potential for substituting one force for another, and the consequences this can have for the degree of public participation in shaping society. To the extent that software is substituted for statute, for example, a form of privately created regulation is being used rather than publicly adopted laws. Such substitution should not be accepted by default, or permitted to go unexamined.

Many Threads Are Intertwined: Technology, Law, Economics, Psychology and Sociology, and Public Policy

The issues are also complex because of the variety of threads intertwined in them, each thread representing a different perspective from which to analyze the problems and evaluate possible solutions. The committee believes that five threads are particularly important: technology, the law, economics, psychology and sociology, and public policy. Each brings a particular conception of and approach to analyzing the problems. As the committee has attempted to employ the perspective, mindset, and vocabulary of each of these viewpoints at various points in this report, a brief characterization of each will prove useful in understanding the remainder of the report.

Consider the issue of someone who wishes to duplicate a copyrighted software program for a friend. This issue will be conceived of differently from each of the five perspectives. For a technologist working for the software vendor, the question is how easy or difficult such reproduction would be (e.g., what technical mechanisms are available to inhibit copying and how easily can they be deployed?). The discussion would consider questions of strength of protection, complexity of development, reliability, and so on.

A lawyer would ask whether the action conforms to the law as it is currently written. Reference may be made to recent cases or the legislative history (e.g., the record of congressional debate when the law was being developed), and there may be appeals to analogies and past precedents, but the focus is on the law as it exists, rather than what it might or ought to say.

For the economist the question typically involves costs and benefits, economic efficiency, and trade-offs. For example, in the economic view, protecting intellectual property is worthwhile to the degree that the net total benefits exceed the cost of protection. The benefits are measured in terms of the societal impact of the IP, determined by the production of new IP information, its value and so on, while costs are those incurred by all parties: IP creators, government (e.g., police and courts), and consumers (including time required to cope with protection measures). In this view, additional IP protection or enforcement is appropriate only if it leads to enough new and valuable IP. This contrasts with the legal and property view, which sees protection and enforcement as an issue of property and ownership rights.

A second, sometimes counterintuitive economic concept used in this report is the notion of economic efficiency. Economically efficient production and distribution require that all consumers who are willing to pay the marginal cost of production and distribution be able to obtain the product. In the case of an information good, the costs of producing the first copy (i.e., cost of its original creation) do not vary with the number of users, so economic efficiency is served when all consumers willing to pay the (typically very low) cost of reproduction and distribution have received the product. This is not to the benefit of the individual rights holder, but it is optimal from an economic cost-benefit perspective for society as a whole (though rights holders do of course need sufficient revenues to recover first-copy costs and costs associated with reproduction and distribution, in order to be willing to create and distribute the works to begin with).

The psychological and sociological view concerns individual behavior, grounded in perceptions of fairness and responsibility, fear, shame, guilt, convenience, and pragmatism, and the perception of the individual as beneficiary, victim, or patron in transactions involving intellectual property. This is perhaps the least well understood factor in developing policy, yet it may be one of the most critical because it affects appraisals of the enforceability of different options. Little is known about what the typical individual believes about IP and how he or she perceives the relationship of actions to broader social and economic interests. In addition, little is understood about the social norms surrounding the use of IP.

For the public policy analyst the question concerns what social goal society is trying to accomplish (e.g., is it better for society as a whole for such copying to be allowed or prohibited?). Here one would include a discussion of what the law ought to be in order to support important policy objectives (e.g., promoting national economic competitiveness). The policy analyst uses the knowledge base and techniques of the economist, lawyer, and social scientist and operates in a multidisciplinary manner.

Clearly, the richness of each of these perspectives can only be hinted at here. The important point is the multiplicity of perspectives that are relevant to the issues and the rather different mind-set, vocabulary, and analytic approach that each offers.

The Problems Are Global, with Differing Views, Laws, and Enforcement Around the World

For the most part this report focuses on circumstances and possible actions within the United States. The conclusions and recommendations must, however, be considered in a worldwide context. Laws and intellectual property practice differ by country and will likely remain different despite efforts at harmonization. Cultural attitudes toward intellectual property and the premises on which law is founded may also differ. As one example, European IP law incorporates the concept of "moral rights" in a work and gives to the author a stronger degree of control than exists in U.S. law, and one that persists beyond the sale of a copy of the work (see Box 1.5). U.S. law and culture also treat intellectual works as property, but this is not universal either (see Box 1.6).

Attitudes toward intellectual property may also depend on perceptions of national interest: Just as in other domains, where IP is concerned what is in one country's economic interest may be inimical to another's. Those interests may also vary over time. In the 19th century, for example, Charles Dickens' serials were widely republished in U.S. newspapers without payment of royalties to English publishers. The change over time in U.S. attitudes toward the IP laws of other countries is widely ascribed to the change in U.S. status in the 20th century to a major producer of IP content (Warner, 1999).

National variations in law and enforcement matter for several reasons. First, IP laws and attitudes obviously have consequences that go beyond national borders, as for example, the U.S. ultimatum to China in 1994 threatening to brand it a copyright pirate and initiate a "Special 301" investigation (a government action that can lead to trade sanctions) unless China cracked down on copyright violations leading to pirated software, CDs, and movies intended for sale in the United States.

Second, the existence of networks with international scope makes the issue of international variation a matter the average user can face daily. Even the unsophisticated computer user can access the information resources of countries around the world without leaving the room. As one consequence, practices required or prohibited in one country may be circumvented by actions taken, over the Internet, in another country.

Third, jurisdictional problems arise in enforcement of laws as cyberspace blurs the concept of the location of an action, both internationally and within the United States. For example, in 1994 two operators of a computer bulletin board in Milpitas, California, were arrested for (among other things) distribution of obscene material, on the basis of an indictment made by a grand jury in Tennessee. This, gave rise to the question of what constitutes "community standards" when the geographical basis for the community is blurred by cyberspace.[30]

The problems are difficult in part because the influence of international data networks is not easily controlled on a national basis. The Web is inherently international and cannot be divided by national lines. One difficulty comes from trying to determine the origin of an information request. When a Web server (the computer providing information) gets a request for some of its information, that request contains the Internet

---

[30]Doyle (1994), p. A12; see also *United States v. Thomas*, F. App. 0032P (6th Cir. 1996).

BOX 1.5
Moral Rights

Many so-called "civil law" countries (e.g., much of Europe) have in their intellectual property law a notion that the creator of an artistic work (and potentially his or her progeny) has an inalienable (i.e., irrevocable) right—variously called a moral right, *droit morale,* or *droits d'auteur.* This right protects both the artistic integrity of the work and the artist's interests against the unauthorized modification or desecration of the work, which would damage the author's reputation. The concept has its origins in the aftermath of the French Revolution, when the tradition arose that "the most sacred and legitimate, the most unchallengeable and personal of all the properties is the oeuvre, the fruits of a writer's thought" (Holderness, 1998). Moral rights are intended to protect an author's name, reputation, and work; these things are seen as integral to the very act of creation, which is why they are regarded as perpetual and irrevocable as long as the work exists.

As a consequence, the right survives any sale of the work: Article 6 *bis* of the Berne Convention indicates: "Independently of the author's economic rights, and even after the transfer of the said rights, the author shall have the right to claim authorship of the work and to object to any distortion, mutilation or other modification of, or other derogatory action in relation to, the said work, which would be prejudicial to his honor or reputation."

The U.S. and British tradition sees works of authorship differently, viewing them as commodities to be freely traded, under the control of whatever person (or corporation) holds current title to it. In consequence, U.S. law allows authors a broad freedom of the right to contract, including the right of an author to divest himself or herself of what would otherwise constitute such moral rights. This is seen frequently in the entertainment industry, where the "all rights" contract in routine use grants the purchaser of the rights to a work the right to adapt, change, modify, add to, subtract from, satirize or otherwise change the creative work in any manner that the purchaser may choose.

In order to qualify for membership in the Berne Copyright Convention in 1989, the United States had to make the case that its local law honored the notion of moral rights; it did so by pointing to a number of laws that create rights similar to moral rights. For example, Section 43A of the federal Lanham Act creates a federal law of unfair competition, part of which prohibits inaccurate descriptions of goods and their origin. A number of U.S. cases have cited this law in finding a violation of author's rights, when, for example, a television program was edited by someone other than its creator, to an extent that so fundamentally changed the nature of the original work that it was a misdescription of origin to continue to exhibit the edited work with the author's credit (see, for example, *Gilliam v. ABC* (538 F.2d 14, 192 U.S.P.Q. 1 (2d Cir. 1976)). The misattribution of credit on a motion picture has been held to be actionable under the same doctrine (see, for example, *Smith v. Montoro* (648 F.2d 602,211 U.S.P.Q. 775 (9th Cir. 1981)).

In 1990 the United States passed the Visual Artists Rights Act, embodied in section 106A of the Copyright Law, giving the author of a work of visual art certain rights of attribution and integrity (e.g., the right to claim authorship of the work, the right to prevent the use of his or her name as the author of any work of visual art which he or she did not create; the right to prevent any intentional distortion, mutilation, or other modification of a work which would be prejudicial to honor or reputation).[1] There are also specific statutes in a number of jurisdictions that protect the creative rights of artists and sculptors against the unauthorized desecration or modification of their works in the hands of subsequent purchasers (see, for example, California Civil Code Section 987).

Despite these provisions, the notion of moral rights is clearly not an established element in U.S. legal tradition, illustrating how even the basic premises for IP law can vary internationally in fundamental ways.

---

[1]Because these rights do not cover works for hire and may be waived, one commentator says of them,"A more grudgingly tokenistic implementation of moral rights is difficult to imagine" (Holderness, 1998).

BOX 1.6
A Copyright Tradition in China?

China offers a second example of how copyright tradition differs globally. Michael Oksenberg, a Stanford University professor of political science and China scholar, commenting on this study, suggested that in China and most Asian cultures, the concept of intellectual property in creative expression is completely foreign, and to some sacrilegious.[1] Acts of individual expression are seen as rooted in the contributions of ancestors and are "in the air"—the person expressing them is simply doing what is in the wind. What is regarded in the United States as an individual act of creativity would be regarded there as one person playing the role of scribe for ancestors and other contemporaries. As a consequence, one who expresses an idea has no right to it—it is a social expression, not an individual one, and part of the process of passing and extending a society's cultural legacy. He believes that it is not realistic to expect China to pay more than lip service to IP rights for the foreseeable future, independent of the political system in place there.[2]

By contrast, Professor Guo Shoukang of The People's University of China, who wrote the chapter on China in Paul Geller's edition of Melville Nimmer's *International Copyright Law and Practice* (Geller, 1993), noted that the Chinese had developed printing "technologies at least as far back as the Tang Dynasty, that is, between 704 and 751 A.D.," and that "legal prohibitions against the unauthorized printing of books followed closely after the invention of printing." Professor Guo also points to a number of instances of more recent vintage in the 13th century of restrictions on reprints and attempts to stop piracy.

---

[1]See, for example, Alford (1995), which concludes that China never had a copyright law or tradition.

[2]Professor Oksenberg also disagreed with the view, held by many economists, that the propensity to adhere to IP rights is a function of the degree of development of a country. This view is based on the observation that poor countries are generally consumers of knowledge but not significant producers. He suggested that this may be true for European offshoot countries (e.g., Latin America), but not Asia. His alternative view is that as a poorer country becomes richer, Western countries have a greater incentive to enforce their IP rights, and put more pressure on the emerging country to comply, and it is this pressure that leads to increased adherence.

address of the requesting computer, and from this the requesting computer's physical location can sometimes be determined.[31] But it is not difficult to keep anonymous the requesting computer's identity and location,[32] so the server may be unable to determine where the request is coming from and hence could not know whose laws were relevant to the request. Of course, even if the server could reliably determine the source of the request, it is unclear why the server should be used to enforce the laws of another country (see Box 1.7).

Equally daunting problems arise when information is "pushed" across a border. There are simply too many routes and too much data flowing for a nation to police its information borders effectively. As a result, attempts by a country to enforce national laws on "its part" of the Internet are difficult, except perhaps where they concern the rights and responsibilities of its own citizens.[33] For better or worse, we are inextricably interconnected and all must deal with the difficulties—and opportunities—that such interconnection brings.

Potential Solutions Have to Be Evaluated from a Variety of Perspectives

The multiplicity of forces at work in these issues have to be kept in mind when evaluating solutions. For example, a technical mechanism may seem promising at a first glance, but later turn out to be intolerably awkward for the average user (as were, for example, early attempts at copy protection for software). Similarly, it is easy to suggest changes in the law, but markets can exert powerful forces that defeat the intent of a law (as seen, for example, in the experience of some cities that instituted rent control to preserve low-cost housing, only to discover that this prompted conversions to condominiums, resulting in a reduced stock of rental housing).

The five perspectives outlined above are useful to keep in mind. Suggested solutions should be technologically feasible (e.g., for a software solution, the desired program will require a plausible amount of time and hardware to run). Solutions have to be evaluated with respect to the specifics of the law as it currently exists, the legislative history and intent

---

[31]The network address is just that, a network address, not a physical location. Although the two are often correlated, there is no guarantee. The computer in question may be physically somewhere other than its Internet address indicates, particularly in the case of laptops connected to the Internet via a phone line.

[32]See, for example, <http://www.anonymizer.com>, a site that offers this service.

[33]As, for example, the Digital Millennium Copyright Act's limitation of liability for Internet service providers.

BOX 1.7
A Server as a Law Enforcer?

Should a server in one country be in charge of enforcing the laws of another country? The question presents a variety of interesting issues. Two examples illustrate the conflicting values that may come to bear on this issue. Consider a nation like Singapore, which is quite concerned about its citizens receiving content via the Internet that it considers undesirable, ranging from political ideas deemed unacceptable by the government, to pornography and "undesirable" Western influences in fashion and music.[1] Should Web servers in the United States help enforce this policy by restricting what they will send to Singapore Web addresses? Given the long tradition here of freedom of speech and expression, the reflexive answer is no. Clearly, U.S. Web servers should not be used to enforce the laws of another country.

On the other hand, consider the existence of Web servers located in nations whose local laws lack any respect for intellectual property rights, or even affirmatively encourage their citizenry to pirate works of the sort protected under U.S. intellectual property laws. Should their Web servers be permitted to enable individuals in the United States to freely access, download and copy anything they wanted, without regard to its U.S. copyright status? Many would say no, but would we not be asking the Web servers of other nations to enforce our laws? Perhaps so.

This second example is more than speculative. There is global interest in fostering cooperation among nations under the auspices of the Berne Convention, the Universal Copyright Convention, and the World Intellectual Property Organization, in an attempt to bring the laws of all nations into general conformity with protection of intellectual property rights. The enforcement of copyright, patent, and trademark law as it exists in United States and other developed nations could be severely undermined if an individual could freely download content from such "offshore" servers.

So unless, and until, there is global uniformity in IP laws, the question will be with us: Should a Web server help to enforce the laws of another country? The answer may depend on whose values are at stake.

---

[1]See, for example, Einhorn (1999), referring to Singapore's requirement that Internet service providers block access to certain sites, and Chapman (1999), "The Internet is tightly controlled here—the City's three Internet service providers all filter network content as instructed, and the schools get even further filtered material." Other countries that have passed legislation to attempt to filter content on the Internet include China, Saudi Arabia, and Australia (see Finlayson, 1999).

(how it got that way), and some understanding of legal context (i.e., how the statute fits into the collection of laws). Solutions have to make economic sense as well. Consideration should be given not only to the costs and benefits but also to who pays the costs and who derives the benefits. Solutions should take account of psychology and sociology; they must ultimately be viewed as fair and pragmatic by the majority of citizens. Finally, solutions should take account of public policy goals, as embodied in, for example, copyright law, antitrust law, and foreign policy.

Given the complexity of the challenges for intellectual property presented by the information infrastructure, and the variety of perspectives that should inform evaluation of any proposed solution, it is certain that any proposed solution will inevitably be imperfect in some manner. The committee believes that it is important, when exploring both the problems and potential remedies, to not just point out that the solution is imperfect; that's a given, and far too easy. The more interesting approach, indeed, the more effective mind-set, is to ask questions such as, Is it good enough? Will it do enough of the task to be worth the cost and effort? How can it be improved? It is in that spirit and mind-set that this report proceeds.

ROAD MAP FOR THE REPORT

The issues described in this report are difficult and contentious, because the stakes are high and the needs and desires of various stakeholders often are in conflict. A more detailed discussion of the stakeholders and their concerns is the subject of the addendum to this chapter.

Chapter 2 addresses many of the pertinent issues of digital intellectual property within the context of digital music. The digital music phenomenon is worthy of discussion both in itself and for what it may portend for other information industries. This case study also serves as a means of introducing topics (such as technical protection services) in a context that will be familiar to many readers.

Chapters 3, 4, and 5 present detailed discussions of the major issues. Chapter 3 considers implications for public access and archiving of the social and cultural heritage, addressing the major effects the information infrastructure has on intellectual property and society. Chapter 4 includes an analysis of individuals and their understanding and behavior with respect to digital intellectual property and addresses specifically the fair-use/private-use issue. Chapter 4 also addresses the key issue of whether "copy" is still an appropriate basis for the protection of digital intellectual property. Chapter 5 describes the technical means available for protecting intellectual property, explores how business models can be used in collaboration with (or at times in place of) technical protection, describes how we might begin to measure the success of using either or both of these approaches to protection, and discusses the impact of granting patents for information inventions.

Chapter 6 presents the conclusions and recommendations of the study. Although some conclusions and recommendations are quite specific, the majority are not. Part of the difficulty in proposing specific solutions is the inherent complexity of the problems; part of it is the rapid pace of technology evolution. It would be imprudent to base legislation or public policy on any particular technology. Instead, the committee attempts to provide guidance at a more general, strategic level. The committee made a concerted effort to make accessible what one needs to know about technology, law, economics, and sociology. And the committee has attempted to provide a clear, objective, and insightful framework in which to address the issues, to help make fruitful the vigorous debate that must occur among all the stakeholders.

ADDENDUM: THE CONCERNS OF STAKEHOLDERS

The issues surrounding digital intellectual property addressed in this report derive much of their complexity from the varied nature of the stakeholders and the wide range of their concerns. This addendum provides additional background on the stakeholders to better illuminate IP issues. Given this report's emphasis on IP, other stakeholder concerns relating to digital information production, distribution, and use are not described in detail. Within each of the broad classes of stakeholders, both coinciding and quite different interests are evinced.

Creators of Intellectual Property

The creators of intellectual property are a heterogeneous group. They range from corporate entities driven largely by economic motivations to individual artists and authors who may create for any number of reasons, including economic gain, prestige, or the desire to share what they do with peers and the public. Categories may of course overlap: Individual creators of IP may be employed in corporate or academic organizations that constrain ownership or use of intellectual property as a condition of employment. Some content creators are entrepreneurs, while others are members of the research community.[34] Notwithstanding their differences, IP producers have several interests in common: a concern with some control over the disposition and dissemination of their work (whether for economic gain or intellectual credit), a concern with the accurate attribution of authorship, concern with the integrity or fidelity of the work that may be associated with their names and reputations, and

---

[34]See "The Research Community" section in this addendum for a discussion of the specific concerns of researchers.

concerns about derivative works (which relate to the unauthorized distribution and reproduction or modification of what they create).

Generally, content creators want the ability to continue to use or otherwise benefit from their work and/or protect its integrity in teaching, in communication with their peers, or in the creation of subsequent works—even if they had to relinquish some or most of the rights to the work as a condition of publication.[35] The creators of some kinds of work may benefit from cheaper distribution methods that could increase their control over their work, as some have begun to do (see Box 1.8). For writers and others continuing to rely on conventional publishing outlets, author control has been diminishing over time. Prior to the early 1990s, the typical contract, particularly in traditional print media (e.g., periodicals), was for first-time print publication only. A work was generally published for use only in a print format, often only in a single publication, and typically the publication was limited geographically (e.g., the norm was first North American serial rights). This has changed, as authors are increasingly expected to transfer all rights in perpetuity to their works without additional compensation (Kaminer, 1997; Manly, 1997), and in some cases, an author's work has been used without the formal transfer of digital rights (Stone, 1998; Phipps, 1998). Where there are few choices among publishers (e.g., specialized journals), publishers are likely to control terms; authors have more negotiating leverage where there is a choice of publishers. (See Box 1.9.)

Much creativity builds on material that already exists, a process protected by three cornerstones of intellectual property law: the public domain, which provides a rich source of materials that can be reexamined, recontextualized, "repurposed," and reinterpreted; fair use (including the rights under fair use that allow a creator to quote for a variety of specified purposes); and limitations on what can be protected by intellectual property rights, in particular, exclusions of ideas, facts, and mathematical algorithms. Increasingly, issues arise in protecting likenesses, characters, imaginary worlds, and, perhaps in the future, factual databases.

Follow-on creation challenges for intellectual property protection are epitomized by digital multimedia works, which bring together visual images, sound recordings, text, programs, and other materials, rarely created completely *de novo*, or out of original material. Instead, they often weave together elements of previous works in a novel arrangement; some-

---

[35]The concept of a continuing connection with creations for which rights have been transferred relates to so-called moral rights—a concept honored in European intellectual property law though relatively new in the United States. See Box 1.5 above in this chapter for a discussion of moral rights.

BOX 1.8
The Possibilities for Self-Publishing

Self-publishing[1] is hardly a new idea, but the widespread use of the Web provides many new opportunities. Examples are easily found of authors self-publishing their work online. For example, Matt Drudge, creator of the Drudge Report, began mailing news items to a small circle of Internet subscribers in 1996, initially charging a modest sum for access. By 1998 Drudge had successfully established himself as one of the most read and discussed journalists of the 1990s. Today, he hosts his own television talk show. The Web facilitates self-publishing of all kinds of media, not just text.

Well-established creators, such as the musician called The Artist Formerly Known as Prince and John Kricfalusi, the creator of Ren & Stimpy (cartoon characters), have taken to the Internet exclusively as a new venue to promote and distribute their work. Although many other artists, including the musical group U2 and pop music celebrity David Bowie, have experimented with prereleasing work over the Internet, The Artist and Kricfalusi are notable insofar as they explicitly turned to digital network distribution in frustration with what they saw as intolerable terms in the contractual status quo of the music and cable television industries, respectively. While both met with only modest financial success in their first online projects, each was able to demonstrate that a relationship could be forged directly with consumers in a manner that hints at the possibility of a viable economic model.

Like the many Internet software companies that have made a practice of releasing free software online and charging for related products and services, some creators have taken to publishing content without charge online and looked for ancillary revenues elsewhere. For the first few years of its existence, Amazon.com's best-selling title was a guidebook on Web site production, David Siegel's *Creating Killer Web Sites* (1996), that both reiterated and was promoted by material supplied by the author for free on his popular Web site. Similarly, Tom and David Gardner—the brothers who founded the online Motley Fool investment site—have proven quite successful in reaping ancillary revenues in books and merchandising through promotion on their free online Web site.

The Internet facilitates low-cost distribution, but it does not necessarily easily attract desired audiences. The ease of distribution means that the challenge of the Internet is to become noticed among many sites competing for attention. These examples notwithstanding, there will likely continue to be an advantage to high-end marketing efforts—underwritten with significant capital—that will remain out of the reach of most authors.

---

[1]In 1999, self-publishing and music attracted a lot of attention; see Chapter 2 for a discussion.

BOX 1.9
The Increasing Use of Broad Contracts

With the advent of the digital era, nearly all contracts take the form of a broad license to use a first-time print publication work in a wide array of electronic formats. The most comprehensive of the new contracts have been all-rights and work-for-hire contracts. An "all-rights" contract implicitly argues that the writer owned the copyright when the work was created and is now licensing its entire use away, whereas under "work for hire," the employer, from a legal standpoint, is considered the original creator of the work. From an economic standpoint, the contracts have similar effects as the creators cannot profit from the exploitation of the works they create once either of these contracts is signed. The problematic issue is not the use of contracting per se, but that additional rights are often transferred for little or no increase in compensation.

The *New York Times* issued a "work-for-hire" agreement in the summer of 1995, which decrees that all articles will be "works made for hire and that, as such, the *New York Times* shall own all rights, including copyright, of your article. As works made for hire, your articles may be reused by the *New York Times* with no extra payment made to you." The *Boston Globe*, after being purchased by The New York Times Company, adopted virtually the same language in May 1996.

More typical of the "all-rights" language is the *Philadelphia Inquirer's* contract, which states that the grant of publication rights "shall include the right to publish the material; to create derivative works; to use, adapt, modify, perform, transmit or reproduce such material and derivatives in any form or medium whether now or hereafter known throughout the world including without limitation, compilations, microfilm, library databases, videotext, computer databases and CD-ROMs." This language was effectively adopted by all of the *Inquirer's* sister newspapers of the Knight-Ridder newspaper chain.

In an attempt to control their rights in the digital age, authors have sought legal assistance. In August 1997, a federal judge ruled in *Tasini et al. v. the New York Times et al.* against publishers' claims that authors had expressly transferred the rights to their articles through contracts, oral or written, or through check legends stamped on the back of publishers' checks.[1] The judge also rejected the publishers' claims that granting rights to first publication automatically extended to other media, and that electronic use was simply an archive.

However, the judge ultimately ruled that publishers could use authors' work under the concept of a "revision" in the meaning of the U.S. Copyright Act. Under Section 201(c) of the Copyright Act, someone who puts together a "collective work" can reproduce or revise it. The ruling found that putting a work online or on a CD-ROM is simply a revision of the original print version of, for example, the *New York Times* because these reproductions include precisely the same selections as the original print version. In September 1999, this ruling was overturned by the U.S. Court of Appeals for the 2nd Circuit (Hamblett, 1999).

---

[1]See <http://www.jmls.edu/cyber/cases/tasini1.html>.

times the term "repurposing" is used to describe some of this activity. The work of the creator concerns selection, arrangement, and linkage as much as in the creation of wholly original content. The use of works still under copyright requires authorization from the rights holder, making a wide range of multimedia works economically impractical because of the costs of clearing rights. There are two components to this cost. The first is the overhead cost of determining whether the existing material is copyrighted and who holds the copyright and then negotiating with that entity. The second component is the actual payment that must be made to the rights holder once he or she has been identified and the license negotiated.[36] Developments such as rights clearinghouses, stock photo archives, and the like may greatly reduce the overhead costs (the cost of the clearinghouse operations can result in lower royalty rates to the artist). Because a considerable amount of material remains protected by copyright, there is a significant amount of work that cannot readily be reused because the overhead costs of clearing rights so greatly exceeds the amount that would be paid for its use. The inclusion of copyright management information with digital objects may facilitate the tasks of rights clearance and payment.

Distributors

There are many kinds of distributors of intellectual property:

• A wide variety of publishers, including mass media (newspapers, magazines, and so on), entertainment enterprises (film and television distributors, music publishers, and performance organizers), think tanks with their own employees and relationships with creators based elsewhere (e.g., the Brookings Institution, Cato Institute, and the National Bureau of Economic Research), scholarly journal publishers (commercial, nonprofit societies, universities, and so on), and the government (e.g., the U.S. Government Printing Office). Some publishers are content producers as well.

• Aggregators that bring together access to the IP products of others for resale, for example, book and music stores, video rental stores, and catalog merchants of packaged software.

• Services that organize information conveniently for the purpose of offering that package of information to a public audience, ranging from

---

[36]The first component is a measure of the inefficiency of the current system of managing IP rights; the second component represents a measure of the real value of the existing content. The unfortunate reality today is that the overhead costs alone in many cases are large enough to discourage projects from being undertaken.

such services as Lexis-Nexis, Yahoo, or MSNBC, to libraries (both hard copy or electronic) and link-filled Web sites.

In addition to the general issues that creators of intellectual property are concerned with, content distributors are concerned about the following issues:

- *Economic viability.* Most distributors are in business: They are concerned about developing a framework that permits them to be economically viable. Digital technology presents additional challenges to the business plans of distributors; if it promotes disintermediation, for example, it is reasonable to expect a change in the number, identity, and size of distributors.
- *Legal uncertainty.* Many of the legal issues surrounding content in the information infrastructure are unclear, and they are being settled slowly and haphazardly through case law. The international nature of networked access generates additional concerns relating to jurisdiction and conflict of laws that raise questions about international legal consistency and international enforcement of intellectual property rights. This type of uncertainty is very dangerous and discouraging to distributors and is reflected in pricing that incorporates a "risk premium."
  In particular, distributors are concerned about the liability that they may incur in distributing intellectual property. The nature of rights associated with intellectual property influences liability. For example, have they been transferred appropriately to enable the intended distribution to proceed lawfully? Distributors are concerned about incurring additional obligations in return for their IP rights. These obligations include permanent access or archiving responsibility, requirements for use tracking, maintenance of the confidentiality of use, censorship, and so on. Finally, distributors may be concerned about liability associated with the content of intellectual property (e.g., the risk of libel or defamation charges).
- *The boundaries of fair use.* Copyright distributors and certain categories of users differ over what kinds of use are fair uses of digital works, with distributors tending toward a more limited interpretation.
- *The boundaries of derivative works.* Derivative works may have a substantial effect on the ability of distributors to distribute their content. As one example, "framing" takes a page from one Web site and surrounds it with other content in another Web site, often obliterating the identity of the original site and blocking out advertising that was placed on the original site. This seriously concerns publishers or distributors who mount the original site as well as to those who want the right to "frame."
- *Protection of trademarks and related brand and corporate identification.*

Because content distributors are often the entities providing the "branding" of the distributed content, they have special interest in protecting their brand. There are stakeholder issues because of the relatively limited availability of Internet domain names as compared to the number of trademarks, as well as the question of whether holders of trademarks have the right to insist on using certain domain names by virtue of the ownership of their trademark.

• *Efficient management of the rights process.* Content distributors frequently bring together the rights of many parties. Although permissions and licenses may be governed by standard processes designed to cover large numbers of content creators under a variety of circumstances (e.g., publishers' policies, standard contract terms and conditions), efficient processing of payments is the distributors' responsibility.

Publishers have been moving to secure more control over the works they publish, given the severe uncertainty about the impact of digital media on revenue, including the effective value of existing content to which they hold rights, potential losses from experiments with new media, consolidation in media industries, and so on. It is generally understood that few publishers can accept agreements covering only print distribution, because digital distribution is increasingly a standard and essential part of their business. However, to the degree that digital rights are sought solely to protect a future option (not a present product), pricing is difficult. It is difficult for both parties to place a value on future rights or future exploitation, as relatively few works will achieve classic or best-seller status, and discounting of future value is as appropriate for information as for other investments.

Many media distributors have argued that the global exploitation of content and the swiftness with which such exploitation must take place in the marketplace requires that companies obtain broad assignment of rights from authors. They argue that it is simply too costly to find an author in order to obtain subsequent rights. One solution to this concern is the use of rights clearinghouses and the ability of digital networks to facilitate rights management and clearances. This would require the creation of digital-age versions of the American Society for Composers, Authors and Publishers (ASCAP) or Broadcast Music, Inc. (BMI)—clearinghouse-type structures for handling author payments and licenses.[37]

The concept is quite simple. An author would assign his or her rights to the licensing entity (or designate the licensing entity as an agent), which

---

[37]Information about ASCAP is available at <http://www.ascap.com>; information about BMI is at <http://www.bmi.com>.

could then grant, in turn, the right to any user—individual or corporate—to use a piece of content for an agreed-upon price, based on the duration and type of use. Technology allows speedy presentation of rights options and price: A page at the clearinghouse's Web site would indicate what an article would cost to purchase for a particular use and allow the user to click on an icon to complete the purchase and receive the content. The system might also allow the user to contact the owner of a work to conduct a direct negotiation.

Two relatively new author-operated models exist in the United States: the Publication Rights Clearinghouse (PRC), a project of the National Writers Union; and the Media Photographers Copyright Agency (MPCA), a project of the American Society of Media Photographers.[38] Author-operated models have also been established abroad, including The Electronic Rights Licensing Agency in Canada and the Authors Licensing and Collecting Society in the United Kingdom.

Schools and Libraries

Historically, schools and libraries have played two special roles with regard to public access to intellectual property. They serve as custodians for the cultural record, and they make intensive and integral use of a wide range of intellectual property in the course of teaching and supporting scholarship. Libraries and schools also share a common tradition of concern for free expression, free inquiry, academic freedom, and unfettered discourse, including freedom of research, commentary, and criticism, all based on a foundation of accurate and authentic information. Schools and libraries are at the center of tensions among competing policy goals relating to intellectual property rights, privacy, and free speech—all of which affect who has access to information and under what conditions. Schools and libraries are also vulnerable to changes in long-standing practices and expectations resulting from broad shifts in the production, distribution, access, and use of information occasioned by networking.

Schools care about the following issues:

• *Fair use.* Educators need to know in direct and simple language under what terms or conditions portions of copyrighted works can be used in classroom presentations, course packs, and electronic reserves. Does fair use differ if the source is digital rather than hard copy or if the

---

[38]See <http://www.nwu.org/prc/prchome.htm> for the PRC, and <http://www.mpca.com> for the MPCA.

distribution is digital? For example, should copying a paper for distribution to students be viewed differently from posting an electronic copy of the paper on a Web site? Do school licenses constrain student and/or faculty fair use in specific ways? What are the limits of fair use in a context in which course-related information can be distributed to networked students anywhere in the world? Fair use relates both to how educators obtain and share information and how they convey the process and the ethics of using intellectual property—the nature of "fairness."

• *Rights management procedures.* Educators would like to adapt what they teach to changing circumstances and discovery of relevant materials. It would help them if requirements for clearance of rights were not time and process intensive.

• *Educational use of research results.* Educators who have produced work protected by copyright (and, in particular, works for which they have transferred copyright to others) want the ability to use those works in their classrooms. Conditions limiting such use have been imposed by some conventional journal publishers. The strict copyright transfer policies of some academic publishers, with few rights retained by authors, have caused a backlash: Scientists in some disciplines have begun to develop their own refereed online publications or to support systems to deposit papers on public servers.

Libraries care about the following issues:[39]

• *Archiving and preservation of the public record and cultural heritage.* Libraries have an interest in long-term access to information; part of their mission is to act as agents on behalf of future generations of students and researchers. How is digital archiving to be organized and what rights does the library and archival community have pertaining to the archiving of copyrighted works?[40]

• *Fair use.* What is fair use in the sharing of digital resources with other libraries or individuals? What does a "loan" mean in the digital environment—is that concept still appropriate? Who is a member of a library's authorized community in a networked environment? What is fair use in gaining access to protected digital materials?[41]

• *Liability.* Will librarians have liability in cases of patron abuse of copyrighted information, carried out on the library's network?

---

[39]For an additional discussion, see Henderson (1998), available online at <http://www.ala.org/washoff/copylib.html>. See also Office of Technology Assessment (1993).

[40]See Chapter 3 for a discussion.

[41]See Chapter 4 for a discussion.

The Research Community

The research community, associated largely with higher education, has a number of specific concerns that are closely related to those of libraries and schools: [42]

• *Access to information.* A very large body of information vital to continued progress in research and development is available in the open research literature. Proposed changes to intellectual property (particularly copyright) law that may strengthen rights holders' control over IP should be evaluated with respect to possible implications for the research community and the nation's research capability. Researchers also value access to information and databases for its educational value and its potential for reuse and development.

• *Impact of technology on current models for dissemination of research results.* Researchers are prolific producers of IP, which typically is published, sold, and distributed primarily by commercial publishers and professional societies, the business models of which depend heavily on existing copyright law. Technological changes challenge the viability of today's publication system through electronic publishing of journals, pre-published material ("preprints"), and direct posting of articles.

• *Web-specific issues.* World Wide Web publishing raises questions about who has rights to create links to which Web pages and who owns the links themselves. A number of researchers have produced Web bibliographies—featuring either links to specific works or to sites containing works and other links. (Some collections of links are large, highly structured, and potentially commercially valuable.) Determining who has the rights to such links could have significant effects on all research communities by the effects on fundamental scholarly practices such as quotation and citation. The technology, for example, supports the production of Web articles that link directly to cited material, but the law may or may not support such links. Universities are increasingly claiming IP rights over the course materials that faculty post on a university Web server. As distance education grows in importance (including its financial contribution), arguments over who owns digital course material can be expected to increase. Changes in electronic distribution and in the way academic research is distributed (e.g., posting preprints and finished papers on personal Web sites) mean that an increasing number of publishers now permit authors to retain more rights, including in some cases the right to

---

[42]See AAU (1994) (available online at <http://www.arl.org/aau/IPTOC.html>) for a discussion of the IP concerns of universities. See Appendix G for a discussion of the specific concerns of cryptography and security researchers.

post their paper on their own Web server for free public access, even though this is in direct competition with the published version.

The General Public

Until fairly recently, the general public remained largely untouched by copyright law and the policy debates that shaped it. Copyright issues have instead been resolved primarily by negotiation among representatives of the entities most directly involved—publishers, libraries, educational institutions, entertainment industry, communications, music companies, and the software industries. The recent policy debates have seen a continuation of that historical pattern: Those parties are invested in, familiar with, and equipped to address the issues. By contrast, most citizens have not seen themselves individually as IP producers or rights holders, a characterization that is changing as the Web enables many people to see themselves as suppliers of information (whether or not they regard it as "property" or appreciate the legal rights). To the extent that the interests of the general public have been represented, the burden of advocacy has often fallen on libraries and universities.

Individual citizens, in their role as information consumers, have the following concerns:

• *Availability.* Individuals want the broadest range of intellectual property to be available with the least impediment to access. Broadness of range includes concern about diversity of views; public concern about the potential for limiting diversity through control of information sources has in the past motivated public policy relating to broadcasting content, and new digital media raise new questions about risks of and antidotes to such control. Impediments to access can include price, procedural difficulty, licensing terms constraining how material can be used, and continuity over time. Historically, public libraries and broadcast media have provided a lot of information without fees; digital media are associated with both fee and no-fee content.

• *Quality.* Individuals often relate the value and usefulness of intellectual property to its "quality." Quality ranges from the fidelity of a copy relative to an original, to such attributes as accuracy or completeness. Authenticity can be associated with quality: Assurances about the source of content can contribute to authenticity.

• *Privacy.* Individuals have legitimate concerns, especially in the context of access to networked information, over the privacy of their use of intellectual property. Because accessing networked information involves an interaction between the consumer and a server run by a distributor, there is the potential that consumers may be revealing informa-

tion about themselves or what they are reading. Media attention to
Amazon.com's publication of the books bought most by employees of
different organizations illustrates some of the potential,[43] and the flaunt-
ing of former U.S. Supreme Court nominee Robert Bork's video rental
records led to legislation protecting the privacy of such records.[44]

• *Simplicity and clarity of the legal regime.* The actions of individuals
are governed by a set of IP laws that they may not understand well and
that, even when understood, they may not easily be able to conform to. In
certain cases, there is a mismatch between the law and common sense
models of information and ownership. The personal use/fair use distinction has been especially problematic in this regard (Chapter 4). Also,
individuals should be able to determine whether they incur liability in
using intellectual property, yet this is often not easy.[45]

Discussion of the general public or citizens at large raises questions
about the concept of public interest as opposed to private interest, which
may be organizational or personal. Public interest is an elusive and
abused concept; it is enhanced by private and public action, and it is
abstract and therefore harder to measure than the costs and benefits of
changes in the treatment of IP on private parties. In a 1998 editorial, the
New York Times stated, "What vexes any discussion of copyright is the
idea of benefit. It is easy to see what the Disney Corporation will lose
when Mickey Mouse goes out of copyright. It is harder to specify what
the public will lose if Mickey Mouse does *not* go out of copyright. The
tendency, when thinking about copyright, is to vest the notion of creativity in the owners of copyright" (*New York Times*, 1998).

The circumstances that gave rise to this situation have changed, however, and the rate of change is increasing. Beginning with the development of inexpensive document copying in the 1950s and 1960s, copyright
law emerged as a matter of direct concern to individual citizens. The
development of successful tape recording formats in the 1960s and 1970s
made the copying and transcription of music and other sound program-

---

[43]See Streitfeld (1999).

[44]The Video and Library Privacy Protection Act of 1988 was enacted on November 5 (P.L.
100-618). For additional information, see Hinds (1988).

[45]IP, as it affects the average citizen, is moving beyond copyright: Web sites raise trademark issues; content placed on Web sites may involve state law about rights of likeness and
publicity; and perhaps most alarming, software authors can unwittingly infringe patents on
a regular basis as they learn to program. If copyright is arcane, these other issues are even
more obscure to the average citizen, and he or she will surely be incredulous as the lawyers
descend with cease and desist orders. Unfortunately, the popular press has not yet recognized these issues in a serious way and drawn public attention to them.

ming accessible to consumers. The emergence of videotape recording in the late 1970s and 1980s raised the stakes for individuals still higher. The widespread diffusion of personal computers in the 1980s and 1990s has made the copying of software and other forms of digital material easy even for unsophisticated consumers.

Other Consumers and Producers of Intellectual Property

Governmental Organizations

Government at all levels—federal, regional, state, and local—is a producer, distributor, and consumer of intellectual property. The federal government differs from other producers in its goals, being concerned with universal accessibility of most of its information and being unable to hold copyright in the works that it creates. Government has not traditionally sought to maximize revenue from its IP assets, but that is changing somewhat with the privatization of certain functions, budget pressures that motivate user-fee charges for certain services, and recognition that information has value for which at least some are willing to pay.[46]

Government agencies are concerned about the following issues:

• *Information integrity.* Documents published by the government in many instances have an important, authoritative status, ranging from their immediate impact (e.g., interest rate changes, major reports) to their archival role (e.g., legislative histories). These documents must be available in verifiably unmodified form.

• *Universality of access.* Government addresses all citizens by definition. Networked information is, in principle, vastly more accessible than print information distributed through physical distribution systems such as the depository library program. In addition, the flexibility of electronic information allows it to be much more accessible to citizens with disabilities, for example. It does require access to appropriate equipment and network services, which imply different costs than those of traveling to a library or government office, and raises questions about differentials in citizens' ability to access government information. (See Box 1.10.)

• *Rights of access.* Repackaging, adding value, and sale of some government information by commercial vendors raises questions about loss of information from the public domain.

---

[46]See Chapter 3 for additional discussion.

BOX 1.10
Digital Intellectual Property and the Digital Divide

According to a report from the National Telecommunications and Information Administration (NTIA), *Falling Through the Net: Defining the Digital Divide* (NTIA, 1999), more Americans than ever have access to telephones, computers, and the Internet. However, NTIA also found that two distinct groups of "haves" and "have-nots," remain and that, in many cases, this digital divide has widened in the past year. For example:

- Households with incomes of $75,000 and higher are more than 20 times more likely to have access to the Internet than those at the lowest income levels, and more than nine times as likely to have a computer at home.
- Black and Hispanic households are approximately one-third as likely to have home Internet access as households of Asian/Pacific Islander descent, and roughly two-fifths as likely as white households.
- Americans in rural areas are lagging behind in Internet access. At the lowest income levels, those in urban areas are more than twice as likely to have Internet access than those earning the same income in rural areas.

The proliferation of digital intellectual property has important implications for the digital divide. In some respects, the digital revolution holds the promise of new and innovative ways to improve information access for both haves and have-nots. The Web provides easy access to an enormous and rapidly growing amount of information—much of it free. It is "easy access" for those with convenient access to computers and Internet connections, with the requisite proficiency in the technology to find and organize the desired information, and the capability to adapt readily to the likely changes in IP mechanisms—licensing, fair use practices, and so on. For those who lack such technology access, proficiency, and ability to adapt readily, the digital divide may become a digital chasm.

Government is also concerned about the role of IP and IP industries in the local, state, regional, and U.S. economies; in international trade; and in reducing piracy, both domestically and abroad.

Private Sector Organizations

Private sector organizations—comprising for-profit companies and not-for-profit organizations—use prodigious quantities of intellectual property in the course of their operations and, consequently, share with other consumers of IP many similar concerns already discussed, such as fair use and information integrity. In addition, private sector organizations, as visible targets for those who conduct IP enforcement programs, are concerned about complying with IP laws in a simple way that can be readily documented (e.g., the purchase and use of site licenses for digital IP).

Journalists

The press has a special role in our society implied by the First Amendment to the Constitution. Journalists have some specific concerns with intellectual property and suffer intensely from the "multiple roles" problem described above. As authors, they are very concerned about control over their work; in particular, the efforts of newspapers and magazines to exercise broad, long-term control over that work—as a condition of publication—and with the incorporation of their own writings into the cultural record. As researchers, they share with educators, libraries, and the research community concerns about the availability of the public record, government information, and factual information; accountability as it is operationalized by the archiving of the cultural record; freedom of speech (including the ability to use copyrighted materials freely for criticism and for news reporting); and fair use.

Standards Organizations

Standards organizations play an important role in the continuing evolution and health of the information infrastructure. They make extensive use of copyright to ensure control over their works, the integrity of these works, and the continuity of access to these works. In this regard, they are much like other producers and distributors of intellectual property. Standards organizations also have some specific concerns related to patents and their interaction with the standards development process; the incorporation of a patented technology into a standard may give an unfair advantage to the patent holder or raise the difficult issue of licensing terms. Standards organizations must also operate under certain expectations of openness of participation and information flow associated with antitrust law.

2

Music: Intellectual Property's Canary in the Digital Coal Mine

Of all the content industries affected by the digital environment, the music industry has, for a variety of reasons, been thrown first into the maelstrom. Events have proceeded at the dizzying pace that has been called "Internet time," with technical, legal, social, and industrial developments occurring in rapid-fire succession.[1] Yet the problems facing the music industry will likely soon be found on the doorstep of other content industries. This chapter presents the developments in the music industry and reviews its early phases of coming to grips with digital information. These developments offer an intriguing case study illustrating the problems, opportunities, possible solutions, and cast of characters involved in dealing with digital intellectual property (IP). The focus is not on the day-to-day specifics, as these sometimes change more rapidly than daily newspapers can track. Instead, the perspective is on the underlying phenomena, as a way of understanding the issues more generally. Not all these issues will play out identically in different industries, of course. But some

---

NOTE: Underground coal deposits are invariably accompanied by methane gas, which is highly explosive, but colorless, odorless, and tasteless. Before more advanced detectors became available, miners would take canaries into the mines with them because the birds were far more susceptible to methane and thus offered advance warning. The concept has since become a metaphor for anything that serves in that role.

[1]For example, according to net measurement firm Media Metrix, an estimated 4 million people in the United States listened to digital music in the month of June 1999, up from a few hundred thousand less than a year ago (*Wired News*, 1999).

of the problems will be widespread, because they are intrinsic to digital information, no matter what content it carries. The problems include distributing digital information without losing control of it, struggles over standards and formats, and evolving the shape of industries as the new technology changes the previous balance of power.

WHY MUSIC?

The problem, or opportunity, has hit music first for a variety of reasons. First, files containing high-fidelity music can be made small enough that both storage and downloading are reasonable tasks. Digitized music on a standard CD requires about 10 megabytes per minute of music; with a format called MP3, that same information can be compressed so that it occupies about one-tenth as much space.[2] As a result, music files currently require about 1 megabyte for each minute of music (or about 45 megabytes for a typical album) yet offer generally acceptable (though not quite CD quality) sound. With multigigabyte disk drives common, dozens of albums are easily stored directly on a hard drive or inexpensively written to writable CDs.[3] Video, by contrast, contains a great deal more information: A digitized 2-hour movie (e.g., on a DVD) contains about 5 gigabytes of information.

Second, access to digitized music is abundant, and demand for it is growing rapidly. Numerous MP3 sites offer free MP3 playback software, songs, and albums. With a 56K modem (which provides a sustained transfer rate of about 5K bytes/second), a 5-minute song takes about 17 minutes to download, an album about 3 hours. With access to high-speed network connections growing more commonplace (e.g., at work and on campuses), sustained download speeds of 50K bytes/second and higher are widely available, making possible the transfer of a song in under 2 minutes and of an entire album in about 18 minutes.

---

[2]MP3 is shorthand for Moving Picture Experts Group (MPEG) 1 Layer 3. Other compression techniques are available that offer even higher compression ratios, making the files smaller still. Although the MP3 format has been the most recent concern of the recording industry, problems involving digital music and IP arose some years ago, with the proliferation on the Web of unauthorized versions of songs written in MIDI format, and recordings in WAV and AU format (see Box 1.2 in Chapter 1 for an explanation of formats and file compression).

[3]A gigabyte of hard drive space can hold 1,000 minutes (roughly 20 albums' worth) of music; desktop machines routinely ship with multigigabyte hard drives and can thus easily hold 50 to 100 albums. The hardware to write one's own CDs was a few hundred dollars in 1999 and is sure to become cheaper, and a blank recordable CD costs about a dollar. Using MP3 encoding, a recordable CD will hold 10 albums' worth of songs. Hence to someone with the equipment, the marginal cost of copying an album is approximately *ten cents* and is sure to decline further as technology improves.

But you need not go to the Web to find digitized music. A very large percentage of the music industry's current content is already available in an unprotected digital form: CDs. Widely available software programs known as "rippers" (or "digital audio extractors," in more polite circles) can read the digital data from CD tracks and rewrite it in a variety of formats, notably as MP3 files.[4] These files are easily shared among friends or posted around the Web.

The third reason that the problem has surfaced first in the music world is that music is popular with a demographic group (students in particular, young people generally), many of whom have easy access to the required technology, the sophistication to use it, and an apparently less than rigorous respect for the protections of copyright law.[5] Students also constitute a well-defined and geographically proximate community, which facilitates the sharing of digital music files.[6]

Fourth, music can be enjoyed with the existing technology: Good speakers are easily attached to a computer, producing near-CD quality sound, and a variety of portable players (e.g., the Rio from Diamond Multimedia) are available that hold 30 minutes to an hour of music. By contrast, even if it were available on the Web, downloading a best-selling novel is not enough; it would still have to be printed before you could enjoy the work.

W(H)ITHER THE MARKET?

What are the consequences for the recording industry? It is facing an age-old question that lurks in the background of most innovations that affect intellectual property: Something is about to happen, but will it be a disaster or an opportunity? New technology and new business models for delivering content are almost always greeted with the belief that they will destroy the existing market. In 17th century England, the emergence of lending libraries was seen as the death knell of book stores; in the 20th century, photocopying was seen as the end of the publishing business, and videotape the end of the movie business (Shapiro and Varian, 1998). Yet in each case, the new development produced a new market far larger than the impact it had on the existing market. Lending libraries gave

---

[4]Rippers copy the digital data from the CD to the computer's hard drive, reading the CD as if it were a (very large) floppy disk.

[5]On college campuses, the use of digital music files has created sufficient levels of network congestion to cause some network administrators to put MP3 servers on their own subnet to minimize the disruption to the main campus network (as reported by a briefer at the committee's meeting on July 9, 1998).

[6]See Gomes (1999).

inexpensive access to books that were too expensive to purchase, thereby helping to make literacy widespread and vastly increasing the sale of books. Similarly, the ability to photocopy makes the printed material in a library more valuable to consumers, while videotapes have significantly increased viewing of movies (Shapiro and Varian, 1998). But the original market in each case was also transformed, in some cases bringing a new cast of players and a new power structure.

Will digital information do the same for music? Some suggest that the ability to download music will increase sales by providing easy purchase and delivery 24 hours a day, opening up new marketing opportunities and new niches. For example, the low overhead of electronic distribution may allow artists themselves to distribute free promotional recordings of individual live performances, while record companies continue to focus on more polished works for mass release. Digital information may also help create a new form of product, as consumers' music collections become enormously more personalizable (e.g., the ability to create personalized albums that combine individual tracks from multiple performers). Others see a radical reduction in aggregate royalties and, eventually, in new production, as pirated music files become widely distributed and music purchases plummet.

The outcome for digital music is still uncertain; there is of course no guarantee that the digital music story will play out in the same way as it has in these other industries. But past experience is worth considering, reminding us as it does that at times innovation has contributed to a resurgence in the market rather than a reduction.

WHAT CAN BE DONE?

There are two major lines of response to the challenges outlined above: find an appropriate business model, and develop and deploy technical protection mechanisms. Each is considered below.

The Business Model Response

"The first line of defense against pirates is a sensible business model that combines pricing, ease of use, and legal prohibition in a way that minimizes the incentives for consumers to deal with pirates" (Lacy et al., 1997).[7] This view nicely characterizes the business model response, suggesting that one way to cope with piracy is to provide a more attractive

---

[7]The paper cited describes a new technical protection mechanism for music, yet (appropriately) begins by acknowledging the power of a good business model.

product and service. The difficult part of this approach is that it may require rethinking the existing business model, industry structure, and more.

Make the Content Easier and Cheaper to Buy Than to Steal

One example of that rethinking is the suggestion by Gene Hoffman, CEO of EMusic, Inc. (formerly GoodNoise, a digital music provider): "We think the best way to stop piracy is to make music so cheap it isn't worth copying" (Abate, 1998).[8] EMusic sells singles in digital form for 99 cents.

But if those singles are in the form of MP3 files, they are unprotected—nothing prevents the purchaser from passing a file on to others or posting it on a Web site (illegal though these actions may be). So how could there be a market in such things, when it appears that all the value in the item could be extinguished by the very first sale?

One answer becomes evident to anyone who has actually tried to download MP3 music from any noncommercial site: The service is terrible and the experience can be extraordinarily frustrating.[9] Search engines can assist in finding songs by title, performer, and so on, but you have to know how to look: Can't find what you're looking for when you type in "Neil Young"? Try "Niel Young." In any collection, quality control is a problem; when the data are entered by thousands of individual amateurs, the problem is worse.[10]

When the links are found, the next question is, How long are you

---

[8]A similar comment came from Steve Grady, vice president of marketing for EMusic, who concluded that "the best way to combat piracy is to make it easier to buy rather than to steal" (Patrizio, 1999a).

[9]Some commentary from the Lycos Web page on downloading MP3 files illustrates these problems:

> In some situations there can be problems downloading the MP3 files. . . . This is due to the unstable nature of the MP3 servers—most of these are run on an amateur basis, and the reliability is not very good.

> On many servers only a very limited number of users can be logged in at a time. The only way around this problem is to repeatedly try to download the files until there are available connections on the server.

> Many servers are up for only a limited time. On others, the MP3 files may have been removed. In both these cases we will have dead links until these changes have been updated in our database . . . it is not usually known whether a server is permanently down, or whether it will be available again. We have therefore chosen to keep the files in the database until we are sure that they have actually been removed, with the cost of displaying a few dead links.

[10]A particularly insidious example of the sites distributing music illegally are the so-called "ratio" sites, which permit downloading of MP3 files only in exchange for the uploading by the user of at least one "ripped" MP3 file.

willing to keep trying, when receiving responses such as "Host not responding," "Could not login to FTP server; too many users—please try again later," and "Unable to find the directory or file; check the name and try again"? The computers containing the files are often personal machines that are both unreliable and overloaded.

Even once connected, the speedy download times cited earlier are ideals that assume that both the computer on the other end and its connection to the Internet are up to the task. The real-world experience is often not so good: Creating a Web site with a few music files is easy; providing good service on a site with hundreds or thousands of songs is not: The hardware and software requirements are considerably more complex.

Where is the business then, if files are unprotected by technical mechanisms (even if legally still protected)? It may be in the service, as much as the content. Why experience 30 minutes or an hour of frustration, if for a dollar or so you can have what you want easily, reliably, and quickly? This is one example of how, in the digital age, content industries may mutate, at least in part, into service companies. The key product is not only the song; it is also the speed, reliability, and convenience of access to it.[11]

There is also a more general point here about the relative power of law and business models. Although legal prohibitions against copying are useful against large-scale pirates (e.g., those who would post MP3 files for sale and hence have to be visible enough to advertise), they are unlikely to be either effective or necessary against individual infringers, where detection and enforcement are problematic. Where such private behavior is concerned, business models may offer a far more effective means of dealing with IP issues.

Use Digital Content to Promote the Traditional Product

Another business model approach sees free online distribution of music as a way to build the market for the traditional product. In March 1999 Tom Petty put a song from his new album online; it was downloaded more than 150,000 times in 2 days. Other groups have made similar efforts, releasing digital versions before the albums were available in stores (Cleary, 1999), all in the belief that distribution of a sample track will increase sales of the traditional product.

---

[11]This is an increasingly common observation. For example, the success of Web sites such as Amazon.com, Barnes and Noble, and others are attributable in part to the speed, reliability, and convenience of the service they offer, as well as the product available for sale.

In October 1999 the rock group Creed made the most popular song from their CD release "Human Clay" available for free download at more than 100 sites. The Creed CD jumped to the number one spot on the Billboard top album sales when it debuted, notwithstanding the free give-away of what was believed to be its first hit.[12] This phenomenon calls into question the conventional wisdom that, although one may choose to give away free songs as a promotional tool, to give away the most popular song will eliminate some motivation to buy the entire album. The utilization of the digital download of free songs in their entirety in order to drive sales of traditional "packaged goods," in the form of CDs and audio cassettes, is perhaps yet another example of the so-called "clicks and mortar" business strategy, under which the Internet does not replace brick and mortar sales, but can be used as an adjunct to traditional brick and mortar retailing methodologies.

Give Away (Some) Digital Content and
Focus on Auxiliary Markets

A more unconventional approach takes the position that most digital content is so difficult to protect that a more sensible business model would treat it as if it were free (Dyson, 1995):

> Chief among the new rules is that "content is free." While not all content will be free, the new economic dynamic will operate as if it were. In the world of the Net, content . . . will serve as advertising for services such as support, aggregation, filtering, assembly, and integration of content modules. . . . Intellectual property that can be copied easily likely will be copied. It will be copied so easily and efficiently that much of it will be distributed free in order to attract attention or create desire for follow-up services that can be charged for.

The value instead is in the auxiliary markets. The classic example is the Grateful Dead, who have long permitted taping of their live shows (and have taken the additional step of permitting fans to trade digitized versions of those recordings on the Web) (Buel, 1999). They gave away their live performances (which were generally not released as records) and profited from the increased draw at concerts, and the income from related merchandise and traditionally produced studio recordings.[13]

---

[12]See Philips (1999).

[13]They were, however, rigorous at pursuing those who sought to profit from selling recordings of the live shows. Trading among fans was fine; selling the material was not. This is also an example of the suggestion above that distribution of live performance recordings might be handled by the artists rather than the record companies, although of course in this case the artists let the fans do it themselves.

The breadth of applicability of this model is of course not immediately clear. The model may be idiosyncratic to a particular band, audience, and tradition. But the thought is at least worth entertaining that the model might be used in other musical genres and perhaps in other publishing businesses, as well.

This variety of approaches illustrates the challenge and the opportunity of finding an appropriate business model in the world of digital IP. One challenge is in determining whether the existing models and existing industry structure can be made to work in the face of the new technology. In some cases it can; for example, movie theaters remain viable in the presence of VCRs and rental tapes. But in other cases the existing business model and industry structure cannot be maintained, no matter how vigorous the legal or technical efforts. In such cases some form of adaptation and creative rethinking can be particularly effective; the business model responses noted above offer a few examples of that type of thinking.

The Technical Protection Response

Many technical protection mechanisms are motivated by the key issue noted earlier, lying at the heart of the difficulties with digital IP: the liberation of content from medium. When content is bound to some physical object, the difficulty of duplicating the physical object provides an impediment to reproduction. However, when digitally encoded, text need no longer be carried by a physical book, paintings by a canvas, or music by a record or CD, so reproduction becomes easier. What options does technology offer for controlling reproduction?

Mark the Bits

One response is to "mark" the bits, that is, add to the content the digital equivalent of a watermark that identifies the rights holder. While it does not prevent the content from being copied and redistributed, this technique can at least make evident who owns the material and possibly aid in tracking the source of the redistribution.

Music can be watermarked by very small changes to some of the digital samples. As explained in Chapter 1, music is typically digitized by measuring the sound intensity 44,100 times a second, using a 16-bit number (0 to 65,535) to indicate intensity level. If the intensity of a sample is actually 34,459, it can be changed slightly and the human ear will never hear the difference. This permits encoding information in the music, by deciding, say, that the last two bits in every 150th sample will not encode the music, but will instead encode information about the music (e.g., the identity of the rights holder). This change to the music will be imperceptible to a person but will be easily read by a computer program.[14]

A variety of such watermarking techniques is commercially available, including those that are "robust" (i.e., difficult to remove without affecting the music) and those that are "fragile" (i.e., distorted by most modifications to a file, as for example compression using MP3). One proposed plan is to embed both a robust and a fragile watermark in newly released CDs and then require that licensed portable players refuse to play digital music that has a robust watermark but lacks the fragile watermark.[15] The presence of the robust watermark indicates that the music is newly released, while the absence of the fragile watermark suggests that the music file is no longer in its original form (and hence may have been copied).

Reattach the Bits

A second, more ambitious approach is to find a way to "reattach" the bits to something physical that is not easily duplicated. A number of technical protection mechanisms are motivated by this basic observation; the description that follows draws on features of several of them as a way of characterizing this overall approach.

One relatively straightforward technique for reattaching the bits is to employ special hardware that enforces copy protection. Digital audio tape players, for example, have a serial copy management system (SCMS) in which the hardware itself enforces the prohibition against making digital copies of copies. The first-generation copy contains an indication that it is a copy rather than the original, and any SCMS-compliant device will not copy this copy.

This technique works well but is limited to single-purpose devices. It will not work on a general-purpose computer, because the user would be able to gain access to the original information and make copies by means other than the SCMS. As a result, the challenge of reattaching the bits becomes more difficult for a general-purpose computer. A succession of increasingly more sophisticated and complex mechanisms can be used to approach this goal:

• *Encrypt the content.* This mechanism provides, at a minimum, that the consumer will have to pay to get a decryption key; without it, a copy of the encrypted content is useless. Buy a song, and you get both an encrypted file and a password for decrypting (and playing) the song.

---

[14]See Chapter 5 and Appendix E for further discussion of watermarking and other technical protection mechanisms.

[15]See Robinson (1999).

- *Anchor the content to a single machine or user.* Simply encrypting the content is not enough, as the purchaser can pass along (or sell) both the encrypted content and the key, or simply decrypt the content, save it, and pass that along.[16]   There are a variety of ways to anchor the content; one conceptually simple technique encodes in the decryption key (or the song file) information about the computer receiving the encrypted file, such as the serial number of its primary disk. The decryption/playback software then checks for these attributes before it will decrypt or play the song.[17]
- *Implement persistent encryption.* The scheme above is still not sufficient, because a consumer might legally purchase content and legally decrypt it, then pass that on (or sell it) to others who can modify the decrypted file to play on their machine. Some technical protection mechanisms attempt to provide additional security by narrowing as much as possible the window of opportunity during which the decrypted information is available. To do this the information must be decrypted just in time (i.e., just before it is used), no temporary copies are ever stored, and the information is decrypted as physically close as possible to the site where it will be used. Just-in-time decryption means that decrypted information is available as briefly as possible and then perhaps only in very small chunks at a time. Decrypting close to the usage site reduces the number of places inside (or outside) the machine at which the decrypted information might be "siphoned off." Persistent encryption is complex to implement in its most ambitious and effective form, because it requires the IP protection software to take control of some of the routine input and output capabilities of the computer. If this is not done, there are a large number of places (e.g., in the operating system) from which decrypted information might be obtained.

A variety of systems in use in 1999 employ one or more of these mechanisms, including a2b from AT&T, the Liquid Audio player from Liquid Audio, the Electronic Music Management System from IBM, ContentGuard from Xerox, the InterTrust system, and others.[18]

---

[16]Encryption by itself does convey to the user that he or she is not supposed to pass along the content casually—thus, it could have some deterrence effect in helping to keep honest people honest, at least for the present, when encryption is still a novel technology for most people. More about this is presented in Chapter 5.

[17]Hard drives are of course sometimes changed (e.g., to upgrade to a larger drive), so this technique is not foolproof. But the claim is that there are enough such stable attributes about a particular computer that this scheme will work most of the time.

[18]Information on a2b (both the company and the technology) is available online at <http://www.a2bmusic.com>; Liquid Audio, at <http://www.liquidaudio.com>; IBM's system, at <http://www.ibm.com/security>; ContentGuard, at <http://www.contentguard.com>; and InterTrust, at <http://www.intertrust.com>.

Another element of technology provides a useful additional capability. Differences among consumers make it useful to offer content on a variety of different terms. For music, for example, one might want to sell the right to a time-limited use, a finite number of uses, or an unlimited time and usage count. A variety of systems have been developed to provide an easy way to specify a wide variety of such conditions. When a music file is downloaded, then, in addition to the music it will contain information indicating the license conditions under which the music may be used. The playback software checks these conditions and enforces them appropriately.

Note that the technology picture outlined here is optimistic, in a sense discussed further in Chapter 5, where the significant difficulties involved in providing secure content handling in the real world are considered in detail. Note, too, that as with any security mechanism, the key question in the real world is not the purely technical issue of whether it can be defeated. All mechanisms can be, eventually. Instead the key questions go back to the three fundamental factors: technology, business models, and the law. The legal system sets the basic rules on what may be controlled; technology and business models then work in tandem: Is the technology strong enough to provide a meaningful disincentive for theft, yet not so expensive (for either the distributor or the consumer) that the added costs drastically curtail demand? At this point in the development of digital content delivery mechanisms, companies have relatively little real-world experience on which to make such judgments.

A SCENARIO

A system using some of the technology described above is easy to imagine; the details used here are intended to convey the general idea rather than describe any particular system. The user downloads and installs software that provides for music playback and assists with online purchasing. He or she then connects to a Web site offering music in that form, selects a song or album, and provides a credit card number to pay for it. As part of the transaction, the vending site is provided with information specific to the computer requesting the file (e.g., the serial number of its disk); this information is embedded in the decryption key supplied with the file. The vending site may, in addition, mark the music file with a unique ID that enables linking back to this transaction, in the event that the decrypted audio file is later found to have been distributed.

The customer now has the song, the vendor has the money, and the consumer can use the song according to the terms of use that are embedded in the file. Redistribution of the file is pointless, as the song won't play on another computer. Yet all the other advantages of digital audio are maintained: The user can store large amounts of music compactly, have random access to any track on any album, and create personalized albums containing selected tracks from selected albums in a particular order.

The same basic model applies to portable players like the Rio: To anchor the bits (i.e., make them playable on only one portable player), a hardware identifier may be built into the file, such as the serial number of the player. Just as on the PC, the player's software can check for this information and refuse to play the song unless the identifier in the file matches the device.

CONSTRAINTS ON TECHNOLOGICAL SOLUTIONS

The scenario above sounds simple, but there are inherent difficulties. First, no protection scheme lasts forever. Any time content is valuable, some people will be motivated to find ways to break the protection mechanism, and some of them will be more than willing to share their techniques. For example, by the end of 1998 a program called a2b2wav that was available on the Web purportedly cracked the protection scheme then in use by both a2b and Liquid Audio, producing an unprotected file of the music in another audio format called WAV. Another example illustrates both the interest in digital music and the speed with which protection mechanisms are subject to defeat. On August 17, 1999, Microsoft released Windows Media 4.0, intended to be a secure format for music and other media files. On August 18, 1999, various Web sites offered a program that reportedly defeated the security features of Windows Media, stripping out the license information and making the files shareable. The new program had apparently been in development for only a month, based on the beta releases of Windows Media (Livingston, 1999).

One plausible countermeasure is to design protection to be renewable (i.e., easily changed so that the new protection scheme can quickly replace the old). This solution protects content from that point forward, limits the profits from piracy, and keeps the protection a moving target for those trying to break it.

A second difficulty in developing technical protection solutions is that consumer devices must be easy to use. Cumbersome content protection schemes may discourage use, particularly as consumers are likely to be impatient with mechanisms that they perceive are intended to protect someone else's interests. This requirement puts stringent performance demands on a system (e.g., decryption must be fast enough to become imperceptible) and requires careful design, to ensure that the system is conceptually simple.

A third difficulty arises with any system that "anchors" content to a specific device: the potential loss of all of that content if the device in question fails or is replaced. Does the consumer have to repurchase every piece of music that he or she owns if the portable player fails, is lost, or is replaced?

A fourth difficulty is the diversity of interests at work here, including the computer owner (i.e., music consumer), computer manufacturers (of both hardware and software), music publishers, and performers. Consumers have expectations about the ability to share and the ongoing use of content, publishers are concerned about the overall market, and performers are concerned about their audience and royalties. Getting significant content protection machinery in place and widely distributed would require a concerted and coordinated effort, yet each of the players has its own goals and aims that may not necessarily align (see, for example, Hellweg, 1999).

A fifth difficulty is the inherent complexity of providing end-to-end protection within a general-purpose computer. PCs have been successful to a significant degree because they have open architectures; that is, components of the machine can be replaced by the user (e.g., replacing a hard drive with a larger one or buying a new sound card[19]). As long as the machine is designed this way—to be accessible to users—decrypted information can be captured in numerous ways as it passes from one place to another inside the machine. One could modify the software used by the sound card, for example, so that it not only generates the signal for the speakers but also stores away the decrypted music samples. Clearly, hardware and software designers could make such steps progressively more difficult, but the effort they must expend and the consequential costs would be substantial.

A sixth difficulty arises from the installed base of PCs. With more than 100 million computers in use, any scheme that requires new hardware faces a significant barrier to acceptance. Clearly the benefits have to outweigh the cost and inconvenience of changing machines. One benefit that may encourage the adoption of new hardware is the interest in electronic commerce. Efforts have been mounted to create new hardware and software with security built in at all levels.[20] If this succeeds, PCs may routinely come with technology that makes possible secure electronic

---

[19]The sound card is the hardware that turns digital samples into the signals necessary to drive the speakers.

[20]The Trusted Computing Platform Alliance, a collaborative effort founded by Compaq, HP, IBM, Intel, and Microsoft, is aimed at providing security standards for computers, trying to create "an enhanced hardware- and operating system-based trusted computing platform." For more information, see <http://www.trustedpc.org>.

commerce, and that may also be usable for enforcing intellectual property rights.

A second mitigating factor here is the relatively short lifetime of computers, at least in the corporate environment, where 3 years is a common figure for turnover. A related opportunity arises with new technology: With portable players, a relatively recent development, there is not a major installed base requiring compatibility, offering the chance to set security measures in place near the outset of the new technology (as indeed the Rio player and others have agreed to do).[21] The installed base problem here is more likely to arise from the need to be compatible with the existing MP3 format, widely and legally used by consumers to make personal copies of CDs they own and who are likely to want players capable of playing those files.

There is also a traditional chicken and egg problem involving technology development and content owners. Investing in new content delivery technology is risky without content to deliver, yet content owners are reluctant to release their content for digital distribution until they feel the delivery system has been tested in real use, is secure, and will be accepted by consumers. For example, in June 1999, Digital Video Express, the company that invented and was marketing Divx technology, ceased operations. Its announcement mentioned as one of the contributing factors its inability to obtain adequate support from studios.[22]

Finally, there is the problem of the digital infrastructure, one element of which is transaction support: Buying content online requires systems for secure transactions, in high volume, possibly involving rather small amounts of money (e.g., $1 for a song). A second element of infrastructure, public-key cryptography, is effective at protecting content but requires a substantial infrastructure to make it easy to use on a wide scale (see Appendix E for a description of public-key certificate authorities). Although some progress has been made on creating the infrastructure to support electronic commerce, these and other elements of this digital infrastructure are not yet in place for routine use and will require both a sizable investment for their creation and a major effort at agreeing on standards.

INDUSTRY CONSEQUENCES OF THE NEW TECHNOLOGY

The digitization of music in general and the availability of an easily used format like MP3 in particular have wide-ranging consequences for

---

[21]See Strauss and Richtel (1999).
[22]See Ramstad (1999).

the industry, consequences that are being played out in a number of struggles.

One consequence is the possibility of a radical shift in power. As noted earlier, one of the fundamental changes brought about by the Web is the availability of an inexpensive publishing and distribution medium with worldwide reach. If composers and performers choose to take advantage of that medium, what is to be the role of traditional music publishers and distributors?

This phenomenon has been called disintermediation, referring to the elimination of middlemen in transactions. In the view of some, traditional publishers are becoming unnecessary, because authors, composers, and performers will be able to publish and distribute their product online themselves. Some performers have already done so, though generally offering their products as free samples. A variety of MP3 Web sites have also emerged, modeled on the notion of artists' cooperatives: Composers and performers can post their work and receive royalties, with no effort on their part other than the original posting.

Even as ease of publication may provide alternatives to and hence reduce the demand for one kind of intermediary—traditional publishers and distributors—it may simultaneously increase the demand for another kind. If anyone can be a creator and publisher, content will proliferate, producing a world of information overload. The consumer's problem will not be obtaining content, but rather wading through it all. This difficulty has long been recognized: Nearly 30 years ago Herbert Simon suggested that "a wealth of information creates a poverty of attention" (Simon, 1971). In the content-rich world, then, information intermediaries may become even more important because, although content may proliferate, attention is on an immutable budget.

But these intermediaries would not be publishers in the traditional sense; hence the phenomenon may not be so much disintermediation in general, as the diminishing need for one variety of intermediary and an increasing role for another. The new role of publishers in an information-rich world may require a different kind of company with a different focus. This scenario presents the possibility of a significant shift in the power structure of the music industry and a significant economic impact. Little wonder, then, that battles are emerging over the future character of the industry.[23]

---

[23]One example of a new kind of publisher is Garageband.com, an Internet start-up company that is attempting to bridge the competing worlds of online digital music and the traditional recording industry. Garageband's focus is on the thousands of small and struggling musicians and groups that have not been able to sign with a major recording studio and have discovered that they cannot make a profit by giving their music away over the

Those battles are sometimes indirect, as for example struggles over music delivery hardware. In October 1998, the Recording Industry Association of America (RIAA) sued Diamond Multimedia, maker of the Rio. The suit sought to block the sale of the device on the grounds that the Rio violated the Audio Home Recording Act (AHRA) because the device did not contain a serial copy management mechanism. The request was denied on the grounds that the Rio was strictly a playback device; there was no way to copy from it to another device. This ruling was subsequently upheld by an Appeals Court in June 1999.[24] The Appeals Court noted that the main purpose of the AHRA was "the facilitation of personal use," pointing out that "[a]s the Senate Report explains, '[t]he purpose of [the Act] is to ensure the right of consumers to make analog or digital audio recordings of copyrighted music for their private, noncommercial use.' S. Rep. 102-294, at § 86." It also mentioned in passing the notion of "space-shifting" as analogous to the "time-shifting" permitted under *Sony Corp. of America v. Universal City Studios:* "The Rio merely makes copies in order to render portable, or "space-shift," those files that already reside on a user's hard drive. . . . Such copying is paradigmatic non-commercial personal use entirely consistent with the purposes of the Act."[25]

A similar sort of lawsuit occurred in March 1999, when the International Federation of the Phonographic Industry (IFPI) sued the Norwegian company FAST Search and Transfer over the use of its MP3 search engine and database, licensed to Lycos. IFPI claimed that the database consists almost entirely of MP3 files that violate copyright. The suit is

---

Internet. Musicians will be able to upload recordings that will be rated by music enthusiasts on the Web. Beginning in the year 2000, Garageband.com plans to begin awarding a recording contract for the most popular music each month. The company plans to attract listeners who are willing to sample unknown recording artists by offering prizes such as backstage passes to concerts and visits to recording studios (Markoff, 1999).

[24]The issue before the court concerned whether the Rio was a digital audio recording device subject to the terms of the AHRA (devices under the purview of the AHRA must prohibit unauthorized serial copying). The court indicated clearly that the Rio was not such a device, affirming in passing that computers are not such devices either, even though:

> The district court concluded . . . the exemption of computers generally from the Act's ambit, "would effectively eviscerate the [Act]" because "[a]ny recording device could evade regulation simply by passing the music through a computer and ensuring that the MP3 file resided momentarily on the hard drive." RIAA I, 29 F. Supp. 2d at 630. While this may be true, the Act seems to have been expressly designed to create this loophole.

Diamond Multimedia later formed an alliance with Liquid Audio, a vendor of audio files in a protected format that makes songs easily playable only on the PC to which they were downloaded.

[25]*RIAA v. Diamond Multimedia*, F.3d (9th Cir. 1999).

novel in part in claiming a search engine firm as a party to copyright violation. The suit illustrates as well the mixed collection of interests involved: One of the backers of the suit is the Swiss company Audiosoft, which offers its own version of a secure container for digital music distribution (Robertson, 1999).

The attempt to ban the sale of the Rio illustrates several interesting aspects of the struggle over digital music. First, attacking the Rio on serial copying grounds is curious, because the device can't copy. Virtually all of the copying of MP3 files is done on computers, but (as noted) computers are not covered by the Audio Home Recording Act (because their ability to do digital recording is not "designed or marketed for the primary purpose of . . . making a digital audio copied recording"[26]). Hence the RIAA aimed at the available target, on the available basis, even if it was a tenuous match.

Second, the battle illustrates the difficulty of drafting law and policy that are technology specific in the age of general-purpose computers. The AHRA goes to some effort to define the particular class of device it attempts to regulate. But general-purpose computers can, by their nature, do anything with digital information, including the very things outlawed by the AHRA. Short of insisting on hardware that limits the behavior of the computer, little can be done to prevent this.

The difficulty is further highlighted by the announcement in October 1998 of a company called Empeg, in Somerset, England, that claimed to be bringing out a portable device that not only played MP3 files but also ran the Linux operating system. In other words, it was a portable MP3 player that was also a computer and hence outside the purview of the AHRA (Patrizio and Maclachlan, 1998).

A third interesting aspect of the battle is the indirection involved in going after the playback device, when the real effort is to discourage the proliferation of illicit MP3 files. The rationale is clear: "The RIAA claims that it is unable to stamp out the proliferation of pirated music download sites. . . . [Cary] Sherman [senior executive vice president of the RIAA] stated that the only viable solution to prevent free downloads is to attack the problem on the receiving end . . ." (Anderson, 1998).

Fourth, the battle over the Rio illustrates the difficulty of large-scale policing of private behavior. Trying to take the devices off the market might make relatively little difference even if it succeeds, as MP3 files were downloaded to personal computers and played in large numbers before the Rio, and would continue to be without it. Removing the Rio (and all the other similar players) might prevent the appeal of MP3 files

---

[26]Title 17, sec. 1001.

from increasing, but would do little to change the personal behavior of individuals using their own personal computers.

The final and perhaps most important point in the struggle over the Rio is that the real battle is over standards, because the future character and structure of the industry will be determined to a significant degree by what becomes the popularly accepted format for digital music. This fact is the major motivation behind the various industry consortia that have been formed to develop formats and standards (e.g., the Secure Digital Music Initiative (SDMI) formed by BMG Entertainment, EMI Recorded Music, Sony Music Entertainment, Universal Music Group, and Warner Music Group; the pairing of Sony and Microsoft; and others). If MP3 wins out, putting the genie back in the bottle will be difficult, and we will all be engaged in an experiment to see what effect unrestricted digital distribution has on the music industry.[27] If one of the more secure formats wins out, the industry character and structure will likely suffer far less upheaval.

Control of the standard may also affect the disintermediation phenomenon, as control of a standard, particularly a proprietary standard, puts some degree of control over publishing into the hands of the standard owners. Control of the standard will have consequences for the consumer as well. A fundamental motivation for the SDMI effort is to produce a secure music format and corresponding devices that would refuse to play files containing unlicensed MP3 tracks. Note, too, that "winning" here means popular acceptance, much as it did in the battle of VCR formats between VHS and Beta. That in turn means that the battle will involve a wide variety of issues, such as technical adequacy, ease of use, amount of content available, and promotion by the various standard bearers.

The struggle to determine the future course of the industry naturally pits sites like MP3.com against conventional publishers. Yet signs of a convergence of interests have also appeared: In June 1999 MP3.com signed an agreement with the American Society of Composers, Authors, and Publishers (ASCAP), the largest of the performing rights societies, to license the streaming broadcast of music from ASCAP's catalog.[28] This agreement allows MP3.com to act as a radio station of sorts, making songs

---

[27]There is also a successor to MP3, called MP4, that provides a framework in which IP protection information can be specified. It does not protect IP but does provide a place and a language for specifying the rights holder's desires concerning protection. Software to play MP4-encoded music would have the task of enforcing those desires, but at least there is a place in the music file itself for such information to be recorded. See Chapter 5 for additional discussion of MP4.

[28]See Thomason and Pegoraro (1999).

available on demand for playback from the site, but the bits are not as easily captured and stored, reducing the opportunity for piracy. This service is new for MP3.com, which previously had focused on allowing users to download MP3 files. The interesting element here is the careful dance of rapprochement engaged in by these two representatives of seemingly opposed interests, an apparent attempt to bridge the gap between the traditional world of music copyrights and royalties, and the proliferation of free music. Yet in the long term the interests are in many ways aligned, as each can benefit from the wide distribution of music and growth in the market.

One additional and possibly wide-ranging consequence of the new technology is unbundling of content. Music is typically downloaded by the track, not the album. This practice has become widespread in part because of size (downloading an album is still slow) but may also represent a challenge to the current marketing model, which emphasizes selling albums, not singles. It is common experience that not every song on a CD is equally appealing, but without a way to pick and choose, consumers had no choice but to buy the package offered. Hence the new technology may be promoting a "new" business model, in which content is more easily unbundled and as a consequence marketed and sold in smaller chunks.[29]

THE BROADER LESSONS

Three general lessons can be learned from the early years of the struggle over digital music. First, what has happened there may happen in other content industries as well, as other products become digitized.

Books and movies have begun to feel the effects. Electronic books are appearing, with several Web sites selling full-length books in digital form, while others offer reloadable book-sized portable display hardware.[30] This development is made possible in part by the creation of more secure forms for the content, of the sort provided, for example, by Adobe's use of ContentGuard technology from Xerox, resulting in a more secure version of its popular PDF document format.

With electronic publishing of books come many of the same issues concerning the role of the publisher: Various sites on the Web offer electronic publication of works, presenting an alternative to traditional pub-

---

[29]At one time, of course, singles were a viable part of the music market (as 45 rpm records); perhaps that time has come again.

[30]Electronic books are available online at sites such as <http://www.fatbrain.com and http://www.1stbooks.com>, and books and a portable reader are available at <http://www.rocket-ebook.com>. For a general discussion, see Clark (1999).

lishers, while also promising 40 to 50 percent royalty rates and author retention of copyright, practices far from common in the print publishing business.[31]

Movies in digital form are currently saved from widespread illegal copying because of their large size, but this barrier is likely to be overcome before too long. A number of sites have begun already to sell full-length movies in digital form,[32] but at upwards of 200 megabytes for a (compressed) movie, and 5 megabytes for even a trailer, the space requirements and download times are still quite substantial. Others are exploring the possibility of Internet distribution of movies.[33] Digital movie piracy has also appeared; in 1999 pirated copies of "The Blair Witch Project," "The Matrix," and "American Pie" were all available online. These copies are relatively low-quality, still sizable to download and store, and not easy to find (they are generally traded in low-profile news groups and chat rooms). But the struggle over digital movies has clearly arrived and will grow worse as storage capacity and transmission speeds increase.

The second lesson is that struggles over protecting intellectual property take many forms and reach into a variety of areas, including battles over technology, standards, industry structure, and business models. Keeping this in mind often makes it easier to decode the disparate agendas and strategies of the many players engaged in the struggle.

The third lesson is that among the various battles, the struggle over standards is often the most intense as it typically has the most far-reaching effects, with consequences for authors, publishers, and consumers alike, as well as the shape and character of the industry.

---

[31]See, for example, Fatbrain.com's e-matter, online at <http://www.fatbrain.com/ematter/home.html> for an example of these rates, effective in September 1999. For background on the struggle over electronic rights to textual works, see Zeitchik (1999).

[32]For example, see Sightsound at <http://www.sightsound.com>.

[33]For example, Metafilmics, an established filmmaker that won an Academy Award for best visual effects in 1998, announced it is making "The Quantum Project" specifically for initial Internet distribution (Pollack, 1999).

3

Public Access to the Intellectual, Cultural, and Social Record

For the past two centuries, copyright law has promoted broad public access to a vast array of informational works. As James Madison once observed, copyright has been characterized by a happy coincidence of public and private interests. The private interests of authors and publishers in creating and selling copies of works have coincided with the interests of members of the public eager to get access to the ideas, information, or expression in the works.[1]

Public access has not, however, been achieved solely by virtue of the incentives arising from the grant of rights to authors and the transfer of some of these rights to publishers; a number of other mechanisms have been at work as well. A variety of public policies promote access, including fair use, the first-sale rule, and the copyright registration requirement of depositing a copy in the Library of Congress; some of these accomplish their ends by placing limitations on the rights of authors and publishers.[2] The emergence of organizations such as public libraries, universities, and cultural heritage institutions has also added significantly to public access, making a wide variety of works available, by acquiring them and making

---

[1]In *Federalist* 43, Madison wrote, "The public good fully coincides in both cases [patent and copyright] with the claims of individuals."

[2]Fair use, the first-sale rule, and the copyright registration requirement are discussed in the Chapter 4 section entitled "The Challenge of Private Use and Fair Use with Digital Information" and in this chapter's sections entitled "Public Access Is an Important Goal of Copyright" and "Archiving of Digital Information Presents Difficulties," respectively.

them available to the public for no direct payment by the user.[3] Public access and use occur in a variety of ways, including purchase (new or used), borrowing (from libraries or friends), educational exposure, and reading of commentary that quotes a work. To date, this collection of mechanisms has worked well, providing protection and thus incentive to authors and rights holders while also ensuring wide public access to work through a variety of routes.[4]

Changes brought about by digital intellectual property (IP) and the information infrastructure are challenging the existing set of policies and practices for public access. This chapter focuses on the implications of those changes for public access, including both the optimistic and pessimistic projections of their possible consequences.

PUBLIC ACCESS IS AN IMPORTANT GOAL OF COPYRIGHT

U.S. courts and commentators have repeatedly emphasized the fundamentally utilitarian nature of copyright, noting that the Constitution provides for intellectual property protection with the pragmatic goal of promoting the public interest in access to knowledge and innovation. This intent is evident from the Constitution's grant to Congress of power "to promote the progress of science and useful arts, by securing for limited times to authors and inventors the exclusive right to their respective writings and discoveries." In *United States v. Paramount Pictures, Inc.*, the Supreme Court's decision considered the purpose of this clause:[5]

> The copyright law, like the patent statutes, makes reward to the owner a secondary consideration. In *Fox Film Corp. v. Doyal*, 286 U.S. 123, 127, Chief Justice Hughes spoke as follows respecting the copyright monopoly granted by Congress, "The sole interest of the United States and the primary object in conferring the monopoly lie in the general benefits derived by the public from the labors of authors." It is said that reward to the author or artist serves to induce release to the public of the products of his creative genius.

Further comment on the constitutional concern with access to infor-

---

[3]Of course, costs are involved in operating these institutions, which are funded by government, universities, and other organizations.

[4]Because the term "public access" has been used historically in a variety of ways, it is useful here to note one way in which the committee is *not* using the term. It does not include in public access any access to specific copies of a work, especially any unique copies of a work (e.g., the originals of artwork). The public has access to versions of a work that have been published and distributed, placed in publicly accessible collections (e.g., libraries), or otherwise made available through normal channels.

[5]334 U.S. 131, 158 (1948).

mation is found in the Court's decision in *Sony Corp. v. Universal City Studios, Inc.*:[6]

> As the text of the Constitution makes plain, it is Congress that has been assigned the task of defining the scope of the limited monopoly that should be granted to authors or to inventors in order to give the public appropriate access to their work product. Because this task involves a difficult balance between the interests of authors and inventors in the control and exploitation in their writings and discoveries on the one hand, and society's competing interest in the free flow of ideas, information and commerce on the other hand, our patent and copyright statutes have been amended repeatedly.

U.S. copyright law has promoted public access to ideas, information, and works of authorship in a number of ways. The grant of exclusive rights to authors promotes public access because it presumes—usually correctly—that authors want to make their works widely available to the public and will, in fact, do so as long as they have the right to stop unauthorized exploitations.

Numerous other means to promote public access have also been used. For example, U.S. copyright law, like the copyright law of many other nations, promotes public access by the "first-sale" rule, a limitation on the right of rights holders to control copies of their works that have been distributed to the public. This rule provides that the first sale of a copy of a work to a member of the public "exhausts" the rights holder's ability to control further distribution of that copy. A library is thus free to lend, or even rent or sell, its copies of books to its patrons. Libraries are, of course, not the only institutions that are reliant on the first-sale rule for their operation. Bookstores, art galleries, and auction houses also depend on it, as does the practice of sharing copies of books or magazines with friends or of giving purchased books to friends.

The Congress recognizes that the preservation of history—the record of social and cultural discourse, scholarship, and scientific debate and discovery—is of fundamental importance to society in innumerable ways. Both while these items are still protected as intellectual property and later when they are part of the public domain, they form an essential base upon which new artistic and scholarly work is developed. As a result, Congress has adopted a number of other rules enabling libraries and archives to promote public access to informational works. These rules have, for example, permitted the making of some copies for specified purposes, such as to replace pages from a damaged book or to preserve materials that are deteriorating, to ensure that future access to these items will be

---

[6]464 U.S. 417 (1984).

available. From time to time, stakeholders have negotiated guidelines to establish standards for such activities (e.g., the Guidelines for Educational Uses of Music (U.S. Copyright Office, 1998)). Congress has also passed rules limiting copyright in order to promote public access to copyrighted materials in classroom settings, including a privilege to perform or display such materials in the course of face-to-face teaching in the classroom.

Nonprofit institutions, such as libraries and schools, are not the only institutions that have public-access-promoting privileges. Congress has occasionally used compulsory licenses to promote new recordings of copyrighted music, jukeboxes of sound recordings, and cable distribution of broadcast network programming, all of which have public access implications.

Fair use rules also promote public access.[7] The most common form of fair use is the ability of an author to quote from previous works (thereby copying a small part of them) in order to comment on them or report news about them. The Supreme Court's decision in the *Sony* case cited public access concerns as favoring a fair use ruling about making copies of television programs for time-shifting purposes and about the impact of the ban on videotape recorders that certain motion picture studios sought.[8]

Finally, public access is also promoted by the limited duration of monopoly rights conferred under copyright law.[9] Every copyrighted work eventually becomes part of the public domain, making the work even more widely accessible to the general public by virtue of its royalty-free public domain status (e.g., books by Mark Twain, operas by Giuseppe Verdi). Unfortunately, public domain status can have an opposite effect as well. A public domain work may go (or stay) out of print if no one is willing to invest in preserving the work without the ability to assert exclusive rights as a way of recouping the investment. Contractual arrangements can attempt, as a matter of contract between two parties, to create protection beyond the date of copyright expiration.[10]

---

[7]See Chapter 4 for an extended discussion of fair use in the digital age.

[8]The reading of the *Sony* case is not uniform among committee members. Some members view *Sony's* fair use analysis as applying to private home taping of programs from cable and other forms of pay television as well as from free broadcast television, whereas other committee members view *Sony* more narrowly, as applying to free broadcast television only, which is what was directly involved in this case.

[9]Although the Constitution restricts the duration of copyright, a specific term is not specified. The last two major amendments, the Copyright Act of 1976 and the Copyright Term Extension Act of 1998, extended the term of copyright. The constitutionality of the 1998 act is challenged in *Eldred and Eldritch Press v. Reno*, filed in the U.S. District Court for the District of Columbia on January 11, 1999.

[10]The proper duration of copyright protection has always been and remains a highly

Access: Licensing Offers Both Promise and Peril

Licensing has been widely used in the software world for some time.[11] Software developers routinely distribute their products under licenses, rather than by the sale of copies. For example, the ubiquitous shrink-wrap license on a box of mass marketed software is just that, a license— one that typically recites as one of its first points that the consumer is purchasing a right to use the software and that the software remains the property of the manufacturer.[12] A variety of other licensing schemes exist, including site licenses, in which an organization purchases the right for all its members at a particular location to use the software. Licenses may also provide a natural mechanism for use-metered payment for software (i.e., paying by how much you use the program, rather than buying a copy for unlimited use). More generally, licenses for software offer the potential to custom tailor the agreement in ways that are less easily done with outright sale. Within the entertainment industry, licensing is often used (e.g., in the exhibition of movies in a theater). Access to online databases also occurs frequently under a licensing arrangement.

Licensing is a newer phenomenon for some other digital information products—particularly for those traditionally delivered in physical form. An increasing amount of the information acquired by libraries, for example, is in digital form, and unlike print materials, which have historically been available on a sale-of-copy basis, digital materials are frequently available only through licenses. Licensing can have advantages: The

---

controversial issue. For the purposes of this report, the question is whether the proliferation of digital IP, networks, and the Web materially change the long-standing debate. The committee believes that the first-order answer is no, because the underlying issues concerning the duration of copyright in the digital world are fundamentally the same as they were in the predigital world.

[11]For an overview of the licensing of electronic content, see Okerson, Ann, "Buy or Lease: Two Models for Scholarly Information at the End of an Era," *Daedalus*, 125(4):55-76. Of course, some forms of software (e.g., video games) are often sold rather than licensed. For a discussion of the licensing of electronic content and the implications for libraries, see "Accessibility and Integrity of Networked Information Collections" (Office of Technology Assessment, 1993).

[12]The law is unsettled on issues concerning the enforceability of shrink-wrap licenses. See, for example, *Step-Saver Data Systems v. Wyse Technology*, 939 F.2d 91 (3d Cir. 1991) (holding that a shrink-wrap license delivered in a package after a telephone contract was not part of the contract); *Arizona Retail Sys., Inc. v. Software Link, Inc.*, 831 F. Supp. 759 (D. Ariz. 1993) (holding that a shrink-wrap license was enforceable where there had been no prior agreement but was not enforceable where there was a prior telephonic agreement); and *ProCD, Inc. v. Zeidenberg* (see Footnote 20). See also *Vault Corp. v. Quaid Software Ltd.*, 847 F.2d 255 (5th Cir. 1988) (a state statute authorizing shrink-wrap licenses was preempted by federal law; the lower court had held that the shrink-wrap license would be an invalid contract of adhesion (a standard contract form offered to consumers without affording the consumer a realistic opportunity to bargain) were it not for the statute).

license may provide clarity on terms and conditions of access (e.g., who is included in the authorized user community and how the material may be used);[13] it may provide for an increase in the rights for the institution that go beyond those provided under copyright (e.g., the ability to make unlimited copies for local use); and it may limit the organization's liability in the case of misuse by patrons. Considering the degree of financial, ethical, and legal exposure, establishing a written agreement is increasingly endorsed by both librarians and publishers.

Licensing, coupled with a rights management and payment infrastructure that is efficient and easy to use, can facilitate access that has been impractical in the past. It does this by providing consumers with a straightforward way to access large collections of information items, often when the individual items are of low value.[14] In the absence of licensing arrangements, publishers likely would not make some of these collections available at all. Such arrangements can also allow consumers to access information at a more specific level; perhaps the day will come when instead of purchasing a book, one can obtain access to the executive summary for a limited period of time (presumably at a much lower price than the purchase of the book).[15] Thus, licensing can increase the options for making information available.

However, there are also concerns about licensing as a model of information dissemination, particularly the impact it may have on public access. The trend toward licensing means that (digital) information is in some ways becoming a service rather than a product.[16] Buy a book and you own it forever; pay for a service and when the period of service is over, you (typically) retain nothing. The increased use of licensing seems to diminish greatly the public access accorded through the first-sale rule. Consider libraries as an archetypal example. In the print world, a library's failure to renew a subscription or buy an updated version of a book has no effect on the availability to patrons of earlier volumes or editions. In

---

[13]A book in a traditional library has a limited user community by virtue of the effort needed to access it (geographic limits) and the inherent restriction on the number of patrons who can read it at once. As noted above, neither of these natural limits exists in the digital world, hence the utility of an agreement on who is to be included in the library's community of users of its digital works.

[14]For example, consumers could be able to view a digitized photograph for the payment of a license fee (which might be very small—perhaps much less than one dollar). Physical delivery mechanisms and standard payment methods (checks/credit cards and billing departments) will cause such transactions to be cost prohibitive. Licensing with an efficient rights management and payment infrastructure can cause such transactions to become economically viable.

[15]See CSTB (1998) for a discussion on pricing models for digital information.

[16]This trend toward licensing is a part of a larger shift from a goods-based to a service-based economy that is enabled in part by information technology (CSTB, 1994).

the world of licensed information, ending a subscription to an electronic journal may mean the end of access to earlier volumes or editions, as well.

A second issue arises from the nature of licenses as contracts. Contracts might not incorporate and indeed may attempt to override the public policy considerations that have been carefully crafted into copyright law. Those who contract for information may find that their access is far more restrictive than what they were accustomed to for print materials, unless fair use and other such considerations are explicitly a part of the agreement.

Some institutions (especially libraries) have worked to negotiate licenses that preserve fair use and other public access features. Publishers are currently experimenting with licensing models to respond to these concerns. Yet the concern remains about the use of a mechanism such as licensing that lacks any of the built-in protections for public access that are embodied in copyright law.

Mass marketed information products raise a more general concern about the proliferation of licenses. Where some institutions are by necessity becoming more sophisticated negotiators, the situation is far less clear for the typical consumer. Do consumers face the prospect of having to scan the tiny print of licenses to see whether fair use and other public policy considerations have been incorporated? Will they even know what such things are? Will contracts simply override some of those considerations? Copyright law currently gives owners of copies of computer software the privilege to make back-up copies; can that privilege be taken away by a shrink-wrap license in the software package that says loading the software onto the computer means one has agreed to the license terms that expressly prohibit making a backup? Does merely designating a transaction a "license" and stipulating that use of the product constitutes the user's agreement to the terms convert a transaction that would otherwise seem to be a sale into a license? If a license term prohibits making disparaging remarks about the licensed information or disclosing flaws to other potential users, should that term be enforced?[17] These and related questions have been the subject of heated debate in the past few years and

---

[17]A restriction in a negotiated contract prohibiting a licensee from disclosing flaws in the licensed information may, as a matter of law, sometimes be enforceable (even if as a practical matter it might be difficult to monitor and enforce). For example, if as a matter of private contract under a software license, a licensee agreed not to make disparaging remarks or not to reveal publicly any flaws in the software, that licensee's publication of a book or a magazine article criticizing the software and revealing its flaws may constitute a breach of contract and under the contract might entitle the licensor to seek money damages or even to seek equitable relief to enjoin the further publication of the disparaging or flaw-revealing information. There is, however, very little precedent for awarding damages, let alone injunctive relief, in such cases except in special circumstances.

are far from resolved.[18]   In the print world, efforts to impose "license" restrictions on mass marketed copies of copyrighted works were generally treated as ineffective.[19] New license models that entirely avoid the foregoing problems could be part of these new paradigms for public access. Some recent case law and proposed legislation would enforce mass market licenses in the case of digital information.[20] Nevertheless, the issue remains unsettled.

The mass market issue raises an additional concern if licensing becomes the dominant model of distribution for works that are considered part of our intellectual and social heritage. One could imagine a world in which novels, poems, and plays, for example, are available only (or mostly) by license rather than sale. The consequences of such a world for

---

The foregoing discussion, regarding the legal enforceability of a contractual waiver of fair use rights by a licensee, supposes a negotiated license agreement at arm's length between sophisticated parties. The use of such prohibitions in mass-market form license agreements, particularly as they may be embodied in shrink-wrap licenses or "point-and-click" licenses on the Internet, may present different issues of enforceability under established doctrines of law regarding contracts of adhesion and unconscionability. Neither the committee report nor the responses indicated above express a view pro or con on whether such waivers would be enforceable in the context of a mass market license.

[18]A model information licensing law (once known as Article 2B of the Uniform Commercial Code and currently known as the Uniform Computer Information Transactions Act (UCITA)), proposes (among other things) to validate mass market licenses of computer information. It regards licenses or license terms as enforceable unless they were unconscionable at the time they were made. In November 1998, this model law was amended to provide expressly for public policy limitations on the enforceability of license terms in mass market or other licensing contexts. Section 105(b) of this model law provides a basis upon which to make an argument that, for example, the public policy favoring a particular user's interest (e.g., in making a back-up copy) is strong enough to override a mass market license provision purporting to deprive the user of that interest. However, section 105(b) will likely require case-by-case determinations about whether the public policy interest asserted should override contractual provisions limiting it. In July 1999, the National Conference of Commissioners on Uniform State Laws (NCCUSL) passed UCITA. State legislatures must approve UCITA before it becomes functional as uniform state law.

[19]License restrictions on the resale of books were found to be unenforceable in *Bobbs-Merrill Co. v. Straus*, 210 U.S. 339 (1908).

[20]See for example, *ProCD v. Zeidenberg*, 908 F. Supp. 640 (W.D. Wis.), rev'd., 86 F.3d 1447 (7th Cir. 1996) (holding a shrink-wrap license surrounding the packaging for a CD-ROM to be an enforceable contract and rejecting a claim that the state law breach of contract claim should be preempted by federal copyright law). Mass market licenses of computer information (including software) would be validated, as a matter of contract law, if states choose to adopt the model licensing law presently known as UCITA, described in footnote 18 above. Section 105(a) of UCITA recognizes that federal law and policy may preclude enforcement of some contract terms that are at odds with federal policy. Section 105(b) recognizes that other fundamental public policies may preclude enforcement of certain contract terms.

public access are far from clear. This underscores the notion that intellectual property should not be viewed solely in terms of economic considerations.

Access and Technical Protection Services

Technical protection services (TPSs) are generally infeasible in the print environment. One can keep a manuscript under lock and key, but when the work has been published, further uses of it generally cannot be controlled by the rights holder. But then the need for control is not so great: Large-scale infringement of the reproduction right of a hard-copy work is generally too expensive or difficult for most members of the public, and most infringements significant enough to undermine the copyright owner's economic interests are public enough that they come to the owner's attention, directly or indirectly.

Technical protection in the digital world is far more practical, despite a somewhat checkered history of success.[21] Recently, there has been great interest in these technologies, especially among publishers concerned about the vulnerability of digital works to inexpensive, rapid copying and distribution. Many rights holders have high hopes that these systems will provide them with control over unauthorized uses, thereby enabling worldwide markets for a variety of digital information products to emerge and thrive on the Internet. And, indeed, without effective TPSs, rights holders may well choose to restrict greatly the availability of their most valuable digital content.

As with licenses, however, there may also be consequences for public access, including issues such as fair use. TPSs can allow content owners to restrict access to and uses of their works in ways not possible in the print world. One method, relying on a combination of cryptography and special hardware and/or software, would make it possible for a vendor to specify that a consumer could read a document but not print it, save an unencrypted copy, or e-mail a part of the unencrypted text to a friend.[22]

---

[21]The first copyrighted works to employ technical protection services in a mass market were copy-protected software programs. These copy-protected programs proved to be sufficiently unpopular with consumers and sufficiently vulnerable to competition by vendors willing to sell competing unprotected products that the industry ceased using them. TPSs are successfully used in the cable and pay-per-view television businesses.

[22]The user could of course capture the protected content through low tech means, such as writing it down or recording it with a camera, then transferring the content to a computer readable format. No TPS is perfect; like any security system, it relies on the deterrence that comes from making some actions sufficiently difficult or expensive (try copying an entire database by hand).

Such a transaction would clearly remove much of the public access that has been an inevitable part of more familiar information access methods.

TPSs raise public policy issues that have been the subject of considerable debate. Because these systems rely in substantial ways on encryption, new tools and technologies for decryption become of special interest. Reflecting this interest, the Clinton Administration's white paper *Intellectual Property and the National Information Infrastructure* (IITF, 1995) argued for new rules to outlaw tools and technologies whose primary purpose was to bypass (or circumvent) technical protection measures used by rights holders to restrict access to or use of their works. The white paper argued that instituting rules against circumvention would have no effect on fair use or on the public domain.

Congress was persuaded otherwise. It heeded the concerns of major copyright industry groups about the dangers to their markets from acts of circumvention and from the availability of technologies to circumvent TPSs. However, it also recognized in the Digital Millennium Copyright Act of 1998 (DMCA) that granting copyright owners the right to control public access to protected works by outlawing the circumvention of technical protection measures was an unprecedented and significant extension of author and publisher rights, one that might have a negative impact on fair use.[23] In addition to carving out several specific classes of circumvention activities that it found, on balance, to be socially useful, Congress provided for a 2-year moratorium on the ban on the circumvention of access controls, during which time the Librarian of Congress is supposed to determine whether the use of technical protection measures is having an adverse effect on the ability to make noninfringing uses of a particular class of copyrighted works. If the librarian finds such an adverse effect, he can identify that class of works or users so that the ban on access control circumvention will not apply to that class of works or users for the next 3 years. The report of the Librarian of Congress on the impact of circumvention legislation, which will be published in the year 2000, is likely to be of considerable importance in addressing public access concerns raised by these systems.[24]

In summary, some people are optimistic about the prospects of technical protection services to enable far greater public access to copyrighted works than has ever been possible previously, and they view the

---

[23]Even if the circumvention of technical protection measures is deemed ultimately to be legal when done in pursuit of access and fair use, such access is then accorded only to those with the technical skill to circumvent effectively or the luck to be interested in content whose protection has already been circumvented.

[24]This report will presumably be written by the staff of the U.S. Copyright Office, a major service unit of the Library of Congress.

anticircumvention regulations as a necessary protection against those who would otherwise bring serious harm to the copyright industries. In contrast, there are those who view TPSs as deeply threatening to public access and to other societal values reflected in copyright law and policy, including promoting freedom of expression, innovation, and competition. This group believes that this threat could be realized as a result of steps toward other desirable goals, such as electronic commerce, yet may have serious consequences nonetheless (this scenario is discussed in Chapter 5). These two views represent the extremes of the spectrum. There is a third group of people who see the likely impact of TPSs to be mixed. This group believes that the implementation of some TPSs will facilitate public access, whereas other uses of TPSs will reduce public access. The perspective of this third group is bolstered by the possibility (see Chapter 5) that TPSs will not be as successful, either technologically or in the marketplace, as their promoters hope.

The committee does believe it naive, however, to think that TPSs, if widely deployed, would have no impact on fair use and, as a consequence, believes that policy makers should continue to monitor that impact and be prepared to act if necessary. How effective TPSs will be and how much acceptance they will achieve with consumers of information remains to be seen. Policy intervention may prove unnecessary, if the market adjusts to provide broad access to consumers while still offering appropriate levels of technical protection.

The New Information Environment Challenges Some Access Rules

The discussion of the exceptions in copyright law for public access is important because there is reason to believe that the change to digital distribution could make a number of those exceptions less applicable and less effective. For example, consider the situation of libraries or other institutions that acquire digital copies of protected works, if the first-sale rule of copyright law no longer provides them with an established legal basis for their lending practices. The white paper expressed the view that the first-sale rule does not apply to digital materials because first-sale limits only the *distribution* right. Lending digital materials to patrons would, however, inevitably involve the making of copies of the materials, copying that the white paper asserts is not privileged by the first-sale rule. Under this view—which some dispute—libraries could not lend digital materials to patrons unless they had negotiated a license to do so.[25] The

---

[25]Some have argued that a temporary copy made to transmit one's copy of a digital work to another person could be justified as a fair use, necessary to enable the first-sale right, as

white paper also predicted a diminishing role for the fair use rule in the digital environment.

In addition, such established practices as interlibrary loans, document delivery services by libraries, and copying for course reserve purposes, particularly if in electronic form, are undergoing serious challenges. Publishers assert that circumstances have changed since guidelines permitting these activities were negotiated, making them unsatisfactory. Putting a paper copy on reserve is one thing, making a (widely accessible and easily copied) version available online is, they argue, something else again.

The New Information Environment Blurs the Distinction Between Public and Private

Public access to informational works has been promoted by the existence of a relatively clear distinction between works that were published (in the legal sense embodied in the copyright law) and those that were not. This distinction is often critical in determining whether the information in the work (e.g., the ideas and facts in it) is public and can be reused and redistributed without permission of the copyright owner.[26] Public access to facts and ideas in published works has been part of the copyright normative structure. As the U.S. Supreme Court said in its venerable *Baker v. Selden* decision:

> The very object of publishing a book on science or the useful arts is to communicate to the world the useful knowledge that it contains. But this object would be frustrated if the knowledge could not be used without incurring piracy of the book. And where the art it teaches cannot be used without employing the methods and diagrams used to illustrate the book, or such as are similar to them, such methods and diagrams are considered as necessary incidents to the art, and given therewith to the public.[27]

In another famous U.S. Supreme Court decision, *Feist Publications v. Rural Telephone Service*, the court explained that copyright could not be used to protect facts, because appropriation of them from published works

---

long as one deleted one's copy after the transmission. Making this argument for the "lending" of digital copies by libraries is, of course, more complicated.

[26] An author's particular *expression* is protected under copyright law, but the facts and ideas in the work are not. Whether published or unpublished, those facts and ideas are in the public domain and may be freely reused under copyright law. See for, example, *Salinger v. Random House*, 811 F.2d 90 (2d Cir. 1987).

[27] 101 U.S. 99 (1879).

was an important means by which the constitutional purpose of copyright—promoting knowledge—could be achieved.[28]

Of course, the choice about whether to publish remains with the author. Some may prefer to commercialize their facts and ideas without publishing them (e.g., by licensing them). Public access is not such an important value in the copyright system that the law imposes on authors a requirement that they must publish their works. Nor is it the case that the distinction between "published" and "unpublished" works has always shimmered with clarity. Nevertheless, the accumulated case law aids in drawing this important boundary line. And the general rule applied: The predominant way that an author could exploit a work commercially was to publish it, and, once published, both the law of copyright and the institutional infrastructure for usage of works meant that public access to information in the work could generally be had. In the print world, the distinctions between public and private, and published and unpublished, have been generally clear. Printed information is typically either one or the other, and where it is public and published, public access expectations arise.

Problems arise in the digital environment because it is not always easy to tell when information has been published and when it has not. The extreme cases may be clear: Placing a copy on a publicly accessible Web page would almost certainly be viewed as a publication,[29] and the information on the Web page would be considered "public."[30] Moreover, this kind of posting is considerably more "public" than a print world act of distributing the same text in a flyer or posting it on a public bulletin board. If instead one loads the only copy of the same text on a password-protected computer system, it is clearly a private and unpublished copy, even if a small number of others are authorized to access it.

But numerous cases also exist in the digital world in which the line between public and private, published and unpublished, is blurred. What if, for example, someone distributes information via a list server (software that distributes information to a specific list of e-mail addressees); is that publication or not? Does it matter how large the list of addressees is or whether outsiders can join the list without permission? What if a user has

---

[28]499 U.S. 340 (1991).

[29]Publication on the Web raises other interesting questions, such as, Where has publication occurred? This is important because copyright laws vary among countries. Suppose a citizen of the United States publishes an article on an AOL-owned (a U.S. company) server based in Germany and that this server has a mirror site in Hong Kong. Which is the country (or countries) of publication? Could one claim that publication on the public Internet is simultaneous publication in all countries of the world?

[30]In this context, "public" is not equivalent to public domain.

to register with a Web site to get access to its contents and use a password; is that site public or private? What about posting information on a multi-firm intranet? What if a scientific organization establishes a system to post draft papers on a site that is intended to be accessible only by members of the appropriate scientific community, but they are from many different institutions; are the papers posted on that site published or unpublished?

A second difficulty arises from the detailed level of control computer systems provide over access to information, making it feasible to provide specific, multiple layers of conditional access. The result is information availability that comes in multiple shades of conditional gray. One class of user may have one level of rights for one set of purposes and another class a different set of rights to other information for other purposes. Whether and under what circumstances each of these situations constitutes publication has significant consequences for public access to the information in the work, but deciding whether the work has been published may be difficult.[31]

Thus, what appears to be a relatively simple distinction in the print world—published vs. private—becomes complex in the digital world, where the boundaries between public and private are blurred. The result is a challenge to our notion of how to make this distinction. The concern also exists that more informational works will be distributed in ways that will be labeled as private, keeping them inaccessible to the public and out of our cultural and intellectual heritage.

Noncopyrightable Databases Present Access Challenges

IP protection of databases is a special case, because in 1991 the U.S. Supreme Court, in *Feist Publications Inc. v. Rural Telephone Service Co.*, restricted protection under copyright so that straightforward collections of facts presented in the obvious way (e.g., an exhaustive listing of phone numbers, presented alphabetically) would not be protected, rejecting the perspective that "sweat of the brow" was grounds for protection. This ruling left database producers with what they deem to be insufficient legal protection for their products,[32] products in which they may have made substantial investments. Legal reforms have recently been pro-

---

[31]Also, by making it much easier to distribute portions of works in varying states of completion, the fluid nature of digital networked information further contributes to the blurring between published and private distribution.

[32]Most databases have a sufficient modicum of originality to be protected by copyright. In addition, contract and unfair competition law, as well as sui generis laws in some countries, provide legal protection to database producers.

posed, both internationally and nationally, to address IP protection of databases, but none has yet been enacted into law in the United States or agreed to by treaty or other international agreements with the United States.

There are many kinds of databases in terms of content, use and source, and the optimum IP policy may differ among them. Government databases, for example, fall into three categories: internal records, reports and background data for reports (e.g., census, budget), and databases created through grants and contracts to nongovernment organizations or individuals. In the nongovernment sectors, numerous types of databases exist, including internal records, records about others (e.g., customers, patients, suppliers), research results, compilations from diverse sources, and so on. The generators of the digital data and information that appear in electronic databases are governments at all levels, the not-for-profit and commercial sectors, and individuals. Some of these entities are both producers and users of databases (e.g., government agencies and researchers), whereas others are primarily producers (e.g., publishers) or primarily users (e.g., private citizens). For both the producers and users of databases, the same advances in information infrastructure that are affecting IP for copyright and patents also provide unprecedented capabilities for disseminating, accessing, copying, and manipulating databases.

In 1996 the European Union issued its Directive on Databases, which requires members to increase significantly the level of protection of databases and explicitly does not provide protection for databases produced in non-E.U. countries that fail to adopt measures affording comparable protection to the E.U. Directive in their laws.[33] In 1997 the World Intellectual Property Organization (WIPO), at the European Union's request, initiated a draft worldwide database treaty largely incorporating the E.U. Directive. The treaty would have given database owners an exclusive right to prevent a temporary or permanent transfer of all or a substantial part of a database to another medium without the rights holder's permission, and it limited the right of treaty countries to provide (public good) exemptions that would interfere with the normal exploitation of the database or the legitimate interests of the rights holder. Because of protests by the scientific community, the United States backed away from signing the treaty after initially favoring it. This action and objections from developing countries halted the treaty negotiations. In 1999, WIPO is reconsidering database protection. Because databases are used globally and access via the Internet cannot easily be restricted to national users, continuing to pursue mutually acceptable international agreements that harmonize the IP protection of noncopyrightable databases is important.

---

[33]See Directive 96/9/E.C. of the European Parliament and of the Council of March 11, 1996, on the legal protection of databases, 39 O.J.L. 77/20, March 27, 1996.

The U.S. Congress introduced legislation in 1996, 1998, and 1999 patterned on the E.U. Directive, although nothing has yet been enacted into law, because of the conflicting views among the stakeholders.[34] The alternatives for IP protection of databases that have been advocated include de minimus changes to existing law, unfair competition/misappropriation models, sui generis models that are close to the E.U. Directive and provide strong property rights to the producers of databases, and the sort of technical protection mechanisms discussed in Chapter 5 of this report.[35]

The Information Infrastructure Is Changing the Distribution of and Access to Federal Government Information

From the earliest days of our nation's history, Congress recognized its responsibility to inform the American public of the work of the federal government. One way it did this was by establishing the Federal Depository Library Program (FDLP) to provide no-fee, geographically dispersed access to government publications. By designating depository libraries in each state and congressional district, Congress ensured that government information from all three branches would be distributed throughout the country and available at no charge to the user. This system reflects a commitment to broad-based democracy and public accountability—principles that are as important today as they have been in the past. The advent of the information infrastructure and the enormous opportunities it offers for citizens to access information has led to both the expected increase in access and, paradoxically, to some situations of sharply diminished access. It has also raised questions about the future role that the FDLP should play.

Online systems such as the Government Printing Office's Access system, the Library of Congress's THOMAS system, and the Security and

---

[34]Efforts to enact legislation to provide protection for databases that do not otherwise qualify for copyright are taking place in the 106th Congress through H.R. 354, the "Collections of Information Antipiracy Act," and H.R. 1858, the "Consumer and Investor Access to Information Act."

[35]A recently released NRC study, *A Question of Balance: Rights and the Public Interest in Scientific and Technical Databases* (NRC, 1999), addresses the scientific, technical, economic, legal, and policy issues regarding the creation, dissemination, and use of scientific and technical data for basic research and other public interest uses, with particular emphasis on the recent developments in IP law in noncopyrightable databases and what IP regime(s) could best accommodate them. An earlier NRC report, *Bits of Power: Issues in Global Access to Scientific Data* (NRC, 1997), provides background for the 1999 report. The recent study addresses database IP issues both in general and specifically as they relate to science and technology from the perspectives of the government, commercial, and not-for-profit sectors and considers the concerns raised by the different science and technology stakeholders.

Exchange Commission's EDGAR system have vastly enhanced public access by providing free, online access to government information. These systems (and many more) are available through the Internet at no extra cost to the user, illustrating how some federal agencies are succeeding in using information technology to enhance public access.

However, in some other agencies, trends are toward the decentralization, privatization, and commercialization of government information. As a consequence, some government information that used to be accessible to the public at no cost through the FDLP might not be accessible any longer (either through the FDLP or the public Internet), with the shift toward government information created or disseminated in digital form. Broad access and use of publicly funded information are substantially impaired when distribution agreements prevent or curtail distribution of information.[36] For example, depository libraries can obtain the *National Criminal Justice Reference Service CD-ROM* for their collections but have to pay subscription costs for access to the Internet database that contains the actual reports.[37]

The response of government agencies has not been homogeneous because different agencies have different relationships and interests with respect to the various kinds of government information. Agencies vary in the extent to which they seek to protect the information that they generate in carrying out their task, with these differences arising from the nature of the responsibility of the agency and its relationship with the rest of the government, especially Congress. The bottom line is that some agencies are charged with the responsibility of disseminating data and new technical knowledge as widely and inexpensively as possible, while others are sometimes given the authority—and a strong financial incentive—to sell information to others. When information is sold, some agencies must pay high prices for data or products that were originally created by other agencies using public expenditures.

Some agencies are explicitly given the job of creating or maintaining important information products. One class of example is the data-based

---

[36]The federal government itself cannot own a copyright to works that it creates. However, private sector organizations can take works of the federal government, add value to them, and own the copyright to the resulting work (assuming that the new work satisfies the requirements for copyright).

[37]Another instance of reduced access occurs when agency CD-ROMs or Web sites are available to depository libraries, but their use is restricted to only one password that must be shared by many people in the congressional district. The Department of Commerce's *STAT-USA* is an example of an information service created by an agency that operates under a cost recovery mandate. Depository libraries are limited to one password to *STAT-USA*, a valuable database that contains thousands of titles that are no longer available in print. Depository libraries that need to provide better access to this information must pay for additional passwords.

agency (e.g., U.S. Bureau of the Census). For the most part, the information products are evaluated on the basis of the quality and utility of the information that they contain. These agencies are extremely interested in disseminating these products as widely as possible, as long as the integrity of their information can be assured. The information that is created from such responsibilities is frequently disseminated widely at little or no cost, and it is sometimes posted on well-managed Web sites that provide unprecedented access. The National Archives and Records Administration is an example of a federal agency chartered to maintain federal government records for historical purposes.

Many agencies are responsible for managing the records of government that reflect the fulfillment of their missions (e.g., the FBI or the courts). For the most part, these agencies either have no particular interest in whether the information they possess is easily accessible or prefer that their activities remain confidential unless they are forced to reveal information through litigation or Freedom of Information Act requests.[38] Sometimes, these agencies allow private parties to copy their internal records for the purpose of packaging and reselling them, usually at high prices, including to other government agencies. Agencies that perform research and development, especially in information technology, can sometimes have some disincentives to be open with their results in order to have valuable knowledge that can be used to attract private partners.

State and local governments are also greatly affected by the shift to electronic distribution. In general, there is far less precedent for viewing information from state and local governments as being in the public domain, and there is a correspondingly wide variability in policies and practices among jurisdictions.

ARCHIVING OF DIGITAL INFORMATION PRESENTS DIFFICULTIES

Historically, most of the materials preserved in archives are in printed form (including printed transcriptions of spoken words) and in the physical arts (painting, sculpture, architecture, etc.). In the last century, the print record has been supplemented and enriched by sound recordings, photographs, moving images, and most recently various forms of digital

---

[38]An example of an important record of government is the contracting information from a federal agency. For the most part, contracts are part of the public record, but agencies often do not facilitate general access to them. In some cases, agencies have allowed others to collect and package contracting information. For example, the quasi governmental, nonprofit RAND Corporation is the sole source of information about defense contracts, which is available to scholars or others undertaking research on defense contracting for a fee in the thousands of dollars.

information, which may have been "born digital" (i.e., created without a physical embodiment, as for example a Web page).

Preservation of these materials has been carried out for society by libraries, archives, and museums and similar cultural heritage institutions, frequently with public support and funding. When published, materials are available for consultation by the public through these institutions even though the publisher may cease to make them available (for any number of reasons, including lack of interest or lack of commercial viability). After they pass into the public domain when copyright expires, materials can be reused or redistributed to make them widely available again.

The keys to making this system work have been the publication and distribution of physical artifacts and the doctrine of first sale. A library, for example, can purchase a book or a journal and then retain possession of that copy of the work. Because of the first-sale rule and other copyright provisions, the institution can preserve it indefinitely and make it available to members of the public. This ability to make the work accessible in turn supports the library, as patrons see the value in supporting the institution.

Overall, the system has worked fairly well for print, despite a number of difficulties. Libraries have had to grapple with the problem of holdings deteriorating because they were printed on acid paper, and they inevitably have to make funding-based decisions about what to preserve.[39] Of course not everything is preserved: Not everything published is acquired by libraries, and not everything acquired is retained indefinitely. Nor are all important printed works published for use by the general public; archives play an important role in such cases, particularly in capturing the processes of government. Yet we have been able to retain the essential core of our printed cultural heritage and enough examples of other materials to provide a good sense of what has not been saved.[40]

---

[39]Unlike earlier printed products, acid paper does not thrive on relatively benign neglect; it deteriorates on the shelves. Libraries have had to spend a great deal of money microfilming or chemically treating deteriorating works to preserve them, and the funds available to support this work have been far less than what has been needed.

[40]It is worth noting, although outside the scope of this report, that the digital environment may place a heavier burden on the government to implement good archival practices. Previously, the FDLP provided a mechanism for the distribution of key government documents to libraries, which then preserved them. Now, with documents available over the Internet, libraries do not add them to their "collections." If these documents are to be preserved, the government will need to assume the archival responsibilities for the electronic documents that it produces, unless other reliable institutions can be found to accept this responsibility.

The archival system has, however, been much less successful with other media of the 20th century. It is important to briefly recount the situation here for the insights it can offer into the fate of the cultural record in the digital era.

Much of the early history of film and sound recordings is lost, although we do have many treasures from this period. Some of the problems were cultural: These new media were not recognized as important parts of the social record until they had achieved a certain level of maturity, popularity, and acceptance and thus weren't saved early on. Part of the problem is technical: Many of the early recordings were made using equipment that is no longer available, and the media on which they were made have deteriorated (in the case of nitrate film stock, it has in some instances spontaneously combusted!). Part of the problem had to do with business models: Until the VCR appeared in the 1980s, most films were simply not available for purchase by cultural heritage institutions. Thus, although they had a huge public impact and were viewed by millions, therefore unquestionably qualifying as part of the social record, they were in effect never published and hence could be preserved only by their owners or by archives if their owners donated them to such institutions.

Historically, copyright deposit has been another important factor in ensuring the archival availability of our cultural heritage, allowing the Library of Congress to amass an extensive collection of printed materials that it has been able to hold in trust for the nation.[41] Sometimes, obtaining voluntary compliance with deposit rules proves difficult. However, the digital world vastly increases the problem. Most publishers will understandably object to an open-ended deposit system for electronic copies of their works, because it could enable anyone, anywhere, at any time of day to access the works in an electronic Library of Congress.[42] However, if the Library of Congress received no deposits of digital works,

---

[41]Although the deposit of a copy of the work in the Library of Congress has been required as a condition for the registration of the copyright, U.S. copyright law does not require such registration as a precondition to copyright protection. Copyright protection automatically inheres in any work that is fixed in a tangible medium of expression, whether it is registered or not. There are, however, a number of practical and legal advantages to registration, including the ability on the part of the rights holder who has registered to receive, under certain circumstances, reimbursement for attorney's fees and statutory damages in a copyright infringement action. Indeed, it is an absolute precondition to bringing a copyright infringement action that the work must first be registered for copyright. So, although copyright registration is not required for protection, it is required if a copyright proprietor wishes to take practical steps to effect such protection, in the form of bringing a copyright infringement action.

[42]That is, publishers would likely object in the absence of technological protection mechanisms that limited the access, copying, and redistribution to a level that would not have an adverse impact on publishers' ability to derive a reasonable profit from their works.

it could offer far less public access to works in the future than it does at present.[43]

Copyright has played an important role in the preservation of many earlier nonprint materials. In the case of films, for example, series of still frames were deposited because film was not yet covered by copyright. These printed strips of still frames turned out to be the only extant copies of some early movies. As these works have moved out of copyright protection and into the public domain making them available to the public again as part of the nation's cultural heritage has been possible—in the case of film, by turning the sequences of still frames back into film. No longer required as a condition of obtaining copyright, copyright deposit can no longer be relied on by society as a means of ensuring the preservation of unpublished materials (which are still protected under copyright). Although it is certainly true that the Library of Congress can, and probably should, be more aggressive in demanding deposit of copies of published works, a tremendous amount of digital material exists that is never "formally published" (to the extent that this definition is understood in the digital environment) and for which it would be difficult to compel copyright deposit.

As the line between publication and limited distribution becomes ever more ambiguous in the digital information environment, the prospect grows of an ever larger mass of technically unpublished material that isn't subject to deposit. In particular, there will be a vast range of ephemera and of new genre material that will not be subject to deposit regulations if the criterion is one of publication. There is also the issue of what can be done with material that is deposited, at least while it is still under copyright—who, if anyone, can view these materials? Finally, in a global world of information, deposit policies tied to the nation of publication are an increasingly awkward proposition.

Fundamental Intellectual and Technical Problems in Archiving[44]

Fundamental technical and intellectual problems are involved in extending the process of archiving beyond printed materials. Once pub-

---

[43]The developments thus far do raise concerns, considering that the deposits of digital materials to the U.S. Copyright Office have been modest at best.

[44]The discussion of archiving in this report is not intended to serve as a comprehensive review of the topic but to highlight that subset of archiving issues germane to a discussion of digital IP. Many challenges exist in the general area of archiving. Some interesting work is currently ongoing, such as the Intermemory Initiative (see <http://www.intermemory.org>), which aims to develop highly survivable and available storage systems composed of widely distributed processors. Discussion of the digital archiving issue will be included in the

lished, printed works are static, and, unless printed on acid paper, they will last a long time in a reasonable environment with little attention or effort. Sound, film, and similar materials are also fixed once published, but they are on media that have variable and often rather short lives. In some cases, works cannot be directly viewed in any practical way because they require equipment that has grown increasingly complex over the years as it has evolved and which, in some cases, is no longer available (e.g., wire recordings and early video recordings for which the recording format is no longer known).[45]

Although offering the possibility of perfect preservation, digital information also raises many pragmatic barriers to long life. It is often stored on media with short life spans; it may require reading equipment that has an even shorter life span; and even transferring the data to another medium is not enough: Software may be needed to interpret and/or view the data. Thus, unless the reading software is also preserved (which may involve in effect preserving entire computing environments), some digital information cannot be meaningfully archived for long periods. Added complexity is associated with some of the new digital works that contain dynamic, interactive digital documents, because they are not fixed in form at publication; they evolve and change. As a result, exactly what the appropriate archival practices are for capturing the essence of these new genres is unclear. Indeed, these practices likely will vary from genre to genre and, in some cases, unless an archival function is designed into the digital object explicitly, fully archiving a record of its evolution over time may be impossible. One effort—The Internet Archive—is capturing wholesale snapshot copies of substantial portions of the Web (on what many view as very shaky copyright grounds), which will undoubtedly become important archival records.[46] Yet the data collection tools used by the Internet Archive to traverse the Web take weeks to capture a single snapshot; so, for some archival purposes, this method may not record

---

forthcoming CSTB report of the Committee on the Information Technology Strategy for the Library of Congress. Additionally, the CSTB proposes to undertake a separate study on digital archiving in the near future.

[45]Of course, building new wire recording players is possible, although at considerable expense. One can also envision a time in the not too distant future when turntables for records will share a similar fate. As players for various recording technologies involve complex magnetic sensors, integrated circuits, embedded software, semiconductor laser technology, and other specialized parts that require an enormous and costly infrastructure to produce, building new players for obsolete media at any reasonable cost becomes less likely.

[46]See <http://www.alexa.com> or "The Digital Attic: Are We Now Amnesiacs? Or Packrats?" available online at <http://www.around.com/packrat.html>. The Internet Archive Web site is at <http://www.archive.org>.

sufficient detail in the evolution of volatile, dynamic Web sites. And there is a great deal of material on the Web or accessible via Web interfaces (e.g., databases) that is not subject to capture by the Internet Archive, either for technical reasons or because the site owners have opted out.[47]

In the case of various types of sound and moving image recordings, one strategy for overcoming media deterioration and the obsolescence and ultimate unavailability of playback devices is to periodically and systematically copy the recordings from older media to newer ones. However, this in turn requires that the owner of the copy have legal authority under the copyright law to take this action for preservation purposes. In fact, the copyright laws have permitted limited copying for preservation purposes, such as the microfilming of books. For the first time, the DMCA permits digitization as a means of preserving printed materials; until the DMCA, digitization was generally considered reformatting, rather than copying, and thus outside of permitted action under the copyright laws.

For digital materials, a similar strategy needs to be followed of copying data from one storage medium to another, as storage media deteriorate and technology changes. This objective is similar to the practices that large data centers have carried out for the last few decades in copying tapes every few years, both to protect against media deterioration and to shift from older to more modern generations of magnetic tape technology. Clearly, if digital information is to be preserved, such copying will need to be permitted under the law.

A number of strategies have been proposed to circumvent the obsolescence of document formats and unavailability of software to interpret older formats. One approach is to reformat documents periodically, converting them from older to newer formats so that they can be read by currently available software. This process, which goes beyond simple copying and might be considered the creation of a derivative work, will also need to be legal if the meaningful preservation of digital information is to occur.

The technical problems of managing digital information into the future are formidable and have been well documented elsewhere; they are not the primary focus of attention here.[48] The key points are that if archives cannot obtain the digital materials that need to be archived and if they cannot obtain clear legal authorization to manage them across time,

---

[47]There is evidence that the search engines used by most people capture a relatively small percentage of the public Internet. For example, Lawrence and Giles (1999) found that no search index captures more than about 16 percent of the Web, which represents a substantial decrease in percentage terms from December 1997.

[48]See, for example, the 1999 NSF Workshop on Data Archival and Information Preservation (Rothenberg, 1999; CPA/RLG, 1995).

then no amount of progress on the technical problems will make any difference. Even if archives can legally obtain digital materials and can manage them (including reformatting as necessary), they will surely find it difficult to obtain the economic support for doing so for a century or more if they cannot provide the benefits that were formerly available to society in terms of limited consultation under the doctrine of first sale.[49]

Intellectual Property and Archiving of Digital Materials

Large-scale archiving of the cultural record requires resolution of two key legal issues—the ability to make copies when migrating from one storage technology to another, and the ability to reformat, thereby creating derivative works when moving from one software technology to the next. But even if these issues could be settled, another issue remains: Libraries and similar cultural heritage institutions continue to be dependent on the framework of publication and first sale for the acquisition of most materials.

This dependence is becoming increasingly unworkable because more and more information is being provided under license. Rights to archive can be negotiated by libraries as part of a license for information; many research libraries are starting this process. With scholarly publishers, who share a common interest with the library community, authors, and readers in ensuring the archival viability of their electronic publications, these negotiations seem to be reasonably successful. With mass market publishers, the likelihood of success is less clear.[50] The information supplier is under no obligation to agree to these terms—compare this to the print world, where one literally could not publish and sell a book commercially without having the material considered for archival preservation in libraries.

Also, as indicated earlier, a significant amount of material is now being made available under apparently ambiguous terms with limited distributions. It is not necessarily published and placed on the market under the first-sale doctrine, but it is available for viewing on the Web

---

[49]The focus of this section is on archiving by cultural heritage and research institutions. However, the issues concerning archiving are also applicable to the consumer realm. For example, consider a consumer's LP record collection today and, perhaps, audio CD collections in the future.

[50]Some mass market information suppliers control the availability of their content to the public to maximize revenue. Disney and others have done so with their films for decades; other mass market content suppliers want to resell the same content repeatedly as popular media change—witness the shift from LP to CD and now to DVD. The replacement of the LP by the CD contributed considerable revenue to the music industry by permitting it to resell essentially the same content.

and on television and for listening on the radio. It is far from clear that libraries or other cultural heritage institutions have the right to capture this material, much less copy it across generations of media, reformat it, and make it available to the public for reference. Unless some process is initiated to permit this material to be collected by our cultural heritage institutions, such material could be omitted from the heritage of our society.[51]

Information distribution through sale and the first-sale doctrine has been an important framework for ensuring the preservation of that information for future generations. The information need not even be sold for it to be preserved: There is a significant body of ephemera that libraries have acquired; once the library had a legal copy, it did not have to worry about who held the rights to the underlying intellectual property but could simply keep and circulate the artifact. In the digital world one must always go to the expense of sorting out and clearing rights, even for ephemera.

In addition, is there a point at which a work becomes sufficiently public, even if not published in the existing legal sense, that it should become part of the collections of our cultural heritage institutions, to ensure both continuous availability to the public (at least on some limited basis, as libraries' holdings are available today) and preservation for access by future generations?

Complex and difficult balances will need to be established, as information increasingly becomes an event to be viewed or experienced, rather than packaged as an artifact to be kept and archived. Rights holders clearly should be able to limit the distribution of their content without running the risk that their works fall victim to some sort of intellectual eminent domain; creators should be able to make works available as "events" and to withdraw access to these events; and rights holders should be able to engage in revenue optimization strategies in marketing materials. This degree of control implies that authors and rights holders will be able to prevent the passage of some works into the cultural heritage. Yet there also needs to be a place for the capture of cultural history and social memory—at some point events are both common enough (widely enough available to the public) and important enough to become part of that record. Social agreements need to be developed about the status of information as event, and these agreements need to be reflected in the law and in practice; the legitimate interests of artists, commerce, and society must all be balanced.

The conflicting forces here are profound. The nightly news on a major television network is seen by millions of people and may play a

---

[51]For an in-depth discussion, see CPA/RLG (1995).

significant role in shaping public opinion. One could argue that such a broadcast should be available for study by future scholars and indeed by the public at large, at least under some constrained circumstances (e.g., viewing in a library). On the other hand, a concert or theatrical performance is a performance (i.e., an event intended to convey a one-time experience to a specific audience). The performers have no obligation to preserve their event for future scholars, and they might reasonably expect that no recording of the event be made without their permission.[52] It is not clear how to draw distinctions and boundaries between these cases: One factor might be the public interest; another might be the scope of the audience (limited attendees versus unlimited availability to the public).

Technical Protection Services and Archiving

As indicated, preservation of digital materials is a difficult and far from fully solved technical problem, and technical protection services for intellectual property make this problem even more challenging.[53] If a digital object can be read only through some type of proprietary, secure content-distribution software, this software must be migrated from one generation of hardware to the next; if the system really is secure, then the ability to preserve the content (even if legally permitted) depends totally on the continued viability and commercial availability of the secure content-distribution software. Even with legal authorization, a library could no longer be confident of its ability to migrate digital content protected by such a system. Similarly, new levels of complexity are added by any type of protection mechanism that tries to limit use of a digital object to a single specific machine, tries to report use of that object over the Internet periodically (e.g., in conjunction with a metering or payment system), or gets authorization over the Internet for each viewing. This complexity is likely to be at odds with the long-term ability to preserve or access the digital object, particularly as it becomes of little or no commercial value, causing the publisher to lose interest in preserving its end of the metering or reporting system. Designing into technical protection services features that facilitate migration (e.g., moving from an older

---

[52]The Vanderbilt University Television News Archive holds more than 30,000 videotapes of individual network news broadcasts and more than 9,000 hours of other news programming. The archive is developing a plan for digitizing its holdings so that researchers can watch recordings via the Internet. The distribution of these broadcasts using the Internet would amount to a retransmission, argues the broadcasters, and therefore would require a license, possibly accompanied with a fee. Of course, the broadcast networks have the discretion to grant a license—or not (Kiernan, 1999).

[53]See the Chapter 5 section entitled "Technical Protection" and Appendix E for an in-depth discussion on technical protection services.

version of secure content-distribution software to newer versions) is certainly possible, but one has to worry about whether software with these capabilities will be made available in a competitive marketplace without a legal requirement.

Existing technical protection services do not, as far as the committee has been able to determine, include an express "self-destruct" mechanism triggered when the copyright on a work expires and it finally enters the public domain. The design of such a mechanism would be tricky, because this date is set at life of the author plus a fixed number of years (or at publication plus a fixed number of years for works of corporate authorship). In the former case, the protection software would in principle have to be able to determine when the author died in order to decide when to cease operating.[54] One can envision situations in which works in the public domain remain entangled with protection software that still attempts to limit access to the work. This largely unforeseen consequence of the recent legislative endorsement of technical protection services may, in the long term, run counter to the public good of a healthy and accessible public domain (Lynch, 1997). An assessment is needed to consider whether the requirements for disengagement of a TPS should be part of the legal and social constraints on the deployment of such protection services in the first place.

Although the focus here has been on the potential difficulties raised by technical protection services, they can also have a positive effect on archiving. For example, the lack of effective TPSs may cause content owners to avoid making their works available to the public in digital formats in the first place. This lack of availability could have serious implications for the public good. The committee acknowledges that the foregoing discussion of archiving and TPSs presumes the existence of works and focuses on the issues for archival institutions and archiving processes. Clearly, authors and publishers need to have appropriate incentives and protections so that they create and distribute content in the first place.

---

[54]The details of authorship are part of the copyright management information protected under the DMCA. Formally recording this information may ease the transition of works into the public domain on several levels. First, it may make determining when a work enters the public domain less costly, and second, it may make automatically disengaging technical protection feasible.

4

Individual Behavior, Private Use and Fair Use, and the System for Copyright

As discussed in Chapter 3, the information infrastructure creates both opportunities for and concerns about public access to information and archiving. Continuing the inquiry into the consequences of the information infrastructure for intellectual property, this chapter considers impacts on individual behavior and discusses the difficult concepts of fair use and private use. In addition, it raises the question of whether the current regime of copyright will continue to be workable in the digital age or whether some of the basic legal models for intellectual property (IP) need to be reconceptualized.

UNDERSTANDING COPYRIGHT IN THE DIGITAL ENVIRONMENT

Earlier sections of the report note how the technology of digital information has vastly increased the ability of individuals to copy, produce, and distribute information, making the behavior of individuals a far more significant factor in the enforcement of IP rights than in the past. Yet we as a society apparently know relatively little about the public's knowledge of or attitude toward intellectual property. The committee found no definitive or widely recognized formal research on this issue, only circumstantial evidence that most people do not have an adequate understanding about copyright as it applies to digital intellectual property.

The General Public

A number of copyright-related myths and urban legends have circulated on the Internet; they are sufficiently widespread that some of the industry trade associations have taken steps to debunk them.[1] In the world of digital music, for example, some misconceptions include the claims that the absence of any copyright notice on a Web site or on a sound file (commonly an MP3 format file) indicates that the recordings or the underlying musical composition have no copyright protection and are freely available for copying; that downloading a copy for purposes of evaluation for 24 hours is not an infringement; that posting sound recordings and other copyrighted material for downloading is legally permissible if the server is located outside the United States because U.S. copyright laws do not apply; that posting content from a CD owned by an individual is not an infringement; that downloading sound recordings is not an infringement, and so on.[2] As discussed in Chapter 2, the extent of the unauthorized copying of copyrighted material posted on the Internet and the apparent ignorance of the rules of copyright are particularly compelling in regard to digital music files.[3]

Other misconceptions concern print, graphics, or other visual content. Some of these are that if the purveyor of the illegal copies is not charging for them or otherwise making a profit, the copying is not an infringement; that anything posted on the Web or on a Usenet news group must be in the public domain by virtue of its presence there; that the First Amendment and the fair use doctrine allow copying of virtually any content so long as it is for personal use in the home, rather than redistri-

---

[1]See, for example, <http://www.audiodreams.com/mtvhits/> (stating, "This page is non-profit and audio files can be downloaded for evaluation purposes only and must be deleted after 24 hours"); <http://www.warezrevolution.com> (indicating "You must delete everything after 24 hours and use it for educational purposes only. All files contained here are only links and are not from our server"); <http://www.fullwarez.com> (indicating, "This page in no way encourages pirated software. This page is simply here to let you TRY the applications before you decide to buy them. Please delete whatever you download after testing them and let the hard working programmers earn their money"); <http://www.escalix.com/freepage/thehangout/disclaim.html> (stating, "MP3's are legal to make yourself and keep to yourself, but illegal to download [sic] publicly and keep them, as they are copyrighted material, that is why you MUST NOT keep an MP3 for more than 24 hours." For an example of a trade association's attempt to debunk some of these myths, see the Software Publishers Association site at <http://www.spa.org/piracy/legends.htm> (the SPA has merged with the Information Industry Association, producing the Software and Information Industry Association, at <http://www.siia.net>).

[2]See <http://208.240.90.53/html/top_10_myths/myths_index.html>.

[3]According to a briefer at the committee meeting of July 9, 1998, "We're bringing up a generation of college students who believe that music should be free because music is free to them in the colleges."

bution to others; that anything received via e-mail can be freely copied and that if the uploading, posting, downloading, or copying does not, in the view of the end user, hurt anybody or is just good free advertising, then it is permissible.[4]

Still other common myths concern software. Among these are the 24-hour rule (i.e., software may be downloaded and used for up to 24 hours without authorization if the ostensible purpose is to determine whether the user wants to continue to use it, at which point he or she would have to delete it or buy it, but there is no limit on the number of times the user can download and reuse the software so long as it is deleted or purchased at the end of each 24 hours) and the "abandonment" rule (i.e., software is available for copying without liability if the copyright owner has ceased actively distributing it for more than some number of years). Some people who copy digital content do not know that they are doing anything illegal and believe that their ignorance of the law should absolve them from liability for copyright infringement. (As a matter of law, it doesn't.)

There is also the question of how well informed the public is about intellectual property more generally, including compliance with the private contracts embodied in shrink-wrap licenses, point-and-click licenses, subscriber agreements, and terms-of-service contracts. The intuitive conclusion is that a relatively small portion of the end-user population can be expected to read and fully comprehend all of the restrictions regarding intellectual property protection by which they may be legally bound, and in that sense the public is not well informed about what constitutes legal behavior.[5]

The committee believes that if, as a matter of legal and social policy, members of the general public are expected to comply with the requirements of intellectual property law, then it is important that the law be set forth in a clear and straightforward manner that the general public can readily comprehend. At face value, those sections of the Copyright Act that relate most directly to the conduct of members of the general public can appear to be somewhat straightforward: the exclusive rights of the copyright owner embodied in section 106, the first-sale doctrine embodied in section 109, and the four factors to be analyzed and balanced in evaluating fair use under section 107.[6] Much less straightforward is the interpretation of these sections in particular instances, which can be very complex and difficult (Box 4.1).

---

[4]See <http://www2.deakin.edu.au/lrs/pub_manual/copyright/Myths/Myths.htm>.

[5]The committee was unable to identify a relevant and authoritative source of data; this conclusion is based on the committee's readings, testimony presented, and deliberations.

[6]These three sections of the Copyright Act are reproduced in the addendum to this chapter.

BOX 4.1
Intellectual Property Law and Common Sense

Ordinary people at times find existing intellectual property law difficult to follow, or they resist following it on the grounds that it violates their common sense. Consider as one example a college professor who develops a set of course notes and posts them on the World Wide Web so that his students can retrieve them. This is a very efficient way of "handing out" notes—the students get the information without the professor's having to photocopy or even e-mail it, and updates, additions, and corrections can easily be linked in as the semester progresses. Because the notes are on a public Web page, any Internet user can download them. Imagine that a second professor teaching a similar course at another college downloads them and uses them in her course.

Under existing law, the second professor may have violated the intellectual property rights of the first.[1] Arguably, she should have obtained his permission before using the notes, even if the Web page contained no instructions about how to contact the professor who wrote them, or about which, if any, uses he wanted to allow. To many World Wide Web users, this does not make sense. The author of the course notes chose to put them on a public Web page and chose not to include with them any instructions about how they were to be used or about how to obtain permission to use them. He had the options of posting them on a password-protected Web page and giving the password only to his students, or of posting them in encrypted form on a public, unprotected Web page and giving the decryption key only to his students. He also had the option of posting them in the clear on a public Web page, along with instructions that they were developed for his course and that anyone who wanted to comment on or use them should contact him at the appropriate e-mail address. Password protection, encryption, and "use-only-as-follows" instructions are standard mechanisms that any Web user can avail himself of, and many Web users do. The fact that the author of these course notes chose not to use these mechanisms is easily interpreted to mean that he thought it was acceptable for anyone who found his notes on the Web to use them.

---

[1]If the professor who created the course notes filed an action against the second professor for copyright infringement, in such an action the second professor would be likely to assert defenses of express or implied consent, waiver, estoppel, and abandonment. Essentially, her argument would be that by posting the course notes "in the clear" on the World Wide Web without any indication of any intent to restrict further dissemination, the plaintiff (the professor who created the course notes) had implied consent to their retransmission and republication. The outcome of such a claimed defense might well turn on whether there is a custom or practice currently observed in academia regarding the posting of such course notes in the clear, whether it was reasonable for the plaintiff professor to expect protection for such a posting in the absence of a notice prohibiting publication, and whether it was reasonable for the defendant to assume or infer from the posting without accompanying restriction that a dedication to the public domain, or at least implied consent to republish the material, was reasonably intended by the plaintiff.

The entire body of written copyright law is voluminous, and many of its subjects (e.g., retransmission royalties) are arcane and complex.[7] This was less problematic when copyright was focused primarily on the behavior of large organizations, but now the behavior and the attitude of individuals can significantly affect markets and industries. Although companies have the resources to analyze, understand, and even help draft legislation, individual consumers do not. Consumers thus face the problem that the law is large, complex, and industry-specific.

The committee favors a greater degree of simplicity, clarity, straightforwardness, and easy comprehensibility for all aspects of copyright law that prescribe individual behavior. This goal might be enhanced by the development of specific interpretive guidelines on those aspects of consumer behavior that raise questions frequently encountered in daily life in dealing with copyright protection.[8] Wherever possible, a clear set of rules would be particularly useful, even if they only outline copying that is assuredly permitted under the law (although there is a risk that such rules could discourage legal copying that is beyond the scope of these rules).[9] This movement toward clarity and specificity must, however, preserve a sufficient flexibility and adaptability in the law so it can accommodate the future evolution of technologies and behaviors.[10] It is important to note that making it easier to comply with the law does not guarantee improved compliance.[11]

---

[7]U.S. copyright law is embodied in several key federal statutes—the 1909 act and the 1976 act, amended by additional statutes from time to time, including, for example, the Digital Millennium Copyright Act of 1998 and the Copyright Term Extension Act of 1998, and interpreted by numerous decisions of U.S. District Courts, U.S. Courts of Appeal, and the United States Supreme Court.

[8]Some efforts in the federal government include the Plain Language Action Network, a group working to improve communications from the federal government to the public. See information available online at <http://www.plainlanguage.gov>.

[9]One attempt at simplification and clarification of the intellectual property law is the set of guidelines for teacher photocopying distributed by the U.S. Copyright Office (1998). Such guidelines can be the product of a negotiated compromise, but they are sometimes issued by an interest group. As desirable as guidelines are, political complexities have thus far made it difficult for truly workable guidelines (for teacher photocopying or other situations) to emerge.

[10]See Box 6.2 in Chapter 6 for a list of principles that the committee recommends for use in the formulation or revision of law related to intellectual property.

[11]In publications and discussions about developing or enhancing the public's awareness and appreciation of the need for the protection of intangible property, a comparison is often made to the stealing of tangible property (e.g., shoplifting), accompanied by the claim that in terms of moral and ethical culpability people (particularly young people) who illegally copy intellectual property would never think of stealing tangible property. There is less comfort in this claim than may be apparent, however. In October 1998, a nationwide survey purportedly showed that nearly half of all high school students admitted that they had

Rights Holders

Although no rigorous study has been done, there is circumstantial evidence suggesting that many rights holders, too, are misinformed about legal behavior with respect to intellectual property, and do not understand the legal limits to their control. For example, a major academic publisher places the following legend on the page bearing the copyright notice for its publications: "No part of this book may be reproduced in any form or by any means, electronic or mechanical, including photocopying, recording, or by any information storage and retrieval system, without permission."

The use of similar legends or legal notices is widespread in the publishing industry. Yet the fair use privilege in the Copyright Act clearly authorizes the reproduction of at least some limited portion of a copyrighted publication for legitimate purposes, including critical commentary, scientific study, or even parody or satire. In that respect the absolute nature of the prohibition above is an overstatement of the copyright owner's rights. Similar observations could also be made with respect to the type of notice prohibiting any form of copying that appears on videocassettes and digital video disks (the so-called "FBI notice"), in the shrinkwrap licenses that accompany mass market software, in point-and-click licenses, and even on some individual Web sites.

Just as it is important for consumers to understand and obey the legal requirements affecting use of copyrighted material, it is also important that rights holders learn about fair use and other limits to their control over intellectual property and that they avoid making overstatements concerning what constitutes legal use of their material. Overstatements may well be counterproductive for rights holders if consumers recognize them as such, judge these statements to be excessive, and ignore them entirely as a result.

---

stolen tangible property from a store at least once within the past 12 months. If the study's results are valid, then a larger deterioration in ethics, particularly among people in the younger age range, concerning all forms of taking of property, whether tangible or intangible, may be taking place and would constrain the effectiveness of IP initiatives based on voluntary compliance.

Results of the study can be found online at <http://www.josephsoninstitute.org/98-Survey/98survey.htm>. Although the study claimed a margin of error of only ±3 percent and drew substantial media attention (see, for example, CNN Newsroom, 1998; Jackson and Suhler, 1998; Ove, 1998; and Thomas, 1998), the committee found no other independent data to validate the methodology, accuracy, or conclusions of the study and believes that they should be viewed with that caveat in mind.

THE CHALLENGE OF PRIVATE USE AND FAIR USE WITH DIGITAL INFORMATION

Perhaps the most contentious current copyright issue concerns the legality of private, noncommercial copying. This is not solely a digital intellectual property issue, but the risk to rights holders from unbridled private copying is especially acute when the information is in digital form and can be copied without loss of quality and disseminated by digital networks. The extremes of the positions on this issue are well established and heavily subscribed to. Some rights holders seem to believe that all, or nearly all, unauthorized reproduction of their works, whether private or public, commercial or noncommercial, is an infringement. Many members of the general public appear to believe that all or virtually all private, noncommercial copying of copyrighted works is lawful.

In the national copyright law of a number of countries, there is a specific private use copying privilege.[12] In the United States, for example, some private use copies are shielded by specific exceptions in the copyright statute. One illustration is the right of owners of copies of copyrighted computer programs to make backup copies of the software under 17 U.S.C. sec. 117. In the main, however, the legality of private use copying will be determined by application of the fair use doctrine and its four-factors test.[13]

The notion of private use copying as a matter of fair use law is most clearly articulated in the Supreme Court's decision in the *Sony Corp. of America v. Universal City Studios, Inc.* case.[14] In this case, the Court stated that courts should presume that private, noncommercial copying is fair. The presumption of fairness should only be overcome by proof of a meaningful likelihood of harm resulting from the private copy. The fair use ruling in the *Sony* case concerned only taping of television programs off the air for time-shifting purposes.[15] It did not discuss, let alone rule on,

---

[12]In some countries, private use copying privileges are being reassessed. In the future, they may be restricted to analog copying.

[13]An appropriate analysis of fair use requires consideration of four factors (the purpose and character of the use, including whether such use is commercial in nature or is for nonprofit educational purposes; the nature of the copyrighted work; the amount and substantiality of the portion used in relation to the copyrighted work as a whole; and the effect of the use on the potential market for or value of the copyrighted work). Evaluating each factor can be a difficult judgment, even for lawyers and judges. See the addendum to Chapter 4 for the full text of the law.

[14]464 U.S. 417 (1984).

[15]The reading of the *Sony* case is not uniform among committee members. Some committee members view *Sony*'s fair use analysis as applying to private home taping of programs from cable and other forms of pay TV as well as free broadcast television, while other committee members view *Sony* more narrowly, and as applying to free broadcast television only, which is what was directly involved in this case.

the legality of taping to build up a library of programs, although it was clear that Betamax machines, the Sony product at issue, could be used for this purpose as well. Given how close the decision was on the fairness of time shifting, it is far from clear that a majority of the Court would have ruled that developing a library of off-the-air tapes was protected under fair use. Yet observed behavior suggests that the general public's views on private copying are closer to the Supreme Court's general pronouncement on private copying, rather than its precise ruling in the case. Among some rights holders, the *Sony* decision remains unpopular. The *Intellectual Property and the National Information Infrastructure* (IITF, 1995) white paper interpreted the *Sony* decision as holding that home taping of programs was fair use because owners of copyrights in these programs had not yet devised a licensing scheme to charge for these uses. Although nothing in the Court's decision provides direct support for this view, subsequent case law developments have considered the availability of licensing as a factor in deciding whether certain educational or research copying was outside fair use.[16] Meanwhile, the Internet and related technology have increased the ease of and demand for private copying, which is likely to influence future case law.

The Wide Range of Private Use Copying

There is much disagreement on the lines separating legal and illegal private uses. One reason it is difficult to make judgments on the legality of private use copying is that the act encompasses a wide range of actions. Closer to the clearly infringing end of the spectrum, for example, is the practice of "borrowing" a software disk from a friend and loading a copy of the software onto one's hard drive without paying for it. This action may be private in the sense that it occurs in one's own home, and noncommercial in the sense that the copy does not enter the market in direct competition with the firm that developed the software. But defending this kind of action as a fair use would be difficult, given the harm that it arguably could have on the market for the software by reducing future sales.[17] Such an instance might present a meaningful likelihood of harm

---

[16]For example, see *Basic Books, Inc. v. Kinko's Graphics Corp.*, 758 F. Supp. 1522 (S.D. N.Y. 1991), which ruled that it was not fair use to photocopy book chapters and articles for academic course packs, and *American Geophysical Union v. Texaco, Inc.*, 60 F.3d 913 (2d Cir. 1995), which ruled that it was not fair use for a commercial research scientist to make archival copies of articles relevant to his research.

[17]Although harm to the market is not necessary to establish copyright infringement, it is an important factor weighing against fair use. The phrase "could arguably" is used in the text because there is not a consensus on the effects of copy protection on the software

to the market that could cause the *Sony* presumption of fairness of private, noncommercial copying to be overcome.

Closer to the middle of the spectrum, at least in the view of much of the public, is making an audiotape of a music CD—for which one has paid the full price—to be able to listen to it as one drives or walks home from work. This view recently found favor with a federal appellate court, which opined that place shifting of this sort was as much fair use as the time shifting in the *Sony* case.[18] Closer to the noninfringing end of the spectrum is making one photocopy of a favorite cartoon or poem to share with a friend or post on an office wall.

In the context of academic research, a wide variety of so-called private use copying occurs. Some of these actions are more clearly fair use than others; some may not, in fact, represent fair uses, even if the copy maker sincerely believes they are. Academic "private use" copying includes hand-copying of quotes from a book or article for research; photocopying of portions of a work for the same purpose; cutting and pasting from an electronic version of a work for the same purpose; making a copy of one's own articles in order to have a complete file of one's writings (a potential act of infringement if the scholar has assigned copyright to the publisher); making copies to send out to reviewers to enable them to assess one's work for tenure; making a copy of an article from one's own copy of a journal so that the copy can be carried into the laboratory and the journal itself kept from possible damage in the laboratory work environment; making a copy of an article to share with a colleague at another institution with whom one is working on a research project; making a similar copy for a graduate student who can't afford to buy his own copy of the journal; doing the same, but instead from a library copy of the journal; and developing an archive of materials on a subject for research purposes, just to name a few. These kinds of uses are widely perceived in the academic community as fair uses. At least some of them are considered by some rights holders to go beyond fair use. Of particular concern to publishers are interlibrary loan, document delivery, and electronic course reserve practices in the academic context (the electronic analogue to materials on reserve in a library reading room).

Also complicating resolution of the private-use-as-fair-use dilemma is the fact that the boundaries of "private use" as a category are far from clear. What does "private" mean in the context of certain acts of copying? What does "noncommercial" mean when paired with "private" in the

---

market. See, for example, "The Economics of Software Copyright Protection and Other Media" (Shy, 1998) and the references contained in the article.

[18]*RIAA v. Diamond/Multimedia*, F.3d (9th Cir. 1999). Also see "In Court's View, MP3 Player Is Just a 'Space Shifter,'" Carl S. Kaplan, *New York Times on the Web*, July 9, 1999.

context of these copies? In the *American Geophysical Union v. Texaco* case, Texaco claimed that the photocopying of articles from certain technical journals by one of its research scientists was a private act, and that it was noncommercial because these copies did not enter the market. The majority court opinion in *Texaco* gave little credence to either claim, although it did not find the commercial nature of Texaco's research projects to be dispositive either. Yet many people seem to consider private use as a very wide privilege, covering everything from photocopies made for one's civic group to a picture of one's favorite celebrity scanned for posting on a Web site.

As a way of providing readers of this report with a better understanding of the competing perspectives about the legality of private use copying, the next two sections present, first, the arguments that private use copying is never or virtually never fair use and, second, the arguments that private use copying is always or virtually always fair use.

Arguments That Private Use Copying Is Not Fair Use

Some rights holders would argue that private use copying is not fair use. A legal argument in support of this position could be based on the fact that copyright law gives rights holders the exclusive right to control the reproduction of their works. Unlike a number of other exclusive rights in the U.S. statute, the right of reproduction is not necessarily restricted to public activities. Rights holders may be entitled to control only "public" performances or displays of protected works, but when the issue of a private use exception to the reproduction right was raised in the legislative debate leading up to the enactment of the Copyright Act of 1976, Congress chose not to pursue this. Although Congress chose to adopt some specific exceptions to the reproduction right to respond to concerns of libraries and archives, it decided that private use should be dealt with in the context of fair use. The fair use provision of U.S. copyright law envisions mainly "transformative" uses of a protected work as fair (e.g., quoting from a work in order to criticize it in a second work). Private use copying, by contrast, is generally regarded as "consumptive" in character (e.g., use of the work so as not to be put to the trouble of paying money for it). In order for either type of use to be "fair," it must meet the four-factors test, and it may be difficult for "consumptive" private use copying to satisfy the fair use standards.[19]

Although there may have been a time when private copying was a minor matter, the widespread use of digital information and networks

---

[19]See footnote 13.

has created increased opportunities for ordinary people to engage increasingly in acts of infringement that are difficult to detect, yet mount up. The availability of this technology has bred a mind-set that seems to regard all copyrighted works as available for the taking without paying compensation. To counteract the seemingly widespread view that private use copies are lawful, it may be necessary to establish a counter-principle that private use copies are not lawful, and, as the argument goes, to use all means necessary within reason to regain control of commercially valuable properties.

An important part of the argument that private use is not necessarily fair use is the position that fair use is a defense against charges of infringement, not an affirmative right possessed by members of the public.[20] This argument maintains that the copyright law is clear in giving authors certain exclusive rights, subject to specific enumerated exceptions. Beyond these specific exceptions, the four factors used to determine fair use become the defense when a use is challenged and do not define affirmative rights. Most who take this position believe that there is no absolute right of public access to material still under copyright. The vagueness of the notion of "publish" and "publication" in an electronic environment adds to the issues of concern. Does the electronic distribution of a work to a controlled list make the work "published"? And, if so, is there then some automatic right for public access? Most rights holders would say no.

Arguments That Private Use Copying Is Fair Use

The views of those who regard private use as always or virtually always fair use stand in stark contrast to the views set forth above. Arguments that private use copies generally fall within the realm of fair use are based on the Supreme Court's finding in the *Sony* case that private, noncommercial copying should be presumed to be fair, as most often it is. Those taking this position argue that many private copies are made for limited purposes, on an occasional basis, and that they do not displace sales of commercial products. As the Supreme Court indicated in the *Sony* case, "A use that has no demonstrable effect upon the potential market for, or the value of, the copyrighted work need not be prohibited in order to protect the author's incentive to create. The prohibition of

---

[20]The idea that fair use could be construed as an affirmative right is very controversial. One of the reviewers of this report calls such a notion "absurd," while another reviewer states that "fair use has evolved strictly as a defense." Some legal scholars, such as William Patry, agree with these reviewers. However, some prominent legal scholars, such as Julie Cohen, David Nimmer, and Pamela Samuelson, do consider fair use to have a more affirmative character.

such noncommercial uses would merely inhibit access to ideas without any countervailing benefit."[21]

The argument favoring private use is based in part on a claim of market failure: The licensing revenue from private use copying is likely to be low compared to the transaction cost of negotiating a license for each such use. If this is true, it may be difficult to establish a meaningful likelihood of harm from private use copying.[22] Hence, there may still be some room for a transaction-cost-based market failure rationale for treating private use as fair use. In addition, many copies made for private use either have no economic significance or can be justified because of special circumstances (e.g., to enable research or to fulfill other valued purposes). Even where private use copies are economically significant and no special circumstances justify the copying, the costs of enforcement against private users may be far greater than can be economically justified. (Rights holders may argue that this is their business decision to make.)

In addition, many private use copies are made by people who have purchased both the work they are copying and many other copyrighted works (e.g., making a tape of music on a CD to play in one's car). Consumers may believe that the price they paid for their copies of copyrighted works implicitly reflects the rights holder's understanding that some consumers will make private use copies of them.[23] A lawsuit against such individuals over private use copying may reduce business both from those customers and from others who may sympathize with them. The publicity from such a lawsuit would be a public relations nightmare. While publishers may again regard this as their business decision, as a practical matter, the public may sense that rights unenforced are rights abandoned.[24]

Some private use advocates also believe that what one does with a copy of a copyrighted work in the privacy of one's own home is simply none of the copyright owner's business. There would be substantial societal costs in establishing an enforcement system that reached into the privacy of people's homes and forced them to show sales receipts or licenses for all the information they possess. The resulting system would

---

[21]464 U.S. 417 (1984).

[22]There is less of a market failure rationale if, instead of negotiating a license, the consumer can buy a second copy easily and inexpensively.

[23]Economists point out that consumers may be willing to pay higher prices for goods if they can be reused, shared, or copied for some purposes. See Shapiro and Varian (1998), p. 98.

[24]There are also, of course, some people who make so-called private copies that they know to be illegal and infringe because they either don't think they'll get caught or they disagree with private ownership rights in intangible works.

likely produce even more negative public reaction than would arise from suing private use copiers. Some leakage of copyrighted works may be part of the price of living in a free society.

Advocates of private use copying are among those who would tend to consider fair use to be a "right" to which the public is entitled once a copyright owner has disseminated her work to the public. Even if the fair use provision is structured as a defense to copyright infringement, once a defense is successful, it may seem to establish a right on the part of the user and other individuals to engage in this sort of act. For example, other software companies benefited from the ruling of the Ninth Circuit Court of Appeals when the court decided that Accolade, Inc., had made fair use of Sega's software when it decompiled the code in order to discern elements of the Sega program interface so that Accolade could make a compatible program.[25] Decompilation to achieve interoperability thus came to be perceived as a right. It is argued that there is a historical lineage to the view that the law grants rights holders only certain limited rights for a limited time, while reserving to the public all other rights, including arguably, the right to make fair uses of works. The Supreme Court's decision in *Campbell v. Acuff-Rose* recognizes that leaving some room for fair use is often necessary if copyright law is to achieve its constitutional purpose of promoting knowledge.[26] Some copyright scholars regard fair use as such a strong right that it overrides contractual provisions or technical protection services aimed at eliminating fair use.[27]

Private Use Copying: The Committee's Conclusions

This report cannot resolve the debate over private use copying. However, it may contribute to a better understanding of the dilemma by articulating the widely divergent positions on this and other issues. The committee as a whole does not endorse the view that all private use copies are illegal or the view that all private use copies are legal. It agrees that private use copying is fair use when, for example, an academic researcher makes a photocopy of a scientific article in the course of her research. However, the committee also agrees that copying a commercial software product from a friend for regular use without payment to the software publisher cannot be justified as fair use. In addition, the committee believes that the view is too prevalent that private use copying is virtually always fair use and agrees that it is often invoked to mask activi-

---

[25]*Sega v. Accolade*, 977 F.2d 1510 (9th Cir. 1992).

[26]114 S. Ct. 1164 (1994).

[27]See, for example, Nimmer et al. (1999) and Cohen (1998).

ties that, in the plain light of day, cannot be justified as fair uses. The committee also believes that copyright education should be undertaken to raise public consciousness about why respect for copyright is a good thing for society, not just for rights holders. Progress would be made if members of the public at least considered the question of whether the copying they do is justifiable or not.[28]

The Future of Fair Use and Other Copyright Exceptions

As with private use copying, diverse views exist about the future of fair use and other exceptions and limitations on copyright owners' rights in digital networked environments. The *Intellectual Property and the National Information Infrastructure* (IITF, 1995) white paper, for example, predicted a diminishment, if not the demise, of fair use in the digital realm because it concluded that rights holders would license uses and copies of digital information, and predicted that fair use would become unavailable when uses could be licensed.[29] Uses that might once have been considered fair, and that have even become customary, may become illegal if the white paper's conclusions become reality and a licensing scheme is put in place to regulate digital uses. Judging whether a use is "fair" involves (among other things) consideration of the impact of the use on the market; where there is no market (e.g., no pragmatic way to negotiate and pay for the right to quote a report), there is no market impact. But where licensing schemes do exist, there is a market and, hence, a chance for market impact, and consequently a potential diminishment of the territory formerly considered as fair use.

The committee came to perceive that fair use and other exceptions and limitations may sometimes have other beneficial functions as applied to digital information. It is obviously for the legislature and the courts to determine how broad or narrow fair use can be and what other exceptions should apply in the digital environment. However, it may further inform debate on these issues for this report to discuss some functions that fair use and other exceptions and limitations might play in the world of digital information. The report briefly discusses seven of the categories into which exceptions and limitations to copyright owners' rights seem generally to fall, and considers how each might arise in the digital environment:

---

[28]Some understanding of copyright is considered to be a basic element of "fluency" with information technology. See *Being Fluent with Information Technology* (CSTB, 1999a).

[29]For example, see *American Geophysical Union v. Texaco*, 60 F.3d 913 (2d Cir. 1995) aff'g 802 F. Supp. 1 (S.D. N.Y. 1992); accord, *Princeton Univ. Press v. Michigan Document Serv.*, 99 F.3d 1381 (6th Cir. 1996) (en banc).

- Those that are based on fundamental human rights;
- Those that are based on public interest grounds;
- Those that arise from competition policy;
- Those that promote flexible adaptation of the law to new circumstances;
- Those that arise from perceived market failures;
- Those that are the fruit of successful lobbying; and
- Those that cover situations in which uses or copying of protected works are de minimis, incidental to otherwise legitimate activities, or implicitly lawful given the totality of circumstances.

A number of fundamental human rights might provide a basis for a limited exception to copyright owner rights, including freedom of speech, freedom of the press, freedom of expression, freedom of information, democratic debate, and privacy or personal autonomy interests. A literary critic for a print magazine, for example, can republish a portion of another author's work in order to develop her critique of that author's work. A reporter for a print newspaper can publish some portions of a politician's speech in order to show the errors it contains. Copyright laws in some countries have specific "rights of fair quotation" to provide for legitimate copying for purposes of criticism and news reporting. In the United States, these concerns are generally dealt with through the fair use doctrine. Given the robustness of criticism, news reporting, and public debate on the Internet, it would seem that fair quotation/fair use rules would likely have some application in the digital world, just as they do in the print world.

Public interest exceptions to copyright vary to some degree from country to country. Among those that could arise under the laws of the United States and some other nations are those that permit performance of copyrighted works in the course of face-to-face instruction in nonprofit educational settings; those that enable libraries and archives to make copies for preservation, replacement, and other legitimate purposes; and those that enable the creation of derivative works for the blind. It is worth noting that the Digital Millennium Copyright Act seeks to maintain an appropriate balance between the rights of rights holders and the needs of others and contains a provision to enable libraries and archives to make digital, as well as print and facsimile, copies for these purposes. It also mandated a study to help Congress consider what copyright rules might be appropriate to promote distance learning.[30] Fair use may sometimes be invoked on public interest grounds to justify some copying of copy-

---

[30]See *Report on Copyright and Digital Distance Education*, a report of the Register of Copyrights (U.S. Copyright Office, 1999).

righted articles in legal proceedings (e.g., as evidence relevant to a contested issue of fact) or to satisfy some administrative regulatory requirements (e.g., to demonstrate the efficacy of drugs).

Competition policy concerns underlie some exceptions to and limitations on copyright owners' rights.[31] Two examples of competition policy-based limitations in U.S. law are rules that impose compulsory licenses on owners of musical copyrights to enable further recordings of those musical works and on owners of rights in broadcast signals for passive retransmissions of the broadcasted material by cable systems. The U.S. fair use defense is sometimes employed to promote competition policies, as in the *Sega v. Accolade* case, which upheld the legality of unauthorized decompilation of computer programs for the legitimate purpose of developing a compatible but noninfringing program. As the *Sega* case demonstrates, competition policy issues may arise at times when information is in digital form.

In times of rapid technological change, it may be difficult for legislatures to foresee what new technologies will arise, how they will be used, and what copyright rules ought to apply. Courts in the United States have often employed the fair use doctrine as a flexible mechanism for balancing the interests of rights holders and of other parties in situations in which the legislature has not indicated its intent. These include not only the *Sony v. Universal City Studios* decision about home taping of television programs for time-shifting purposes, but also a number of cases involving digital information. These include the *Galoob v. Nintendo* case, which upheld Galoob's right to distribute a "game genie" that enabled users to make some temporary alterations to the play of certain Nintendo games,[32] and the *Religious Technology Center v. Netcom* case in which automatic posting of user-initiated Internet messages by an online service provider was found to be fair use.[33]

Exceptions and limitations may arise from a perceived possible market failure. One argument for fair use may be that a market cannot effectively be formed when the transaction costs of negotiating a license far outweigh the benefits derivable from the transaction (whether they be licensing revenues or some other benefit, such as enhanced reputation or goodwill). To the extent that a fair use defense arises, at least in part, from market failure considerations, the scope of the fair use may be affected by changed circumstances that enable new markets to be formed effectively. As noted above, this view is expressed in the *Intellectual Property and the*

---

[31] In some circumstances, ensuring that the public policy underlying the free marketplace is effective may compel restrictions on the rights of rights holders.

[32] *Lewis Galoob Toys, Inc. v. Nintendo of America, Inc.*, 964 F.2d 965 (9th Cir. 1992).

[33] 907 F. Supp. 1361 (N.D. Cal. 1995).

National Information Infrastructure (IITF, 1995) white paper, specifically in relation to the digital environment. To some degree, the exception aimed at promoting publication of works for visually impaired persons reflects market failure, as well as public interest, considerations.

Finally, some exceptions and limitations to copyright owners' rights would seem to be the result of successful lobbying or of a legislative perception of the de minimis or incidental character of a use. In Italy, for example, military bands are exempt from having to do rights clearances for music they perform in public. In the United States, a number of exceptions, such as those creating special copyright privileges for veterans' groups, are the result of successful lobbying. Whether they can be justified on de minimis grounds or are pure pork barrel politics is perhaps debatable. Better examples of de minimis or incidental uses for which special copyright exceptions have been created are those in the Digital Millennium Copyright Act, which provides some "safe harbor" rules for certain copies in the digital environment, such as those made in the course of a digital transmission from one site to another where the transmitting intermediary (e.g., a telephone company) is merely a conduit for the transmission and not an active agent in it.[34] It is conceivable that other such exceptions will need to be devised for incidental digital copying in the future, or that fair use law will be used to exempt incidental or de minimis copying.

During the course of its deliberations, the committee became aware that exceptions to and limitations of the rights of copyright owners may be an area of contention in coming years, particularly in light of the successful conclusion of the Agreement on Trade-Related Aspects of Intellectual Property Rights (TRIPS) in 1994. Among other things, Article 13 of the TRIPS agreement obligates member states of the World Trade Organization not to adopt exceptions or limitations except "in certain special cases that do not conflict with a normal exploitation of the work or otherwise unreasonably prejudice the interests of the rights holders."[35] The U.S. Congress will need to keep TRIPS obligations in mind when it contemplates adopting new exceptions and limitations to copyright law, including those that might apply to digital works. Judicial interpretations of exceptions and limitations will also need to be consistent with TRIPS obligations.

---

[34] 17 U.S.C. sec. 512.

[35] Final Act Embodying the Result of the Uruguay Round of Multilateral Negotiations, Marrakesh Agreement Establishing the World Trade Organization, signed at Marrakesh, Morocco, April 5, 1994, Annex 1C, Agreement on Trade-Related Aspects of Intellectual Property Rights (TRIPS), article 13.

IS "COPY" STILL AN APPROPRIATE FUNDAMENTAL CONCEPT?

All of the preceding discussion accepts a fundamental perspective that underlies copyright, namely, the concept of copying as a foundational legal and conceptual notion. As the very name of the law indicates, the right to control the reproduction of works of authorship is central to the law. Deciding whether a work has been copied has, as a result, been a fundamental question underlying much of copyright history and analysis.

The committee considers here whether the notion of copy remains an appropriate mechanism for achieving the goals of copyright in the age of digital information, exploring two reasons why it might not be.[36]   One reason is that so many noninfringing copies are routinely made in using a computer that the act has lost much of its predictive power: Noting that a copy has been made tells far less about the legitimacy of the behavior than it does in the hard-copy world. A second reason is that, in the digital world copying is such an essential action, so bound up with the way computers work, that control of copying provides, in the view of some, unexpectedly broad powers, considerably beyond those intended by the copyright law.

This issue is clearly controversial, with some on the committee noting that, precisely because the digital world facilitates making an unlimited number of copies without degradation in quality, protection against copying must remain a sine qua non of copyright protection. In raising the issue, the committee is attempting to open the discussion to serious consideration of whether this foundational concept ought to be reconsidered in the world of digital information; the committee makes no recommendation other than that such discussion would be useful.

Control of Copying

The control of copying provided for in copyright law is a means to an end. The ultimate goal of intellectual property policy and law is promoting progress of science and the arts, achieved in large measure by providing incentive to authors and inventors, in the belief that society as a whole will in time benefit from their efforts. Historically, control of copying has been a key means of providing incentive, because it has given authors and publishers control over the use of their work. Control of copying provides rights holders who wish to profit from their work the means to require payment for its use, and offers valuable incentive to those whose reward comes from seeing their work published and distributed in ways they deem appropriate, even if economic reward is not their aim.

---

[36]The committee acknowledges the reviewer who suggested that the committee should deliberate this question.

But control of reproduction is the mechanism, not the goal. To see this, note that loss of control, for example, unauthorized reproduction, is in principle neither necessary nor sufficient to erode an author's incentive. Unauthorized reproduction is not by itself a sufficient disincentive. Consider the situation in which someone makes 500 copies of a work but does nothing with them other than store them or use them for fireplace kindling. Would this have any effect on the rights holder's interest in and incentive for creating subsequent works?

Nor is unauthorized reproduction in principle necessary for disincentive. Disincentive can arise without reproduction, as implicitly recognized in the copyright law's provision for an exclusive right to public performance (e.g., for movies, music, plays, dance). Even assuming no audio or video copy is ever made, unrestrained public performance can have adverse consequences on the market for the work and consequently produce disincentive for the rights holder. The right of public display is similar: In the absence of this right, incentive would be damaged even if no copies were made.

Hence, unauthorized reproduction, by itself, is neither necessary nor sufficient to discourage authors. Its significance lies in what it enables— misuse of a work (e.g., unauthorized distribution)—not in the action of reproduction itself. In addition, the very fact that copyright law specifies the rights holder's exclusive control over performance and display, as well as copying, underscores the central role of copying as one mechanism for protection of intellectual property through copyright, not its goal.

Note that in analyzing an act of unauthorized reproduction under current law, we need not ask what consequences that reproduction has for the rights holder in order to determine its legality; the law specifies an exclusive right to reproduction (within limits such as fair use). The issue addressed here is why the law was written to make it so. The claim is that control of reproduction is the means, not the end. If this is so, then it makes sense to ask whether the means is still appropriate in the world of digital IP, and, if it is not, whether there is a viable alternative that can accomplish the same goals more effectively.

Is Control of Copying the Right Mechanism in the Digital Age?

Control of reproduction may not be an appropriate mechanism to support copyright where digital information is concerned. Consider that control of copying has worked so well in practice through much of the history of copyright because of two important properties of copying a physical object. First, in the world of physical works, copying is an explicit and overt act, carried out with specific intent; one does not accidentally or incidentally copy an entire book. Second, copying is a prerequisite to distribution; before you can sell copies, you have to make them.

In the world of physical artifacts, making a copy, of a book for example, requires an explicit, carefully selected action that has one goal and one obvious result—a copy of the book. There are, in addition, few reasons in the physical world to reproduce an entire work, other than to make a copy that can substitute for an original and, hence, potentially harm the rights holder. In the physical world, the focus on reproduction is thus effective and appropriate, because there is an intimate connection between reproduction and consequences for the rights holder, namely, substituting the copy for the original.

One important consequence of these observations is that, in the world of physical artifacts, reproduction is a good predictor: The act is closely correlated to other actions, such as distribution, that may harm the rights holder and reduce incentive. A second consequence is that, because reproduction is routinely necessary for distribution (and thus exploitation of the work), control of reproduction is an effective means, a convenient bottleneck by which to control exploitation of the work. Finally, because reproduction is not necessary for ordinary use of the work (e.g., reading a hard-copy book), control of reproduction does not get in the way of intended consumption of a work (i.e., reading it).

All of these consequences are false in the digital world: Reproduction is a far less precise predictor of infringement, control of reproduction is a problematic means of controlling exploitation that in some circumstances has important side effects, and reproduction is necessary for ordinary consumption of the work.

Reproduction is not a good predictor of infringement in the digital world because there are many innocent reasons to make a copy of a work, copies that do not serve as substitutes for the original and hence have no impact on the rights holder. For example, as noted in Chapter 1, digital works are routinely copied simply in order to access them. Code must be copied from the hard disk into random access memory in order to run a program, for example, and a Web page must be copied from the remote computer to the local computer in order to view it. More generally, in the digital world, access requires copying.

The numerous ways in which copies get made in the digital world also cloud the question of whether a copy (in the legal sense) has been made. Arguments have arisen, for example, as to whether the copyright in a work can be infringed by the two actions noted above—copying a program from the disk into random access memory to run it and accessing a Web page from another computer. In both cases the information has been copied in the technical sense, but it is unclear whether this constitutes legal infringement.

One consequence of this difference in the digital world is the many discussions that have occurred concerning what kinds of copies to care about, as, for example, the calls for distinguishing between ephemeral and permanent copies, and the discussions that attempt to determine what is sufficiently ephemeral not to matter. Although it has some technical justification, this approach to dealing with the ubiquity of copying in the digital world is not particularly reliable, because, among other reasons, ephemeral copies are at times easily captured and turned into persistent copies, and because, as technology evolves, the list of ephemeral types will become obsolete very quickly.

In the digital world, then, reproduction of many sorts is a common, indeed technologically necessary, action with fundamental technical justification, frequently innocent of any infringement intent or effect. As making those technologically necessary copies is the means of gaining access to the work, such copying is in fact the fundamental mechanism by which a rights holder exploits the work in the digital world.

As a result, the action loses much of its predictive quality. Reproduction is far less tightly linked to a loss of incentive for rights holders. And if this is true, controlling reproduction in the digital world may fail to serve nearly as well as it does in the hard-copy world as a means of achieving the ultimate goals of intellectual property policy. The focus on reproduction also produces debates that may not be particularly revealing or useful, for example, the question posed above about whether accessing a Web page involves making a copy and hence can constitute an infringement of copyright.

But the problem is larger than that. Control of reproduction in the digital world may not be an appropriate means of protecting rights holders. In that world, reproduction is so bound up with any use of digital information that true control of reproduction would bring unprecedented control over *access* to information. In the world of physical works, once a work (e.g., a book) has been published, the rights holder cannot in any pragmatic sense control access to the copies distributed. Social institutions (such as bookstores and libraries) and individuals with copies enable any motivated reader to gain access to the information in the work.[37]

But when access requires reproduction, as it does for works in digital form, the right to control reproduction is the right to control access, even the access to an individual copy already distributed.[38] We do not expect

---

[37]Note that access here is meant in the limited sense of reading; rights holders clearly maintain control over the distribution and sale of copies of the work.

[38]For digital works, access requires reproduction even for an individual copy that has already been published, i.e., reproduced and distributed. For example, an authorized individual copy of a musical work (e.g., an MP3 file) must be reproduced (once again) in order to be heard.

that authors would routinely deny access to their published digital works; why else would they have written and published them, if not to offer some form of access? The point is that while in the hard-copy world the rights holder can control access to the work as a whole by delaying publication or restricting the number of copies that may be made, in the digital world access requires reproduction, so control of reproduction provides control of access to individual published copies. Because control of access to individual published copies was not conceived of as part of copyright, this control is not to be embraced lightly, whether or not routinely exercised by authors or other rights holders.

To summarize, two points are important here. First, in the digital world reproduction loses much of its power as a predictor of important consequences, and hence the question of whether a protected work has been copied may be considerably less important. Second, in the digital world control of reproduction is a blunt instrument whose impact reaches considerably beyond the original intent, bringing into question its use in accomplishing the goals of intellectual property law.

What Can Be Done?

Rather than engage in epicycle-like debates over how ephemeral a copy has to be before it fails to matter, the committee suggests asking whether it might be appropriate to replace copying as a benchmark concept. Considering that control of reproduction is a means, not the goal, can we find some other means of control that is more tightly connected to the goal, whether in the digital or analog world? This will not easily be done, but the committee makes an attempt here solely for the sake of opening the discussion and promoting serious consideration of the issue; the committee makes no recommendation other than that discussion should be initiated.

As above, the committee suggests viewing control of reproduction as a mechanism aimed at preventing uses of a work that would substantially reduce an author's incentive to create. The committee speculates that it may be useful to start from what the law is attempting to achieve—ensuring progress in the sciences and arts—and ask whether a use being made of a work is substantially destructive of a common means of achieving that goal, namely, providing incentive to authors. This approach is similar in overall spirit to the concept of fair use, which requires consideration of (among other things) the impact on the market for the work or on the value of the work. But it is somewhat broader in scope, as incentive arises for authors in more than the marketplace alone, coming as well from, for example, the ability to control the time, place, and manner of publication.

This view would not conflict with all of the (other) traditional exclusive rights in copyright. Creation of derivative works, distribution, public performance, and display of the work can all be conceived of and protected on grounds independent of whether a copy has been made. They also have evident impact on incentive, whether via economic effects in the marketplace or other factors, and hence would be consistent with an incentive-based analysis.

Any such substantial change would of course also bring problems. There would have to be a substantial period of familiarization, as individuals and courts become knowledgeable about the concept and better able to define its currently quite vague boundaries. There would also be the tension between trying to make such a law easier to follow by drawing a sharp line defining what constitutes incentive-destroying use, and keeping the criteria more general (as is the case for fair use) in order to permit the concept to handle new, unanticipated situations in the future.[39]

As the committee discovered, tackling this task would also be difficult because of how highly charged the issue is. The thought of basing copyright on something other than the notion of control of reproduction is viewed with dismay by some in the legal community, given how long and largely successfully it has stood as a central pillar of copyright law and scholarship.[40] Finally, the focus on reproduction also offers a simple mechanism—the first-sale rule—that underlies a number of important social institutions (e.g., libraries); removing the reliance on copy would require a new legal framework for their operation. The needs of these and other stakeholders would have to be carefully considered in any change.

Despite the expected difficulty of the task, the committee believes that it may prove revealing theoretically as well as useful pragmatically in the digital age to undertake this exercise in rethinking copyright protection without the notion of a copy.

ADDENDUM: SECTIONS 106, 107, AND 109 OF THE U.S. COPYRIGHT LAW

The three sections of the U.S. copyright law that relate most directly to the conduct of members of the general public are reproduced in this addendum: the exclusive rights of the copyright proprietor embodied in section 106, the four factors to be analyzed and balanced in evaluating fair

---

[39]While this would be no small undertaking, the legal system has successfully struggled with such tasks in the past, as, for example, elucidating over time the distinction between "idea" and "expression."

[40]For example, committee deliberations on this subject were extremely intense.

use under section 107, and the first-sale doctrine embodied in section 109.[41]

106 Exclusive rights in copyrighted works*

Subject to sections 107 through 120, the owner of copyright under this title has the exclusive rights to do and to authorize any of the following:

(1) to reproduce the copyrighted work in copies or phonorecords;

(2) to prepare derivative works based upon the copyrighted work;

(3) to distribute copies or phonorecords of the copyrighted work to the public by sale or other transfer of ownership, or by rental, lease, or lending;

(4) in the case of literary, musical, dramatic, and choreographic works, pantomimes, and motion pictures and other audiovisual works, to perform the copyrighted work publicly;

(5) in the case of literary, musical, dramatic, and choreographic works, pantomimes, and pictorial, graphic, or sculptural works, including the individual images of a motion picture or other audiovisual work, to display the copyrighted work publicly; and

(6) in the case of sound recordings, to perform the copyrighted work publicly by means of a digital audio transmission.

106A Rights of certain authors to attribution and integrity†

(a) RIGHTS OF ATTRIBUTION AND INTEGRITY.— Subject to section 107 and independent of the exclusive rights provided in section 106, the author of a work of visual art—

(1) shall have the right—

(A) to claim authorship of that work, and

(B) to prevent the use of his or her name as the author of any work of visual art which he or she did not create;

(2) shall have the right to prevent the use of his or her name as the author of the work of visual art in the event of a distortion, mutila-

---

[41]This material was obtained from the Web site of the U.S. Copyright Office at <http://www.loc.gov/copyright/>. It is intended for use as a general reference, and not for legal research or other work requiring authenticated primary sources.

*Section 106 was amended by the Digital Performance Right in Sound Recordings Act of November 1, 1995, Pub. L. 104-39, 109 Stat. 336, which added paragraph (6).

†A new section 106A was added by the Visual Artists Rights Act of 1990, Pub. L. 101-650, 104 Stat. 5128. The act states that, generally, it is to take effect 6 months after the date of its enactment, that is, 6 months after December 1, 1990, and that the rights created by section 106A shall apply to—(1) works created before such effective date but title to which has not, as of such effective date, been transferred from the author, and (2) works created on or after such effective date, but shall not apply to any destruction, distortion, mutilation, or other modification (as described in section 106A(a)(3)) of any work that occurred before such effective date.

tion, or other modification of the work which would be prejudicial to his or her honor or reputation; and

(3) subject to the limitations set forth in section 113(d), shall have the right

(A) to prevent any intentional distortion, mutilation, or other modification of that work which would be prejudicial to his or her honor or reputation, and any intentional distortion, mutilation, or modification of that work is a violation of that right, and

(B) to prevent any destruction of a work of recognized stature, and any intentional or grossly negligent destruction of that work is a violation of that right.

(b) SCOPE AND EXERCISE OF RIGHTS.—Only the author of a work of visual art has the rights conferred by subsection (a) in that work, whether or not the author is the copyright owner. The authors of a joint work of visual art are co-owners of the rights conferred by subsection (a) in that work.

(c) EXCEPTIONS.—

(1) The modification of a work of visual art which is a result of the passage of time or the inherent nature of the materials is not a distortion, mutilation, or other modification described in subsection (a)(3)(A).

(2) The modification of a work of visual art which is a result of conservation, or of the public presentation, including lighting and placement, of the work is not a destruction, distortion, mutilation, or other modification described in subsection (a)(3) unless the modification is caused by gross negligence.

(3) The rights described in paragraphs (1) and (2) of subsection (a) shall not apply to any reproduction, depiction, portrayal, or other use of a work in, upon, or in any connection with any item described in subparagraph (A) or (B) of the definition of "work of visual art" in section 101, and any such reproduction, depiction, portrayal, or other use of a work is not a destruction, distortion, mutilation, or other modification described in paragraph (3) of subsection (a).

(d) DURATION OF RIGHTS.—

(1) With respect to works of visual art created on or after the effective date set forth in section 610(a) of the Visual Artists Rights Act of 1990, the rights conferred by subsection (a) shall endure for a term consisting of the life of the author.

(2) With respect to works of visual art created before the effective date set forth in section 610(a) of the Visual Artists Rights Act of 1990, but title to which has not, as of such effective date, been transferred from the author, the rights conferred by subsection (a) shall be coextensive with, and shall expire at the same time as, the rights conferred by section 106.

(3) In the case of a joint work prepared by two or more authors, the rights conferred by subsection (a) shall endure for a term consisting of the life of the last surviving author.

(4) All terms of the rights conferred by subsection (a) run to the end of the calendar year in which they would otherwise expire.

(e) TRANSFER AND WAIVER.—

(1) The rights conferred by subsection (a) may not be transferred, but those rights may be waived if the author expressly agrees to such waiver in a written instrument signed by the author. Such instrument shall specifically identify the work, and uses of that work, to which the waiver applies, and the waiver shall apply only to the work and uses so identified. In the case of a joint work prepared by two or more authors, a waiver of rights under this paragraph made by one such author waives such rights for all such authors.

(2) Ownership of the rights conferred by subsection (a) with respect to a work of visual art is distinct from ownership of any copy of that work, or of a copyright or any exclusive right under a copyright in that work. Transfer of ownership of any copy of a work of visual art, or of a copyright or any exclusive right under a copyright, shall not constitute a waiver of the rights conferred by subsection (a). Except as may otherwise be agreed by the author in a written instrument signed by the author, a waiver of the rights conferred by subsection (a) with respect to a work of visual art shall not constitute a transfer of ownership of any copy of that work, or of ownership of a copyright or of any exclusive right under a copyright in that work.

107 Limitations on exclusive rights: Fair use*

Notwithstanding the provisions of sections 106 and 106A, the fair use of a copyrighted work, including such use by reproduction in copies or phonorecords or by any other means specified by that section, for purposes such as criticism, comment, news reporting, teaching (including multiple copies for classroom use), scholarship, or research, is not an infringement of copyright. In determining whether the use made of a work in any particular case is a fair use the factors to be considered shall include—

(1) the purpose and character of the use, including whether such use is of a commercial nature or is for nonprofit educational purposes;

(2) the nature of the copyrighted work;

(3) the amount and substantiality of the portion used in relation to the copyrighted work as a whole; and

(4) the effect of the use upon the potential market for or value of the copyrighted work. The fact that a work is unpublished shall not itself bar a finding of fair use if such finding is made upon consideration of all the above factors.

---

*Section 107 was amended by the Visual Artists Rights Act of 1990, Pub. L. 101-650, 104 Stat. 5089, 5128, 5132, which struck out "section 106" and inserted in lieu thereof "sections 106 and 106A." Section 107 was also amended by the Act of October 24, 1992, Pub. L. 102-492, 106 Stat. 3145, which added the last sentence.

109 Limitations on exclusive rights: Effect of transfer of particular copy or phonorecord*

(a) Notwithstanding the provisions of section 106(3), the owner of a particular copy or phonorecord lawfully made under this title, or any person authorized by such owner, is entitled, without the authority of the copyright owner, to sell or otherwise dispose of the possession of that copy or phonorecord. Notwithstanding the preceding sentence, copies or phonorecords of works subject to restored copyright under section 104A that are manufactured before the date of restoration of copyright or, with respect to reliance parties, before publication or service of notice under section 104A(e), may be sold or otherwise disposed of without the authorization of the owner of the restored copyright for purposes of direct or indirect commercial advantage only during the 12-month period beginning on—

(1) the date of the publication in the *Federal Register* of the notice of intent filed with the Copyright Office under section 104A(d)(2)(A), or

(2) the date of the receipt of actual notice served under section 104A(d)(2)(B), whichever occurs first.[†]

(b)(1)(A) Notwithstanding the provisions of subsection (a), unless authorized by the owners of copyright in the sound recording or the owner of copyright in a computer program (including any tape, disk, or other medium embodying such program), and in the case of a sound recording in the musical works embodied therein, neither the owner of a particular phonorecord nor any person in possession of a particular copy of a computer program (including any tape, disk, or other medium embodying such program), may, for the purposes of direct or indirect commercial advantage, dispose of, or authorize the disposal of, the pos-

---

*Section 109 was amended by the Act of October 4, 1984, Pub. L. 98-450, 98 Stat. 1727, and the Act of November 5, 1988, Pub. L. 100-617, 102 Stat. 3194. The 1984 Act redesignated subsections (b) and (c) as subsections (c) and (d), respectively, and inserted after subsection (a) a new subsection (b).

The earlier amendatory Act states that the provisions of section 109(b) "shall not affect the right of an owner of a particular phonorecord of a sound recording, who acquired such ownership before . . . [October 4, 1984], to dispose of the possession of that particular phonorecord on or after such date of enactment in any manner permitted by section 109 of title 17, United States Code, as in effect on the day before the date of the enactment of this Act." It also states, as modified by the 1988 amendatory Act, that the amendments "shall not apply to rentals, leasings, lendings (or acts or practices in the nature of rentals, leasings, or lendings) occurring after the date which is 13 years after . . . [October 4, 1984]."

Section 109 was also amended by the Computer Software Rental Amendments Act of 1990, Pub. L. 101-650, 104 Stat. 5089, 5134, 5135, which added at the end thereof subsection (e). The amendatory Act states that the provisions contained in the new subsection (e) shall take effect 1 year after the date of enactment of such Act, that is, 1 year after December 1, 1990. The Act also states that such amendments so made "shall not apply to public performances or displays that occur on or after October 1, 1995." See also footnote 22, Chapter 1.

†Section 109(a) was amended by the Uruguay Round Agreements Act of December 8, 1994, Pub. L. 103-465, 108 Stat. 4809, 4981, which added the second sentence.

session of that phonorecord or computer program (including any tape, disk, or other medium embodying such program) by rental, lease, or lending, or by any other act or practice in the nature of rental, lease, or lending. Nothing in the preceding sentence shall apply to the rental, lease, or lending of a phonorecord for nonprofit purposes by a nonprofit library or nonprofit educational institution. The transfer of possession of a lawfully made copy of a computer program by a nonprofit educational institution to another nonprofit educational institution or to faculty, staff, and students does not constitute rental, lease, or lending for direct or indirect commercial purposes under this subsection.

(B) This subsection does not apply to—

(i) a computer program which is embodied in a machine or product and which cannot be copied during the ordinary operation or use of the machine or product; or

(ii) a computer program embodied in or used in conjunction with a limited purpose computer that is designed for playing video games and may be designed for other purposes.

(C) Nothing in this subsection affects any provision of chapter 9 of this title.

(2)(A) Nothing in this subsection shall apply to the lending of a computer program for nonprofit purposes by a nonprofit library, if each copy of a computer program which is lent by such library has affixed to the packaging containing the program a warning of copyright in accordance with requirements that the Register of Copyrights shall prescribe by regulation.

(B) Not later than three years after the date of the enactment of the Computer Software Rental Amendments Act of 1990, and at such times thereafter as the Register of Copyrights considers appropriate, the Register of Copyrights, after consultation with representatives of copyright owners and librarians, shall submit to the Congress a report stating whether this paragraph has achieved its intended purpose of maintaining the integrity of the copyright system while providing nonprofit libraries the capability to fulfill their function. Such report shall advise the Congress as to any information or recommendations that the Register of Copyrights considers necessary to carry out the purposes of this subsection.

(3) Nothing in this subsection shall affect any provision of the antitrust laws. For purposes of the preceding sentence, "antitrust laws" has the meaning given that term in the first section of the Clayton Act and includes section 5 of the Federal Trade Commission Act to the extent that section relates to unfair methods of competition.

(4) Any person who distributes a phonorecord or a copy of a computer program (including any tape, disk, or other medium embodying such program) in violation of paragraph (1) is an infringer of copyright under section 501 of this title and is subject to the remedies set forth in sections 502, 503, 504, 505, and 509. Such violation shall not be a

criminal offense under section 506 or cause such person to be subject to the criminal penalties set forth in section 2319 of title 18.*

(c) Notwithstanding the provisions of section 106(5), the owner of a particular copy lawfully made under this title, or any person authorized by such owner, is entitled, without the authority of the copyright owner, to display that copy publicly, either directly or by the projection of no more than one image at a time, to viewers present at the place where the copy is located.

(d) The privileges prescribed by subsections (a) and (c) do not, unless authorized by the copyright owner, extend to any person who has acquired possession of the copy or phonorecord from the copyright owner, by rental, lease, loan, or otherwise, without acquiring ownership of it.[†]

(e) Notwithstanding the provisions of sections 106(4) and 106(5), in the case of an electronic audiovisual game intended for use in coin-operated equipment, the owner of a particular copy of such a game lawfully made under this title, is entitled, without the authority of the copyright owner of the game, to publicly perform or display that game in coin-operated equipment, except that this subsection shall not apply to any work of authorship embodied in the audiovisual game if the copyright owner of the electronic audiovisual game is not also the copyright owner of the work of authorship.

---

*Section 109(b) was amended by the Computer Software Rental Amendments Act of 1990, Pub. L. 101-650, 104 Stat. 5089, 5134, in the following particulars: a) paragraphs (2) and (3) were redesignated as paragraphs (3) and (4), respectively; b) paragraph (1) was struck out and new paragraphs (1) and (2) were inserted in lieu thereof; and c) paragraph (4), as redesignated by the amendatory Act, was struck out and a new paragraph (4) was inserted in lieu thereof. The amendatory Act states that section 109(b), as amended, "shall not affect the right of a person in possession of a particular copy of a computer program, who acquired such copy before the date of the enactment of this Act, to dispose of the possession of that copy on or after such date of enactment in any manner permitted by section 109 of title 17, United States Code, as in effect on the day before such date of enactment." The amendatory Act also states that the amendments made to section 109(b) "shall not apply to rentals, leasings, or lendings (or acts or practices in the nature of rentals, leasings, or lendings) occurring on or after October 1, 1997." However, this limitation, set forth in the first sentence of section 804 (c) of the amendatory Act [104 Stat. 5136], was subsequently deleted by the Uruguay Round Agreements Act of December 8, 1994, section 511 of which struck the above mentioned first sentence in its entirety. See Pub. L. 103-465, 108 Stat. 4809, 4974. See also footnote 20, Chapter 1.

†The Act of November 5, 1988, Pub. L. 100-617, 102 Stat. 3194, made technical amendments to section 109(d), by striking out "(b)" and inserting in lieu thereof "(c)" and by striking out "coyright" and inserting in lieu thereof "copyright."

5

Protecting Digital Intellectual Property: Means and Measurements

Recent years have seen the exploration of many technical mechanisms intended to protect intellectual property (IP) in digital form, along with attempts to develop commercial products and services based on those mechanisms. This chapter begins with a review of IP protection technology, explaining the technology's capabilities and limitations and exploring the consequences these capabilities may have for the distribution of and access to IP. Appendix E presents additional technical detail, attempting to demystify the technology and providing an introduction to the large body of written material on this subject.

This chapter also addresses the role of business models in protecting IP. Protection is typically conceived of in legal and technical terms, determined by what the law permits and what technology can enforce. Business models add a third, powerful element to the mix, one that can serve as an effective means of making more digital content available in new ways and that can be an effective deterrent to illegitimate uses of IP.

The chapter also considers the question of large-scale commercial infringement, often referred to as piracy. It discusses the nature of the data concerning the rates of commercial infringement and offers suggestions for improving the reported information.

The chapter concludes with a discussion of the increasing use of patents to protect information innovations such as software and Internet business models, and explores the question of whether the patent system is an appropriate mechanism to protect these innovations.

TECHNICAL PROTECTION

The evolution of technology is challenging the status quo of IP management in many ways. This section and Appendix E focus on technical protection services (TPSs) that may be able to assist in controlling the distribution of digital intellectual property on the Internet.[1] The focus here is on how technical tools can assist in meeting the objectives stated throughout the report, as well as what they cannot do and what must therefore be sought elsewhere. Appendix E explores how the tools work, details what each kind of tool brings to bear on the challenges described throughout the report, and projects the expected development and deployment for each tool. For ease of exposition, the presentation in this chapter is framed in terms of protecting individual objects (texts, music albums, movies, and so on); however, many of the issues raised are applicable to collections (e.g., libraries and databases),[2] and many of the techniques discussed are relevant to them as well.

A number of general points are important to keep in mind about TPSs:

• Technology provides means, not ends; it can assist in enforcing IP policy, but it cannot provide answers to social, legal, and economic questions about the ownership of and rights over works, nor can it make up for incompletely or badly answered questions.

• No TPS can protect perfectly. Technology changes rapidly, making previously secure systems progressively less secure. Social environments also change, with the defeat of security systems attracting more (or less) interest in the population. Just as in physical security systems, there are inherent trade-offs between the engineering design and implementation quality of a system on the one hand and the cost of building and deploying it on the other. The best that can be hoped for is steady improvement in TPS quality and affordability and keeping a step ahead of those bent on defeating the systems.

---

[1]Note that the phrase "technical protection services" is used deliberately. Although it is tempting to talk about technical protection *systems*—packages of tools integrated into digital environments and integrated with each other—the committee believes that such systems are difficult to implement reasonably in the information infrastructure, an open network of interacting components, lacking boundaries that usefully separate inside and outside. In this environment it is better to talk about technical protection services; each service will be drawn on by information infrastructure components and will generally interact with other services.

[2]For example, as reported by a committee member, in February 1999 the special assistant to the director of Chemical Abstracts Service (CAS) indicated that there were one to three "hacking" attempts per day to get into the CAS database.

• While technical protection for intellectual property is often construed as protecting the rights of rights holders to collect revenue, this viewpoint is too narrow. Technical protection offers additional important services, including verifying the authenticity of information (i.e., indicating whether it comes from the source claimed and whether it has been altered—either inadvertently or fraudulently). Information consumers will find this capability useful for obvious reasons; publishers as well need authenticity controls to protect their brand quality.

• As with any security system, the quality and cost of a TPS should be tailored to the values of and risks to the resources it helps protect: The newest movie release requires different protection than a professor's class notes.

• Again, as with any security system, there are different degrees of protection. Some TPSs are designed to keep honest people honest and provide only a modest level of enforcement; more ambitious uses seek to provide robust security against professional pirates.

• As with any software, TPSs are subject to design and implementation errors that need to be uncovered by careful research and investigation. Professional cryptologists and digital security experts look for flaws in existing services in order to define better products.

• TPSs almost invariably cause some inconvenience to their users. Part of the ongoing design effort is to eliminate such inconvenience or at least to reduce it to tolerable levels.

• The amount of inconvenience caused by a TPS has been correlated historically with its degree of protection. As a result, in the commercial context, overly stringent protection is as bad as inadequate protection: In either extreme—no protection or complete protection (i.e., making content inaccessible)—revenues are zero. Revenues climb with movement away from the extremes; the difficult empirical task is finding the right balance.

• Protective technologies that are useful within special-purpose devices (e.g., cable-television set-top boxes or portable digital music players) are quite different from those intended for use in general-purpose computers. For network-attached general-purpose computers, software alone cannot achieve the level of technical protection attainable with special-purpose hardware. However, software-only measures will doubtless be in wide use soon.

Here (and in more detail in Appendix E) the committee provides a layman's description of the most important technical protection mechanisms, suggesting how each can be fit into an overall protection scheme, describing the limitations of each, and sketching current research directions. There are several classes of mechanisms:

- *Security and integrity features of computer operating systems* include, for example, the traditional file access privileges enforced by the system.
- *Rights management languages* express in machine-readable form the rights and responsibilities of owners, distributors, and users, enabling the computer to determine whether requested actions fall within a permitted range. These languages can be viewed as an elaboration of the languages used to express file access privileges in operating systems.
- *Encryption* allows digital works to be scrambled so that they can be unscrambled only by legitimate users.
- *Persistent encryption* allows the consumer to use information while the system maintains it in an encrypted form.
- *Watermarking* embeds information (e.g., about ownership) into a digital work in much the same way that paper can carry a watermark. A digital watermark can help owners track copying and distribution of digital works.

For effective protection, the developer of an IP-delivery service must choose the right ingredients and attempt to weave them together into an end-to-end technical protection system. The term "end-to-end" emphasizes the maintenance of control over the content at all times; the term "protection system" emphasizes the need to combine various services so that they work together as seamlessly as possible.

Protecting intellectual property is a variant of computing and communications security, an area of study that has long been pursued both in research laboratories and for real-world application. Security is currently enjoying renewed emphasis because of its relevance to conducting business online.[3] While security technology encompasses a very large area, this discussion is limited to describing generally applicable principles and those technical topics relevant to the management of intellectual property.[4]

As cryptography is an underpinning for many of the other tools discussed, the following section begins with a brief explanation of this technology.[5] Next, the techniques that help manage IP within general-

---

[3] As the technology needed for IP may not be affordable for IP alone, there is the possibility of a useful coincidence: The technology needed for IP may be largely a subset of what will be needed for electronic commerce. One concrete example is the Trusted Computing Platform Alliance discussed below.

[4] For example, the committee passed silently over a concern closely related to IP—the effect of the digital world on personal privacy—because, although there is some intersection of the two sets of issues, they are sufficiently separable and sizable that each is best addressed in its own report.

[5] A closely related topic, the Public Key Infrastructure—a set of emerging standards for distributing, interpreting, and protecting cryptographic keys—is primarily of technical interest and is discussed in Appendix E.

purpose computers are described. Finally the discussion turns to technology that can help in consumer electronics and other special-purpose devices.[6]

Encryption: An Underpinning Technology for Technical Protection Service Components

Cryptography is a crucial enabling technology for IP management. The goal of encryption is to scramble objects so that they are not understandable or usable until they are unscrambled. The technical terms for scrambling and unscrambling are "encrypting" and "decrypting." Encryption facilitates IP management by protecting content against disclosure or modification both during transmission and while it is stored. If content is encrypted effectively, copying the files is nearly useless because there is no access to the content without the decryption key. Software available off the shelf provides encryption that is for all practical purposes unbreakable, although much of the encrypting software in use today is somewhat less robust.

Many commercial IP management strategies plan a central role for what is called "symmetric-key" encryption, so called because the same key is used both to encrypt and decrypt the content. Each object (e.g., movie, song, text, graphic, software application) is encrypted by the distributor with a key unique to that object; the encrypted object can then be distributed, perhaps widely (e.g., placed on a Web site). The object's key is given only to appropriate recipients (e.g., paying customers), typically via a different, more secure route, perhaps one that relies on special hardware.

One example of an existing service using encryption in this way is pay-per-view television. A program can be encrypted with a key and the key distributed to paying customers only. (The special hardware for key distribution is in the set-top box.) The encrypted program can then safely be broadcast over public airwaves. Someone who has not paid and does not have the key may intercept the broadcast but will not be able to view it.

There is, of course, an interesting circularity in symmetric-key encryption. The way to keep a message secret is to encrypt it, but then you also have to send the decryption key so the message recipient can decrypt the message. You have to keep the key from being intercepted while it is

---

[6]Where the text that follows identifies specific commercial products and services, it is solely for the purpose of helping to explain the current state of the art. The committee does not endorse or recommend any specific product or service.

being transmitted, but if you have a way to do that, why not use that method to send the original message?

One answer is hinted at above: speed. The key (a short collection of digits) is far smaller than the thing being encrypted (e.g., the television program), so the key distribution mechanism can use a more elaborate, more secure, and probably slower transmission route, one that would not be practical for encrypting the entire program.[7]

Another answer has arisen in the past 20 years that gets around the conundrum—a technique called public-key cryptography.[8] This technique uses two different keys—a public key and a private key—chosen so that they have a remarkable property: Any message encrypted with the public key can be decrypted *only* by using the corresponding private key; once the text is encrypted, even the public key used to encrypt it cannot be used to decrypt it.

The idea is to keep one of these keys private and publish the other one; private keys are kept private by individuals, while public keys are published, perhaps in an online directory, so that anyone can find them. If you want to send a secret message, you encrypt the message with the recipient's public key. Once that is done, only the recipient, who knows the corresponding private key, can decrypt the message. Software is widely available to generate key pairs that have this property, so individuals can generate key pairs, publish their public keys, and keep their private keys private.

As public-key encryption is typically considerably slower (in terms of computer processing) than symmetric-key encryption, a common technique for security uses them both: Symmetric-key encryption is used to encrypt the message, then public-key encryption is used to transmit the decryption key to the recipient.

A wide variety of other interesting capabilities is made possible by public-key systems, including ways to "sign" a digital file, in effect providing a digital signature. As long as the signing key has remained private, that signature could only have come from the key's owner. These additional capabilities are described in Appendix E.

Any encryption system must be designed and built very carefully, as there are numerous and sometimes very subtle ways in which information can be captured. Among the more obvious is breaking the code: If

---

[7]The most basic form of "separate mechanism" to send the key is having a codebook of keys hand-carried to the recipient, as has been done for years in the intelligence business. This is not feasible where widescale distribution is concerned.

[8]The technique was first brought to practical development by R.L. Rivest, A. Shamir, and L.M. Adelman in Rivest et al. (1978). RSA Security (see <http://www.rsa.com>) produces software products based on this development.

the encryption is not powerful enough, mathematical techniques can be used to decrypt the message even without the key. If the key-distribution protocol is flawed, an unauthorized person may be able to obtain the key via either high technology (e.g., wiretapping) or "social engineering" (e.g., convincing someone with access to the key to supply it, a widely used approach). If the system used to read the decrypted information is not designed carefully, the decrypted information may be left accessible (e.g., in a temporary file) after it has been displayed to the user. The point to keep in mind is that cryptography is no magic bullet; using it effectively requires both considerable engineering expertise and attention to social and cultural factors (e.g., providing incentives for people to keep messages secret).[9]

Access Control in Bounded Communities

Perhaps the most fundamental form of technology for the protection of intellectual property is controlling access to information (i.e., determining whether the requester is permitted to access the information). A basic form of such control has been a part of the world of operating systems software almost from the time operating systems were first implemented, offering limited but useful security. In its simplest form, an access control system keeps track of the identity of each member of the user community, the identities of the data objects, and the privileges (reading, altering, executing, and so on) that each user has for each object. The system consults this information whenever it receives a service request and either grants or denies the request depending on what the privilege indicates.

Existing access control, however, offers only a part of what is needed for dealing with collections of intellectual property. Such systems have typically been used to control access to information for only relatively short periods such as a few years, using only a few simple access criteria (e.g., read, alter, execute), and for objects whose owners are themselves users and who are often close at hand whenever a problem or question arises.

In contrast, access control systems for intellectual property must deal with time periods as long as a century or more and must handle the sometimes complex conditions of access and use. A sizable collection—as indeed a digital library will be—also needs capabilities for dealing with hundreds or thousands of documents and large communities of users (e.g., a college campus or the users of a large urban library).

Such systems will thus need to record the terms and conditions of access to materials for decades or longer and make this information acces-

---

[9]See, for example, CSTB (1996).

sible to administrators and to end users in ways that allow access to be negotiated. This raises interesting questions of user authentication: For example, is the requester who he says he is? Does he have a valid library card? It also raises issues of database maintenance: For example, collections change, rights holders change, and the user community changes as library cards expire. Many other questions must be addressed as well so that systems work at the scale of operation anticipated. Some work along these lines has been done (e.g., Alrashid et al., 1998), but a considerable amount of development work is still needed.

Some attempts have also been made to represent in machine-readable form the complex conditions that can be attached to intellectual property. This is the focus of what have been called rights management languages, which attempt to provide flexible and powerful languages in which to specify those conditions.[10] DPRL (Ramanujapuram, 1998), for example, attempts to offer a vocabulary in which a wide variety of rights management terms and conditions can be specified.

An important characteristic of these languages is that they are machine-readable (i.e., the conditions can be interpreted by a program that can then grant or deny the desired use). This is superficially the same as a traditional operating system, but the conditions of access and use may be far more complex than the traditional notions used in operating systems. In addition, as will be shown below, these languages are quite useful outside the context of bounded communities. Finally, although large-scale systems have yet to be deployed, rights management language design is not perceived as a roadblock to more robust TPSs.

Enforcement of Access and Use Control in Open Communities

Access control systems of the sort outlined above can be effective where the central issue is specifying and enforcing access to information,

---

[10]MPEG-4 offers a general framework of support for rights management, providing primarily a structure within which a rights management language might be used, rather than a language itself. It is nonetheless interesting, partly because it represents the growing recognition that rights management information can be an integral part of the package in which content is delivered. The standard specifies a set of IP management and protection descriptors for describing the kind of protection desired, as well as an IP identification data set for identifying objects via established numbering systems (e.g., the ISBN used for books). Using these mechanisms, the content providers can specify whatever protection strategy their business models call for, from no protection at all to requiring that the receiving system be authorized via a certified cryptographic key, be prepared to communicate in an encrypted form, and be prepared to use a rights management system when displaying information to the end user. For additional information on MPEG-4, see Konen (1998) and Lacy et al. (1998).

as is typically true in bounded communities represented by, for example, a single corporation or a college campus. In such communities much greater emphasis is placed on questions of original access to information than on questions of what is done with the information once it is in the hands of the user. The user is presumed to be motivated (e.g., by social pressure or community sanctions) to obey the rules of use specified by the rights management information.

A larger problem arises when information is made accessible to an unbounded community, as it is routinely on the Web. The user cannot in general be presumed to obey rules of use (e.g., copyright restrictions on reproduction); therefore, technical mechanisms capable of enforcing such rules are likely to be needed.

A variety of approaches has been explored. The simpler measures include techniques for posting documents that are easily viewed but not easily captured when using existing browsers. One way to do this uses Java routines to display content rather than the standard HTML display. This gives a degree of control over content use because the display can be done without making available the standard operating system copy-and-paste or printing options. A slightly more sophisticated technique is to use a special format for the information and distribute a browser plug-in that can view the information but isn't capable of writing it to the disk, printing, and so on. Knowledgeable users can often find ways around these techniques, but ordinary users may well be deterred from using the content in ways the rights holder wishes to discourage.

There are also a number of increasingly complex techniques for controlling content use that are motivated by the observation made earlier, that digital IP liberates content from medium—the information is no longer attached to anything physical. When it is attached to something physical, as in, say, books or paintings, the effort and expense of reproducing the physical object offers a barrier to reproduction. Much of our history of and comfort with intellectual property restrictions is based on the familiar properties of information bound to physical substrates. Not surprisingly, then, some technical protection mechanisms seek to restore these properties by somehow "reattaching" the bits to something physical, something not easily reproduced. The description that follows draws on features of several such mechanisms as a way of characterizing this overall approach.

Encryption is a fundamental tool in this task. At a minimum, encryption requires that the consumer get a decryption key, without which a copy of the encrypted content is useless. Buy a digital song, for example, and you get both an encrypted file and a password for decrypting and playing the song.

But this approach secures only the original access to the content and its transit to the consumer. Two additional problems immediately become apparent. First, the content is still not "attached" to anything physical, so the consumer who wished to do so could pass along (or sell) to others both the encrypted content and the decryption key. Second, the consumer could use the key to decrypt the content, save the decrypted version in a file, and pass that file along to others.

There are several ways to deal with the first problem that involve "anchoring" the content to a single machine or single user. One technique is to encode the identity of the purchaser in the decryption key, making it possible to trace shared keys back to their source. This provides a social disincentive to redistribution.[11] A second technique is for the key to encode some things about the identity of one particular computer, such as the serial number of the primary hard drive, or other things that are unlikely to change.[12] The decryption software then checks for these attributes before it will decrypt the content. A third technique calls for special hardware in the computer to hold a unique identifier that can be used as part of the decryption key. Some approaches call for this hardware to be encased in tamper-resistant cases, to discourage tampering even by those with the skill to modify hardware. One form of tamper resistance involves erasing the key if any attempt is made to open or manipulate the chip containing it.

Whatever the approach, the intended result is the same—the content can be decrypted only on the machine for which the decryption has been authorized.

But even this protection alone is not sufficient, because it is not persistent. The consumer may legally purchase content and legally decrypt it on her machine, then (perhaps illegally) pass that on to others who may be able to use the information on their machines. The final technological step is to reduce the opportunities for this to happen. Two basic elements are required: (1) just-in-time and on-site encrypting and (2) close control of the input/output properties of the machine that will display the content. Decrypting just in time and on site means that the content is not decrypted until just before it is used, no temporary copies are ever stored, and the information is decrypted as physically close to the usage site as possible. An encrypted file containing a music album, for instance, would not be entirely decrypted and then played, because a sophisticated pro-

---

[11]This also has privacy implications that consumers may find undesirable.

[12]Hard drives typically have serial numbers built into their hardware that can be read using appropriate software but cannot be changed. However, because even hard disks are replaced from time to time, this and all other such attempts to key to the specific hardware will fail in some situations. The idea of course is to select attributes stable enough that this failure rarely occurs.

grammer might find a way to capture the temporary decrypted file. Instead, the file is decrypted "on the fly" (i.e., as each digital sample is decrypted, it is sent to the sound-generation hardware), reducing the ease with which the decrypted sample can be captured. On-site decryption involves placing the decryption hardware and the sound-generation hardware as physically close as possible, minimizing the opportunity to capture the decrypted content as it passes from one place to another inside (or outside) the computer.[13]

Some playback devices are difficult to place physically near the computer's decryption hardware. For example, digital camcorders, digital VCRs, digital video disk (DVD) movie players, and so on all require cables to connect them to the computer, which means wires for interconnection, and wires offer the possibility for wiretapping the signal.

One approach to maintaining on-site decryption for peripheral devices is illustrated by the Digital Transmission Content Protection (DTCP) standard, an evolving standard developed through a collaboration of Hitachi, Intel, Matsushita, Sony, and Toshiba (see Box 5.1). The computer and the peripheral need to communicate to establish that each is a device authorized to receive a decryption key. The key is then exchanged in a form that makes it difficult to intercept, and the content is transmitted over the wire in encrypted form. The peripheral device then does its own on-site decryption. This allows the computer and peripheral to share content yet provides a strong degree of protection while the information is in transit to the decryption site.

But even given just-in-time and on-site decryption, a sophisticated programmer might be able to insert instructions that wrote each decrypted unit of content (e.g., a music sample) to a file just before it was used (in this case sent to the sound-generation hardware). Hence, the second basic element in providing persistent encryption is to take control of some of the routine input and output (I/O) capabilities of the computer. There are a number of different ways to attempt this, depending partially on the degree to which the content delivery system is intended to work on existing hardware and software.

The largest (current) market is of course for PCs running off-the-shelf operating systems (such as Windows, Mac, and Linux). In that case the content delivery system must use the I/O routines of the existing operating system. The difficulty here is that these routines were not designed to hide the information they are processing. As a result, using an existing operating system opens another door to capturing the decrypted content.

---

[13]Information may be captured by physically wiretapping the cables that route signals inside and outside the computer.

BOX 5.1
Characteristics of the DTCP Copy Protection Standard

• *Copy control information (CCI).* Rights holders need a way to specify how their content can be used. The system offers three distinct copy control states included in the data signature—no copies permitted, one copy permitted, and data not protected. Compliant copy control devices must be able to extract the CCI field from the copyrighted material and act in accordance with the contained instruction. Note that viewing of time-shifted content using a digital recorder is not possible for material marked as "no copies permitted." The one-copy state has been specifically created to allow digital recorder time shifting.

• *Device authentication and key exchange.* Before sharing valuable information, a connected device must first verify that another connected device is authentic. This layer defines the set of protocols used to ensure the identity, authenticity, and compliance of affected devices prior to the transfer of any protected material.

• *Content encryption.* Protected data is encrypted for transmission to reduce the opportunity for unauthorized access to the material. Encryption is necessary because data placed on the wire is (often) simultaneously available to all connected devices, not just the one device for which it is intended. Encrypting the data with keys known only to the sending and receiving devices protects the data while it is in transit.

• *System renewability.* System renewability ensures long-term integrity of the system through the revocation of compromised devices.

NOTE: See <http://www.dtcp.com> for additional information.

Content delivery systems that wish to work in the environment of such operating systems attempt, through clever programming, to reduce the opportunities to capture the decrypted information while the operating system is performing output. But given existing operating systems, abundant opportunities still exist for a sophisticated programmer.

More complex proposals call for replacing parts of, or even the entire, operating system, possibly right down to the BIOS, the basic input/output routines embedded in read-only memory in the computer hardware. Such computers would instead use specially written routines that will not read or write without checking with the decryption hardware on the computer to ensure that the operation is permitted under the conditions of use of the content. This more ambitious approach faces the substantial problem of requiring not only the development of a new and complex operating system but the widespread replacement of the existing installed base as well. This clearly raises the real possibility of rejection by consumers.

The final problem is the ultimate delivery of the information: Music must be played, text and images displayed, and so on. This presents one final, unavoidable opportunity for the user to capture the information. The sophisticated owner of a general-purpose computer can find ways to copy what appears on the screen (e.g., screen capture utilities) or what goes into the speakers (connect an analog-to-digital converter to the speaker wires). As is usual in such matters, the expectation is that this will be tedious enough (capturing a long document screenful by screenful), complex enough (hooking up the converter), or of sufficiently low quality (the captured speaker signal is not identical to the digital original) that all but the most dedicated of thieves will see it as not worth the effort. Nevertheless, those who place substantial faith in elaborate TPSs should keep in mind the necessity of presenting information to the user and the opportunity this provides for capture.

More generally, because all protection mechanisms can eventually be defeated at the source (e.g., as it was with a2b encoding and Windows Media; see Chapter 2), the key questions concern trade-offs of cost and effectiveness. A good mechanism is one that provides the degree of disincentive desired to discourage theft but remains inexpensive enough so that it doesn't greatly reduce consumer demand for the product. A good deal more real-world experience is needed before both vendors and consumers can identify the appropriate trade-offs.

Currently, any system aiming to provide substantial technical protection will rely on encryption, anchoring the bits to a specific machine, and making encryption persistent through just-in-time decryption and low-level control of I/O. Systems using one or more of these ideas are commercially available, and others are under active development. Music delivery systems such as AT&T's a2b and Liquid Audio's Liquid Player, for example, are commercially available. InterTrust, IBM, and Xerox are marketing wide-ranging sets of software products aimed at providing persistent protection for many kinds of content.[14] Similar efforts currently under development include the Secure Digital Music Initiative (discussed in Chapter 2) aimed at providing a standard for protecting music.

Copy Detection in Open Communities: Marking and Monitoring

When a valuable digital object is not encrypted and is outside the sphere of control of its rights holder, the only technical means of hinder-

---

[14]See <http://www.a2bmusic.com> for information about a2b, <http://www.liquidaudio.com> for information about Liquid Audio, <http://www.ibm.com/security/html/cryptolopes.html> for information on the IBM products, <http://www.intertrust.com> for information on InterTrust offerings, and <http://www.contentguard.com> for information on Xerox's offerings.

ing misuse is to change it in ways that discourage wrongdoing or facilitate detection. A variety of approaches have been used to accomplish these goals. One technique calls for releasing only versions of insufficient quality for the suspected misuses. Images, for example, can be posted on the Web with sufficient detail to determine whether they would be useful, for example, in an advertising layout, but with insufficient detail for reproduction in a magazine.

Another technique embeds in the digital document information about ownership, allowed uses, and so on. One of the simplest and most straightforward ways to do this is by labeling the document in a standard way (so the label can be found) and in a standard vocabulary (so the terms of use may be widely understood). In its simplest format, a digital label could take the form of a logo, trademark, or warning label (e.g., "May be reproduced for noncommercial purposes only"). Labels are intended to be immediately visible and are a low-tech solution in that they are generally easily removed or changed, offering no enforcement of usage terms.

Labels could, nevertheless, ease the problem of IP management, at least among the (fairly large) audience of cooperative users. Consider the utility of having every Web page carry a notice in the bottom right corner that spelled out the author's position on use of the page. Viewers would at least know what they could do with the page, without having to guess or track down the author, allowing cooperative users to behave appropriately. Getting this to work would require spreading the practice of adding such information, so that authors did it routinely, and some modest effort to develop standards addressing the kinds of things that would be useful to say in the label. There is an existing range of standard legal phrases.

A second category of label attached to some digital documents is a time stamp, used to establish that a work had certain properties (e.g., its content or the identity of the copyright holder) at a particular point in time. The need for this arises from the malleability of digital information. It is simple to modify both the body of a document and the dates associated with it that are maintained by the operating system (e.g., the creation date and modification date).

Digital time stamping is a technique that affixes an authoritative, cryptographically strong time stamp to digital content; the label can be used to demonstrate what the state of the content was at a given time. A third-party time-stamping service may be involved to provide a trusted source for the time used in the time stamp. Time-stamping technology is not currently widely deployed.[15]

---

[15]Some products do exist, including WebArmor (see <http://www.webarmor.com>) and Surety's Digital Notary Service (see <http://www.surety.com>).

Where the labels noted above are separate from the digital content, another form of marking embeds the information into the content itself. Such digital alterations are called watermarks and are analogous to watermarks manufactured into paper. An example cited earlier described how a music file might be watermarked by using a few bits of some music samples to encode ownership information and enforce usage restrictions. The digital watermark may be there in a form readily apparent, much like a copyright notice on the margin of a photograph; it may be embedded throughout the document, in the manner of documents printed on watermarked paper, or it may be embedded so that it is normally undetected and can be extracted only if you know how and where to look, as in the music example.[16] Visible watermarks are useful for deterrence, invisible watermarks can aid in proving theft, and a watermark distributed through a document can by design be difficult to remove, so that it remains detectable even if only part of the document is copied.

The objectives, means, and effectiveness of marking technologies depend on a number of factors. Designing an appropriate watermark means, for instance, asking what mix is desired of visibility (Should the mark be routinely visible?), security (How easy is it to modify the mark?), and robustness (What kinds of modifications, such as printing a picture and rescanning it, can the mark survive?). The nature and value of the information clearly matters. A recent hit song needs different treatment than a Mozart aria. Modality also matters. Sheet music is watermarked differently than an audio recording of a performance. Some things are difficult to watermark. Machine code for software cannot be watermarked in the same way as music, because every bit in the program matters; change one and the program may crash. Identifying information must instead be built into the source code, embedded in a way that the information gets carried into the machine code but does not adversely affect the behavior of the program.[17] Watermarking digital text also presents challenges: How can, say, an online version of *The Grapes of Wrath* be marked to include a digital watermark, without changing the text? One trick is to change the *appearance* of the text. The watermark can be encoded by varying the interline and intercharacter spacing slightly from what would be expected; the variation encodes the information.

Marking a document is of course only half the battle; monitoring is

---

[16]Embedding IP ownership information in documents in subtle ways has a long history and had been used much before the arrival of digital information. One of the oldest and simplest techniques is the mapmaker's trick of inserting nonexistent streets or roads. Similarly, text has been "marked" by distributing versions with small changes in wording.

[17]This is not difficult technologically, but it adds another step to the marking process and can present significant additional overhead.

needed in order to detect the presence of unauthorized copies. A number of efforts have been made in this direction, many of which rely on "Web crawlers," programs that methodically search the Web looking for documents bearing a relevant watermark. An IP management system that watermarked images, for example, would also have a Web searching routine that examined publicly available image files for that system's watermarks. This is an active area of work; systems have been developed in both the commercial and academic world.[18]

Marking and monitoring technologies do not attempt to control users' behavior directly. In particular, they do not attempt to prevent unauthorized copy and modifications. Rather, they attempt to make these actions detectable so that rights holders can seek legal redress when infringements have been detected. Frequently their intent is simply to indicate that copying is prohibited; the utility of these technologies relies on the fact that many people are honest most of the time.

Trusted Systems

The preceding discussion of technical protection mechanisms points out that the strongest intellectual property protection requires embedding protection mechanisms throughout the computer hardware and software at all levels, right down to the BIOS. In one vision of the future, security will become a major influence on the design of computing and communications infrastructure, leading to the development and widespread adoption of hardware-based, technologically comprehensive, end-to-end systems that offer information security, and hence facilitate creation and control of digital IP. There has been some research (and a great deal of speculation and controversy) about these so-called "trusted systems," but none is in widespread use as of 1999.

One example of this vision (Stefik, 1997b) seeks to enable the world of digital objects to have some of the same properties as physical objects. In these systems, when a merchant sells a digital object, the bits encoding that object would be deposited on the buyer's computer and erased from the merchant's computer. If the purchaser subsequently "loaned" this digital object, the access control and rights management systems on the lender's computer would temporarily disable the object's use on that computer while enabling use on the borrower's computer. These changes

---

[18]Digimarc at <http://www.digimarc.com> is one example of a commercial watermarking and tracking system; Stanford's Digital Library project at <http://www-diglib.stanford.edu/diglib/pub/> has produced systems for detecting copying of text and audio files, using feature extraction techniques to enable fast searching and detection of partial copies.

would be reversed when the object is returned by the borrower to the lender.

The published literature (see, e.g., Stefik, 1997a,b) is fairly clear on what trusted systems are supposed to accomplish, but it does not spell out in technical detail how they are supposed to accomplish it. Stefik, for example, is clear on the need for some sort of hardware component (Stefik, 1997b) to supplement the Internet and PC world of today,[19] but he says little about how that component would work or how it would be added to today's infrastructure. Here, we explore two general ways in which trusted systems might be implemented, then consider the barriers they face.

One way to increase control over content is to deliver it into special-purpose devices designed for purchase and consumption of digital content, but not programmable in the manner of general-purpose PCs. For example, game-playing machines, digital music players, electronic books, and many other types of devices could be (and some are) built so that each one, when purchased, contains a unique identifier and appropriate decoding software. The devices could then be connected to the Web in much the same way as general-purpose computers and download content encrypted by distributors. Legitimate devices would be able to (1) verify that the content came from an authorized distributor, (2) decrypt and display the content (the meaning of "display" depending on whether the content is text, video, audio, and so on), and (3) force the device owner to pay for the content (perhaps by checking before decrypting that the subscription fee payment is up-to-date).

It is expensive to design, manufacture, and mass market such a special-purpose device, and an entire content-distribution business based on such a device would necessitate cooperation of at least the consumer-electronics and content-distribution industries, and possibly the banking and Internet-service industries as well. A particular business plan could thus be infeasible because it failed to motivate all of the necessary parties to cooperate or because consumers failed to buy the special-purpose devices in sufficient numbers. The failure of the Divx player for distribution of movies is perhaps an instructive example in this regard. [20]

Hardware-based support for IP management in trusted systems could also be done using PCs containing special-purpose hardware. Because such machines would have the full functionality of PCs, users could con-

---

[19]For example, a tamperproof clock to ensure that rights are not exercised after they expire or to secure memories to record billing information (Stefik, 1997b).

[20]Production of Digital Video Express LP, or Divx, was terminated by Circuit City in June 1999 (Ramstad, 1999). Although it was not designed to download content from the Web, it was in many other respects the sort of device suggested above.

tinue to use them for everything that they do today. The intent would be that because they had secure hardware, content distributors and their customers could conduct business just as they could in the information-appliance scenario, but without customers having to buy a separate special-purpose device. One problem here, suggested above, is that the content must, eventually, be presented to the user, at which point it can be captured. The capturing may be a slow and perhaps painful process, but, if the content in question is of sufficient value, pirates may well be motivated to go to the effort or to write software that will automate the effort.

The trusted systems scenario faces substantial challenges, in part because accomplishing it would require changes to the vast installed base of personal computers, changes that the marketplace may reject. The need for specialized hardware would require buying new machines or retrofitting existing computers with hardware ensuring that the computer user was able to do with the digital object exactly those actions specified by the rights management language. The tight control of input and output, for example, if universally enforced, would be experienced by the user as an inability to do print redirection, the ability that permits the personal computer user to save into a local file anything he or she can see on the screen or print. The committee finds it questionable whether computer owners would accept the inconvenience, risk, and expense of retrofitting their machines with a device that makes them more expensive and in some ways less capable.

The case is less obvious where purchasing new machines is concerned, but even here there is a substantial question of what will motivate buyers to purchase a machine that is more expensive (because of the new hardware and software) and, once again, less capable in some ways. Note, too, that although consumers might benefit from access to content that would not have been released without trusted systems in place, significant benefit from such systems would accrue to content originators, while the costs would be borne principally by content users.[21]

There are two plausible scenarios for the adoption of such an approach: the "clean slate" scenario and the "side effect" scenario. The clean slate scenario involves the introduction of new technology, which avoids the problem of an installed base and offers opportunities to mandate standards. DVD offers one such example: The hardware and software for a player must use certain licensed technology and obey certain protection standards in order to be capable of playing movies. Such requirements can be set in place at the outset of a new technology, before there is an installed base of equipment without these capabilities. Given

---

[21]For a more thorough discussion of this issue, see Gladney and Lotspiech (1998) and the works it cites.

the size of the installed base of computers and their continuing utility, it is not clear what would provide the analogous clean slate opportunity for trusted systems.

The "side effect" scenario involves technology that is introduced for one reason and turns out to be useful for a second purpose.[22] In this case, the initial reason is business-to-business electronic commerce; the second purpose is IP protection. The Trusted Computing Platform Alliance, a collaborative effort founded in October 1999 by Compaq, HP, IBM, Intel, and Microsoft, is aimed at "building confidence and trust of computing platforms in e-business transactions."[23] It plans to provide security at the level of the hardware, BIOS, and operating system, i.e., thoroughly integrated into the system in ways that would make it transparent to the user. This is a very ambitious undertaking that will require a considerable, coordinated effort among several manufacturers, and its success is far from guaranteed.

Nevertheless, should the alliance make substantial progress, it would offer a foundation for business-to-business e-commerce and would also mean that PCs would likely come equipped with hardware and software that provided a natural foundation for TPSs aimed at IP protection. This report noted earlier that the marketplace for electronic information might benefit from the marketplace infrastructure built for electronic commerce; it may be the case that the computer hardware and software built for secure electronic commerce will turn out to be a useful foundation for IP protection on individual computers.

An alternative version of the trusted system notion envisions creating software-based IP management systems whose technical protection arises from a variety of software tools, including encryption, watermarking, and some of the technologies discussed above. Although this would not provide the same degree of protection as systems using both software and special hardware, it may very well offer sufficient strength to enable an effective marketplace in low- to medium-value digital information. For a variety of nontechnical reasons discussed at length in Gladney (1998), the integration phase of such systems is proceeding slowly, with end-to-end systems not nearly as well developed or well understood as the individual technical tools.

---

[22]In fact there is some reason to believe that general-purpose computers are being used increasingly for entertainment purposes, such as stereo listening, recording studios, and high-tech photo albums, as well as for traditional PC entertainment applications such as computer games, based on a "totally unexpected surge in U.S. home PC buyers," according to industry analyst Roger Kay commenting in McWilliams (1999).

[23]See <http://www.trustedpc.org>.

Protection Technologies for Niches and Special-Purpose Devices

As the discussion above makes clear, there are substantial challenges in creating technical protection services capable of working effectively in the context of a general-purpose computer. However, with more specialized devices, or in contexts of limited uses of the computer, additional techniques may be employed.

For example, for high value, specialized software with smaller, more narrowly defined markets, hardware-based copy protection schemes have had some success. In the computer-aided design software market, for instance, products are distributed with "dongles," simple physical devices that plug into the printer port; the software does not function unless the dongle is installed. But dongles have been tried and have proven impractical for mass market software: Consumers rapidly became frustrated with the need to keep track of a separate dongle for each application and each of its upgrades.

For specific devices, like CD players, copy protection can be based on hardware built into the device. This hardware makes it difficult to use CD-ROM recorders to create unauthorized copies of disks with commercially valuable music, software, or other content. For example, Macrovision's SafeDisc technology uses digital signature, encryption, and hardware-based copy protection in a TPS that is transparent to the user of a legitimate disk.[24] The content of the CD-ROMs is encrypted and digitally signed. The physical copy protection technology prevents CD-ROM readers and other professional mastering equipment from copying the digital signature. This in turn prevents unauthorized copying, because the content can be decrypted only when the digital signature can be read and verified.

Digital video disks provide a second example of hardware-based copy protection for special-purpose devices, in this case for use by the entertainment industry (see Box 5.2).

Technical Protection Services, Testing, and the Digital Millennium Copyright Act of 1998[25]

Understanding the interaction of intellectual property and technical protection services requires an understanding of how technical protection methods and products are developed. One key feature of the technology underlying TPSs is that its creation proceeds in an adversarial manner.

---

[24]SafeDisc is targeted at the software publishing market. See <http://www.macrovision.com> for more information.

[25]These issues are discussed at greater length in Appendix G.

BOX 5.2
Digital Video Disks

Developed by studios and consumer electronics companies in late 1995, digital video disks (DVDs) are used in the entertainment industry to distribute movies and other content. DVDs are compatible with CDs and are of the same size and thickness as CDs, but they have much more capacity—up to 25 times as much as a CD.

Content on DVDs can be protected by a variety of mechanisms:

- Data on the DVD can be encrypted using a system called the content scrambling system (CSS).
- Each disk can indicate whether the contents can be copied, enabling serial copy management. For example, a device getting information from a disk marked "one copy" must change the information on its version to indicate "no [more] copies."
- The DTCP protocol described in Box 5.1 can be used to encrypt information for transmission from the DVD player to other devices.
- Analog copy protection is inhibited by a Macrovision circuit; this adds a signal to the analog video output that will (typically) not distort the display of the video but will cause a recording device to record a significantly degraded copy. This inhibits copying DVDs to videotape.

The DVD technical protection system is useful for keeping honest people honest, but from a security point of view it has defects in its design that prevent it from being a major deterrent for skilled pirates. For example, the effectiveness of the CSS encryption scheme depends on the secrecy of the cryptographic algorithm, not just on the secrecy of the cryptographic key; this is a violation of a well-known cryptographic design principle. CSS has not been adopted elsewhere, partly due to this weakness.

In November 1999, the CSS encryption scheme was apparently broken, due in part to this very issue. Two programmers examined the software used by one DVD player, whose manufacturer had neglected to encrypt its decryption key. Examining the software enabled them to break the scheme for that one specific player, which then provided them with a window into the encryption keys used by any of the other 400-odd licensed players (Patrizio, 1999b).

One member of the community of cryptography and security researchers proposes a protection mechanism, and others then attack the proposal, trying to find its vulnerabilities. It is important that this process go on at both the theoretical and experimental levels. Proposals for new ideas are often first evaluated on paper, to see whether there are conceptual flaws. Even if no flaws are evident at this stage, the concept needs to be evaluated experimentally, because systems that have survived pencil-and-paper attempts may still fail in actual use. This can happen either because flaws were simply not discovered in the theoretical analysis or because a sound proposal was implemented badly. Fielded implementations, not abstract designs, are what customers will use; hence, real implementations must be tested in real use.

A crucial part of the development of good technical protection mechanisms is thus the experimental circumvention, or attack, on hardware and software that are claimed to be secure. Before the device is relied on to protect valuable content, vigorous, expert attacks should be carried out, and they should be done under conditions that are as close as possible to those in which the secure hardware or software will be used.

This process is not merely good in theory; it is how good security technology and products are created, both by researchers and in commercial practice. Vendors, for example, assemble their own "tiger teams" that try to circumvent a security mechanism before it is released in the marketplace. The results of this practice validate its use. The history of the field is replete with good ideas that have been tested by the community, found to be flawed, improved, retested, and improved again. The process continues and the technology constantly gets better.[26]

This in turn has policy significance: Regulating circumvention must be done very carefully lest we hobble the very process that enables the development of effective protection technology. If researchers, vendors, security consultants, and others are unsure about the legal status of their activities, their effectiveness may suffer, and the quality of the resulting products may decline.

This issue arises in part as a consequence of the Digital Millennium Copyright Act of 1998 (DMCA), the U.S. Congress's implementation of the World Intellectual Property Organization (WIPO) treaty. See Box 5.3 for a brief background on the interaction of the development of technical protection mechanisms and the policy in the DMCA.

What Makes a Technical Protection Service Successful?

Whether a TPS is successful begins with its inherent technical strength but depends ultimately on both the product it protects and the business in

---

[26]One ongoing example is the evolution of the Sun Microsystems Java programming system. When Sun launched this innovative system, one of the most important claims that it made was that server-supplied code could be run from any Java-enabled Web browser "safely." This soon turned out to be false, as shown by Dean et al. (1996), whose analysis of the underlying problem led to improvements in Sun's next release of its Java Development Kit. This process continues, as real use and experimentation uncover additional defects, leading to further repairs and improvements.

BOX 5.3
Technical Protection Mechanism Development, Public Policy,
and the Digital Millennium Copyright Act

Congress included in the DMCA two kinds of anticircumvention regulations. The first kind—the access-control provision—generally outlaws circumventing technical protection measures used by rights holders to control access to their works. Simply put, it is illegal to "break" (i.e. circumvent) the technical measures, such as encryption, that rights holders use to control access to their work.[1]

The second kind of anticircumvention regulation—the "antidevice" provisions—generally outlaw devices that are designed or produced primarily for purposes of circumventing technical protection measures, have no commercially significant uses other than circumvention, or are marketed to circumvent technical protection measures. One of the antidevice rules outlaws devices that circumvent access controls; the other outlaws devices that circumvent use or copying controls ("access" concerns whether you can read the document, "use" focuses on what you do with it, for example, print or make a copy of it).

These provisions are, on their own terms, plausible steps providing prophylactic measures aimed at protecting intellectual property. The access-control provision does its part by defining a new legal wrong—breaking the protection mechanism—a step quite distinct from any illegal copying or other use of the content being protected. The antidevice provisions are analogous to similar laws concerning cable television descramblers, working on the presumption that it is inappropriate to manufacture devices whose intended purpose is to enable people to break the law.

As Congress realized, however, problems emerge from the details. First, Congress recognized that circumvention can be done for entirely legitimate purposes, such as encryption research, computer security testing, and achieving interoperability for computer systems. In recognition of this, the access-control provision is subject to seven rather complicated—and at times ambiguous—statutory exceptions that permit circumvention for purposes of the sort noted. These exceptions may not, however, exhaust the full range of legitimate purposes for bypassing technical protection systems, as Appendix G explains. The DMCA as written is inconsistent and unclear as to whether circumvention is permitted to enable fair use, though legislative history suggests that Congress intended the preservation of fair use. Future revision of this law should fix this inconsistency.

Second, Congress was apparently concerned about the potential for technical protection mechanisms to disrupt fair use and other noninfringing uses. The concern is simple: If you can't get access to content, you clearly can't make fair use of it. As a result Congress tasked the Librarian of Congress with a kind of watchdog role. The DMCA requires the Librarian of Congress to determine:

> whether persons who are users of a copyrighted work are, or are likely to be in the succeeding 3-year period, adversely affected by the prohibition under subparagraph (A) in their ability to make noninfringing uses under this title of a particular class of copyrighted works. In conducting such rulemaking, the Librarian shall examine:

---

[1]Relevant excerpts from the DMCA are reprinted in the addendum to Appendix G.

(i) the availability for use of copyrighted works;
(ii) the availability for use of works for nonprofit archival, preservation, and educational purposes;
(iii) the impact that the prohibition on the circumvention of technological measures applied to copyrighted works has on criticism, comment, news reporting, teaching, scholarship, or research.

If such an adverse effect is found, the Librarian can exempt certain classes of users or works from the access-control ban.

Third, there is a significant ambiguity in the DMCA about whether there is an implied right to get access to the tools needed to do circumvention for fair use or other legitimate purposes. It is a hollow privilege indeed to be allowed to circumvent in order to make fair use and then to be told that all the tools necessary to effect that circumvention are outlawed. Some of the exceptions to the access-control provision specifically allow the development of circumvention technologies necessary to accomplish the lawful circumvention, but others do not. As a result, it is somewhat unclear from the statute whether there is an implicit right to develop or purchase a tool to engage in a lawful circumvention. This is an important question that will apparently be left to the courts to address.

Fourth, both the access-control provision and the antidevice provisions are insufficiently clear in their explanation of key concepts and their use of technical terms. Most strikingly, while the provisions indicate that "No person shall circumvent a technological measure that effectively controls access to a work protected under this title" [sec. 1201], they do not adequately explain what is meant by the phrase "effectively controls access."[2] The lack of guidance on this key concept means that ordinary computing professionals cannot reasonably know whether a particular technology will be covered by the statute or not.

The DMCA anticircumvention regulations represent the U.S. implementation of a more general provision in the WIPO Copyright Treaty that requires "adequate protection" against and "effective remedies" for circumvention of technical protection measures used by rights holders to protect their works. The anticircumvention provisions of the DMCA were the subject of considerable controversy during the legislative debate on WIPO treaty implementation, and, as adopted, they bear the imprint of lobbying and political compromise. Rather than specifying a few general principles, the rules are instead very complicated, while at the same time ambiguous in important respects (as discussed in some detail in Appendix G). They adopt, moreover, a copyright-centric view of what is, in fact, a more general public policy issue: When should the circumvention of TPSs used by anyone for *any* purpose be permissible?

All of these difficulties illustrate the complexities of writing regulations for relatively uncharted areas involving complex and fast-moving technology.

---

[2]Section 1201(e)(3)(B) does attempt a definition of "effectively controls access": "a technological measure 'effectively controls access to a work' if the measure, in the ordinary course of its operation, requires the application of information, or a process or a treatment, with the authority of the copyright owner, to gain access to the work." This is inadequate to permit even experienced computing professionals to know what the statute covers (see Appendix G).

which it is deployed. The vendor interested in protecting content is only partly concerned with whether a TPS satisfies an abstract technical definition of security. Indeed, most of the techniques discussed in this section can be circumvented by people who are sufficiently motivated and knowledgeable. Vendors have more concrete concerns: Does the TPS deter enough potential thieves and facilitate enough use by paying customers to produce a profitable content-distribution business?

Some of the properties that bring a TPS in line with a business model include:

- *Usability.* A protection system that is cumbersome and difficult to use may deter paying customers. If that happens, it is a failure, no matter how successful it may be at preventing theft.
- *Appropriateness to the content.* The cost of designing, developing, and deploying the protection system has to be in harmony with the market for the content. For content that is inexpensive or already available in a reasonably priced, non-Internet medium, there is no point to an expensive TPS that drives up the price of Internet delivery.
- *Appropriateness to the threat.* Preventing honest customers from giving copies to their friends may require nothing more than a reasonably priced product, a good distribution system, and a clear set of instructions. At the other end of the spectrum, preventing theft of extremely valuable content that must at some point reside in a networked PC requires a very sophisticated TPS, and even the best available with current technology may not be good enough.

The cost-benefit analysis needed to design or choose an appropriate TPS—if indeed there is one—is difficult, but necessary. Distributors can lose in the marketplace because they choose a TPS that is too sophisticated or too expensive, just as easily as they can because they choose one that is too weak.

THE ROLE OF BUSINESS MODELS IN THE PROTECTION OF INTELLECTUAL PROPERTY

Intellectual property protection is frequently viewed in terms of two forces—law and technology. The law articulates what may legally be done, while technology provides some degree of on-the-spot enforcement. In the early days of the software market, for example, the copyright on some programs was enforced by distributing floppy disks that had been written in a nonstandard way, making them difficult to copy.

But law and technology are not the only tools available for grappling with the sometimes difficult task of distributing intellectual property without losing control of it. In the commercial setting, a third powerful factor in the mix is the business model. By selecting appropriately from the wide range of business models available, a rights holder may be able to influence significantly the pressure for and degree of illegal copying or other unauthorized uses. By thinking creatively about the nature of the product and the needs of the customer, rights holders may be able to create new business models that are largely unaffected by the properties of digital information (e.g., the ease of replication and distribution) that are problematic in the traditional model of selling content. They may even be able to find business models that capitalize on those very properties. Hence, in addition to its traditional role of specifying the nature of the commercial enterprise, the business model may also play a role in coping with the IP difficulties that arise with products in digital form. This section explores a variety of models and their impact on the need for technical protection mechanisms and considers the interaction of law, technology, and business models.

The Impact of the Digital Environment on Business Models

As noted in Chapter 1, the introduction of digital media changes the business environment in a number of important ways. The focus here is on the impact of digital media on the intellectual property issues involved in the commercial distribution of content.[27]

Most business models for traditional copyrighted works involve the sale of a physical item that becomes the property of the customer. The economics of the transaction include the costs associated with creating the initial content and first copy of the work (first-copy costs), the costs of reproduction, marketing, distribution, and other overhead costs. Although copyright does not protect subsequent redistribution of the physical copy, in many cases further reproduction and distribution is protected de facto by the costs associated with creating or re-creating a physical copy nearly equal in quality to the original.

Digital information is of course not the first technology to challenge this business model. Photocopying permits the reproduction and distribution of protected works, and although the quality may not be equal to the original, if made available at a low enough price some customers will find photocopies to be acceptable substitutes. Videotapes and audiotapes are similarly vulnerable.

Digital media disrupt the traditional business model by drastically lowering the cost and effort of reproduction and distribution and by pro-

---

[27]Shapiro and Varian (1998) have studied digital media in detail, exploring a wide variety of issues encountered in doing business in this new environment.

ducing copies indistinguishable from the original. While rights holders and consumers benefit from this, so of course may infringers. Additional impacts of the digital medium include the ability to reproduce material in private, increasing the difficulty of detection, and the ability to copy and distribute material very quickly, often before an intellectual property owner can even detect the offense, let alone seek injunctive relief. Natural barriers to infringement are thus eroded in the digital environment. This erosion may be sufficiently extreme at times that rights holders may be wise to reevaluate their fundamental business model. In some cases digital information may be simply unprotectable, at least in practice if not in law and in principle.

Digital media have other impacts on business models as well. Licensing, rather than sale, is becoming increasingly popular for digital media, in part because of the difficulty of retaining control of it after a sale. In this model the customer becomes a user rather than an owner, buying access to a service rather than a physical good. This raises important issues: In a world of distribution by paper, the customer owns a physical copy of the work. What is "owned" in a service offered over a network? If a library discontinues a subscription to an online journal, for example, what rights, if any, does it have to the intellectual property it had been accessing? While networked services are far from new—Dialog and Lexis-Nexis are now more than 20 years old—the nature of access rights has become a major concern with information products and must be factored into the business model.

Those distributing intellectual property in digital form over networks find they are in a business environment changed by customer perceptions and expectations. The perception is that distribution costs are lower, so customer expectations are that prices will be lower than for analog equivalents. In some cases this is true, as with, for example, the replacement of printed software manuals with online or ondisk help; here the economics clearly favor digital formats over paper. In many other cases, however, first-copy costs are higher with digital products, partly because consumers have come to expect more from digital information (e.g., indexing, searching, hyperlinks, multimedia). There are, in addition, new costs associated with digital distribution that offset at least some of the decreased traditional manufacturing costs (e.g., the cost of keeping up with the rapid evolution in browser capabilities and in Web languages).

This pressure for low-priced goods is exacerbated by the fact that on the Web, by far the largest single supply of digital information, free information currently predominates, creating expectations that content will be available free or for low prices. There is also the misperception that "free" equates to "public domain," leading some to believe that if it can be downloaded freely, it is unprotected by intellectual property law. Traditional business models are thus stressed in a number of ways by digital information; of particular significance here are the erosion of natural barriers to infringement and the pressure for inexpensive goods.

Business Models for Handling Information

Traditional business models include a wide variety of possibilities, including goods paid for solely by the buyer, goods totally or partially subsidized by advertisers, and goods given away at no charge, as well as mixes of these models. These are reviewed briefly here, to indicate how they are used in the digital environment and to set the stage for exploring less traditional business models in the next section.

Traditional Business Models

Some traditional business models are outlined below:

1. Business models based on fees for products or services:
   a. *Single transaction purchase.* Examples: Videos, books, some software, music CDs, some text CD-ROMs, and article photocopies (document delivery).
   b. *Subscription purchase.* Examples: Newsletter and most journal subscriptions.
   c. *Single-transaction license.* Examples: Some software and most text CD-ROMs.
   d. *Serial-transaction license* (usually where there is a flat fee for unlimited use). Example: Electronic subscription to a single title (this is different from item 1c above in that the license will often be renewed from year to year upon payment of fees).
   e. *Site licenses* (these are generally also flat fees for unlimited use, but with a broader licensed community). Examples: Software licenses for whole companies, a package containing all electronic journals from a publisher for all members of a university community.
   f. *Payment per electronic use.* Examples: Information resources paid for per search, per time online, or per article accessed.

2. Business models relying on advertising[28]
   a. *Combined subscription and advertising.* Examples: Newspapers,

---

[28]One of the emerging controversies about the Internet is the relative long-term viability of advertising-supported business models as compared with transaction-based business models. See Caruso (1999).

consumer and business-to-business magazines, Web sites such as the *Wall Street Journal*, and America Online.

   b. *Advertising income only.* Examples: Many Web sites and controlled circulation magazines.

   3. "Free" distribution business models[29]
   a. *Free distribution* (no hidden motive). Examples: Scholarly papers on preprint servers and software like Apache, available for free.
   b. *Free samples*—the traditional notion of providing an introduction to the product. Example: A demonstration version of a software package, in the expectation that the customer will want a full, or more up-to-date, version.
   c. *Information goods for those who buy something else or have another income-producing relationship with the information provider.* Example: Free browser software offered to increase traffic on an income-producing Web site.
   d. *Government information or other information in the public domain.* Examples: Standards, economic data, statutes, and regulations.
   e. *Prestige/vanity/some start-ups.* Example: Garage band wanting to get publicity for other services.

Intellectual Property Implications of Traditional Business Models

   Models in the first category derive all revenue from fees for the product or service. Here revenues depend on the number of copies sold or licenses signed, making the rights holder more sensitive to illegal copying, piracy, and even fair use, to the degree that any of these replace the purchase of a copy. Success of a business model of this type depends, in part, on the producer's ability to control postsale copying.

   Specifically, Models 1a (single transaction purchase) and 1b (subscription purchase) are outright purchases, with all of the first-sale and existing copyright implications as to fair use described in Chapters 3 and 4. Models 1c (single transaction license), 1d (serial transaction license), and 1e (site license), as licenses, are attempts to remove any ambiguity in the copyright law by creating an enforceable contract between the rights holder and the user. Such contracts may attempt to impose other desires

---

[29]The term "free" in quotation marks is used as a way of acknowledging that there are always costs associated with the products and services being given away, costs that must be paid somehow. Free software is paid for by the time of the individuals who create it; the cost of producing and distributing free samples is often built into the price of the full product; government information is paid for by taxes, and so on.

of the rights holder through the terms in the contract. While nominally clearer, many licenses are frequently ignored, not understood, not known about by the end user, or otherwise fail to satisfy all parties. Model 1f (pay per use) is a fee for service that may be implemented through either sale or licensing models.

Business models that include advertising (Models 2a and 2b) add more balance to the revenue stream. Subscription prices are held down or eliminated because a large number of qualified recipients helps to ensure advertising revenues. Intellectual property concerns may be more related to illegal reproduction and framing—for example, it is important to ensure that users come to the rights holders' Web site so that advertisements are viewed by users as intended by the rights holders. There is less concern about unauthorized access when the sole income is from advertising. Many Web sites of this type require user registration as a way to identify viewers to advertisers but, for many others, simply counting page views or some other measure of traffic is sufficient.

In the free distribution business models (Models 3a through 3e), reproduction is generally not an issue: Except for the case where use of intellectual property is tied to the purchase of some other product, the information owner is clearly seeking as widespread a dissemination as possible by giving free access. The principal intellectual property concerns here relate to preservation of the integrity of the information, proper citation if someone else uses the information, and the prevention of commercial use of the material by unauthorized users.

Less Traditional Business Models

A variety of other business models have been explored in an attempt to confront the IP difficulties encountered in the digital world. Some of these are derived from models used for traditional products, while others appear to be unique to the world of information products. Eight of these less traditional business models are described below:

1. *Give away the information product and earn revenue from an auxiliary product or service.* Examples of auxiliary products: Free access to an online newspaper in exchange for basic demographic data; the revenue-generating auxiliary product is the database of information about readers. Free distribution of (some) music because it enhances the market for auxiliary goods and services associated with the artist (attendance at concerts, T-shirts, posters, etc.). Example of auxiliary service: The Linux operating system is distributed for free; the market is in service—support, training, consulting, and customization.

2. *Give away the initial information product and sell upgrades.* Example:

Antivirus software, where the current version is often freely downloadable; the revenue-generating product is the subsequent updates (along with support service).

3. *Extreme customization*—Make the product so personal that few people other than the purchaser would want it. Examples: Search engine output, personalized newspapers, and personalized CDs. MusicMaker will create a CD containing the tracks exactly in the sequence specified by a customer.[30]

4. *Provide a large product in small pieces, making it easy to browse but difficult to get in its entirety.* Examples: Online encyclopedias, databases, and many Web sites.

5. *Give away digital content because it complements (and increases demand for) the traditional product.* Examples: The MIT Press and the National Academy Press make the full text of some books and reports available online; this has apparently increased sales of the hard-copy versions.

6. *Give away one piece of digital content because it creates a market for another.* Examples: The Netscape browser was freely distributed in part to increase demand for their server software; Adobe's Acrobat Reader is freely distributed to increase demand for the Acrobat document preparation software.

7. *Allow free distribution of the product but request payment* (perhaps offering additional value in the paid-for version). Example: Shareware. Where shareware versions have time-limited functionality or are incomplete demonstration versions, this is quite similar to the "free sample" model above.

8. *Position the product for low-priced, mass market distribution.* Examples: Microsoft Windows ('95 and '98).

These examples illustrate that far more than immediate production costs enter into pricing decisions. They also demonstrate the trend of relating pricing and other decisions to efforts to develop relationships with customers.

Intellectual Property Implications of Less Traditional Business Models

These less traditional models all reduce the need for enforcement of intellectual property protection against reproduction. The first two do it by foregoing any attempt to generate revenue from the digital content, using it instead as a means of creating demand for services or physical

---

[30]MusicMaker is available at <http://www.musicmaker.com>.

products, neither of which are subject to the replication difficulties of digital products. Giving away digital content as a complement to a traditional product works because reading information online is still awkward and because most people are not willing to print out a multi-hundred-page book.[31] Selling upgrades relies on the relatively short shelf life of the original product; antivirus software is typically upgraded every 3 months. Extreme customization renders moot any need for enforcing IP protection, because only the original purchaser is interested in the product. Parceling out the product in small pieces simply makes it difficult to copy the entire product, in part restoring a barrier to infringement that comes naturally with physical products.

Giving away one digital product to promote another reduces the need for IP enforcement for the product given away, but it of course does little to reduce the need for IP enforcement for the charged-for product. One related strategy is to differentiate individuals from organizations. For example, Netscape and Adobe give away programs that individuals use, in order to sell the (more expensive) programs purchased by organizations. This approach takes advantage of the comparative ease of enforcing IP rights against organizations as compared to detecting and prosecuting infringement by individuals. It also capitalizes on the expectation that organizations may generate use that is valuable enough for them to pay for the product and recognizes that organizations also have processes and resources to comply more easily with IP laws and license agreements.

Free and low-cost mass market distribution are in the spirit of making the product cheaper to buy than it is to steal. It is worth noting that stealing an information product or service typically comes at a cost. An individual needs to expend the cost, time, and effort to obtain the product or service through infringing means and faces possible downstream costs such as refusal of technical support. When costs (i.e., the price to buy versus the total costs to steal) converge, the need for IP enforcement clearly diminishes.

Business Models as a Means of Dealing with Intellectual Property

As the variety of models above illustrates, business model design and selection can play a significant role in grappling with questions of IP protection. The choice of a model has significant consequences for the role that IP rights enforcement will play and, importantly, models are available that require far less enforcement. Hence, one approach to IP

---

[31]This situation, of course, may eventually change as a result of technological developments that improve the usability of online reading. One such ongoing effort is the activity surrounding electronic books.

rights in a world where digital content is difficult to control entails selecting a business model that does not require strict control.

Relying on a business model rather than a technical protection mechanism may also offer some leverage with the difficulty described in Chapter 1. With the emergence of computers and networking into the mainstream of daily life, attempting to enforce IP rights increasingly involves the difficult task of controlling private behavior. Where IP enforcement has historically been an issue between corporations, and where it has historically regulated public acts, the vastly increased means and opportunity for using (and abusing) IP in the hands of individuals has led to increased concern with the private actions of individuals. Where such private actions of individuals are concerned, detection and enforcement is more difficult, making the law a less effective tool. Technical protection mechanisms may help in such circumstances by making illegal or unauthorized actions more difficult, but the selection of an appropriate business model can reduce the motivation for those actions in the first place.

There are, of course, limits to the applicability of these models. Some properties, such as first-run movies, are unlikely ever to be given away, simply because of their high value. In that case other means of dealing with IP issues become more relevant, such as technical protection mechanisms (e.g., as in DVD) or perhaps not making any digital versions of the intellectual property available to consumers.

In formulating a plan for the commercial distribution of intellectual property, then, the rights holder is well advised to consider all three factors: exploring what boundaries are set by the law, what technical protections are practical and cost-effective, and how the business model will produce revenue. The law sets the foundational context in which the other two must function, drawing the boundaries that specify both what legal protection exists against unauthorized reproduction, distribution, and use, and what limits there are on the rights holder's monopoly (e.g., provisions for public access or time limitations on the term of protection). Technical protection mechanisms and business models can then play complementary roles in grappling with the difficulties of distributing IP content in digital form, each capable of reducing the degree of "leakage" of the product.

All three factors interact. Technology influences the selection of a business model: Any technical protection mechanism has both cost and benefit; it costs the producer to implement and may produce nuisance value for the customer (as for example nonstandard floppies), but may also pay off in lower rates of illegal copying. IP law also influences business model selection, as, for example, the limited lifetime of copyright protection must be considered in developing the business model.

And, in some cases, the selection of a business model may obviate the need for technical protection.

Selecting a business model for an information product is difficult in part because of the curious economics of information products. Appendix D discusses this in some detail; this section summarizes a few observations that have consequences in the marketplace and for the selection of a business model.

The duration of economic value varies over an extraordinary range, from a stock market quote (one minute or less) to a classic play (e.g., timeless Greek tragedies), but generally the economic value of most works is far shorter than the standard period of copyright protection. However, while duration of value is often short (sometimes fleeting), changes in value over time can be quite unpredictable. The novel *Moby Dick* was valueless when published; today's best-sellers may soon be. Today's news is valuable, yesterday's nearly worthless, while the news of 100 years ago is valuable again.

Curiously, there is value in both scarce information and in widely disseminated information. Scarcity confers obvious value. Consider the stock tip known by few others. But wide dissemination of information can produce value as well. Consider network effects in software, where the value of a program increases as more people use it, particularly as it approaches the status of a standard.

For digital information products, there are large first-copy costs and almost negligible production and distribution costs. Particularly in the absence of IP protection, this can produce a very sharp decline in product value over time, as it becomes an easily copied commodity.

A few generalizations are available about selecting business models to deal with IP issues. As a general observation, business models in which intellectual property can be widely disseminated at low cost are more successful in addressing intellectual property problems than are businesses that rely on higher prices and a small number of units distributed. The reasons are straightforward: If the cost of reproduction or piracy is high relative to the cost of acquiring the work legitimately, intellectual property problems will be less serious. Examples include newspapers, magazines, and paperback books.

More interesting, perhaps, is finding ways to permit low-cost distribution. The mass communication media have been the most successful because they make use of advertiser support to cover most or all of the cost of production and distribution, a model widely adopted on the Web. The use of rental markets as in videos (and formerly books) works well where such markets are feasible. The use of intellectual property to promote the use of other products (e.g., free browsers to promote Web traffic) is one of the few successful models available of widespread distribution of a digital information product for (very) low cost.[32]

It is no wonder that the ones most concerned about protection are the producers of intellectual property of high value that is distributed in relatively small quantities. Many high-end professional software packages, for instance, still require a dongle for use, and providers of specialized business information frequently use intranets and extranets protected by passwords in order to keep control of their content.

There is reason to approach doing business on the Web or in other electronic forms with some optimism, for there are a variety of business models to consider. As pointed out by Shapiro and Varian (1998), the goal for commercial information creators and owners is to maximize revenues, not protection. Business models will continue to evolve with the maturation of digital products; their careful design and selection may help to create effective ways to do business in the information world.

ILLEGAL COMMERCIAL COPYING

The U.S. industries that produce and sell copyrighted products constitute a significant part of the U.S. gross domestic product (GDP) and trade with other nations. It is therefore not surprising that the affairs of these industries and the issues they are concerned with attract considerable attention. One of these issues is the definition of illegal copying and the enforcement of the rules against it. A particular focus is illegal commercial copying—piracy.[33] Extensive data are collected and information reported under the auspices of the International Intellectual Property Association (IIPA) and its constituent member trade associations, including

---

[32]The model has not been uniformly successful of course. Despite vast numbers of MP3 files that have been downloaded and traded, there is as yet little systematic evidence that this has benefited more than a very few of the artists in question through increased CD sales.

[33]A briefer at the committee's July 9, 1998, meeting cited her favorite example:

This very cheesy CD, which costs $10 in Hong Kong, includes Windows '95, Office Professional, Microsoft Project, Microsoft Money, Microsoft Works, Norton Antivirus, Norton PCAnywhere, Omni Page, Quark Xpress, Photoshop, Pagemaker, Netscape Communicator, Front Page, Internet Explorer, NetObjects Fusion, Internet Phone, Eudora, McAfee Virus Scan, WinFaxPro, and Winzip Final. All for *ten bucks*. The street price [the selling price that most customers pay for a product] for this collection of software was $4,415 at the Egghead Online Store.

Even though some of the programs on this list are available for free (e.g., WinZip, Netscape Communicator, Internet Explorer), the prices quoted by a major discount software supplier for the rest still totaled over $4,000 in July 1999.

the Association of American Publishers (AAP), American Film Marketing Association (AFMA), Motion Picture Association of America (MPAA), National Music Publishers Association, (NMPA), Business Software Alliance (BSA), Interactive Digital Software Association (IDSA), and Recording Industry Association of America (RIAA), as well as industry trade associations outside of the IIPA, such as the Software and Information Industry Association (SIIA).[34]

These trade associations and other groups representing rights holders publish figures intended to demonstrate the huge dollar cost of infringement of U.S. IP rights that occurs both domestically and abroad. These figures are invariably impressive and are often cited as authoritative in newspaper articles or congressional hearings. The total contribution to the GDP of the United States represented by the copyright industries and the potential economic significance of global copyright infringement that detracts from these industries are substantial and disturbing. For example, in one widely publicized IIPA report published in 1998, it was estimated in 1996 that the total (not just core) copyright industries accounted for $433.9 billion in value added, or 5.68 percent of the U.S. GDP, and that those industries employed over 6.5 million people, or 5.2 percent of the U.S. workforce.[35] In 1996, the core copyright industries alone were estimated to account for $278.4 billion of the U.S. GDP, 3.65 percent of the total.[36] The IIPA estimates U.S. losses due to foreign piracy for the core copyright industries to be approximately $12.4 billion annually; losses resulting from domestic piracy make the total even greater.[37]

Notwithstanding the extensive amounts of data and information made available by these trade associations, some committee members believe there are reasons to question the reliability of some of the data claiming to measure the size of the economic impact of piracy. The com-

---

[34]The Web sites of many of these organizations contain both extensive statistical data on copyright infringement and a description of the methodology utilized to generate their estimates of piracy rates and economic loss (e.g., IIPA data, at <http://www.iipa.com/homepage_index.html>; AAP data, at <http://www.publishers.org/home/issues/index.htm>; AFMA data, at <http://www.afma.com>; MPAA data, at <http://www.mpaa.org/anti-piracy/content.htm>; NMPA data, at <http://www.nmpa.org>; BSA data, at <http://www.bsa.org>; IDSA data, at <http://www.idsa.com/piracy.html>; RIAA data, at <http://www.riaa.com/newtech/newtech.htm>; and SIIA data, at <http://www.siia.net>.

[35]The "core" copyright industries are those that create copyrighted products, whereas the wider set of "total" copyright industries includes those that distribute or depend on copyrighted products. Some members of the committee have expressed skepticism about the accuracy of balance of trade statistics, fearing they may be influenced by political agendas. However, even if somewhat overstated, there is no doubt that the copyright industries' export value makes a significant positive contribution to the balance of trade.

[36]See <http://www.publishers.org/home/press/iipa.htm>.

[37]See <http://www.iipa.com/html/piracy_losses.html>.

mittee considers here some of the issues that arise in collecting and analyzing such data, in part to inform the reader about those issues, and in part as the basis for a recommendation about how such information might reliably be assembled and analyzed. In exploring these issues, the committee takes a strictly economic view, focusing on profits lost from piracy. Lost profits are not the only cost of piracy, nor are the economic consequences the only rationale for enforcing antipiracy laws. But the figures widely circulated by trade organizations are intended to be economic analyses of profits lost, hence it is appropriate to explore these figures and their methodology from a strictly economic viewpoint.

One concern is that those who read figures of the sort found in the IIPA report may infer that all or most of the copyright industries' contribution to the GDP depends on copyright policy that protects works to at least the current degree and that perhaps the contribution now requires still greater copyright protection as a consequence of digital information and networks. However, within the economics community, the specific relationship between the level of IP protection and revenue of a firm in the copyright industries is unclear.[38]

A second problem is the accuracy of estimates of the costs of illegal copying. A number of difficulties arise here. One difficulty is that the needed data have to be based on extrapolation from very limited information, because illegal sales and distribution are frequently private acts. A second difficulty is determining the extent to which illegal copies are displacing sales. One widely quoted study, by the SIIA, estimates the number of illegal copies and then derives the net loss to the industry by assuming that each illegal copy displaces a sale at standard market prices[39] (other studies appear to rely on more complex formulations). This approach is problematic because it is unlikely that each illegal copy displaces a sale at the market price (some people will buy at the pirate price but not the legal market price) and because it estimates reduction in gross revenues rather than net loss to the copyright industries (see Box 5.4). Consequently, these estimates may be taken to represent an upper bound on the reduction in gross revenues by these industries.

There is, as shown in Box 5.4, disagreement about the economic impact of piracy on the copyright industries. It is clear, however, that there

---

[38]See, for example, Shy (1998) and the works cited in that article.

[39]Testimony at the July 9, 1998, committee meeting by a representative of the Software Publishers Association (now SIIA) indicated that they did this when calculating piracy estimates. In addition, the BSA/SIIA's *1999 Global Software Piracy Report* indicates that this is still the approach: "By using the average price information from the collected data, the legal and pirated revenue was calculated. This is a wholesale price estimate weighted by the amount of shipments within each software application category" (p. 12). Pirate revenue is then taken to be equivalent to a loss to the legal sellers.

BOX 5.4
Estimating Losses from Piracy

The economic significance of pirating to rights holders is appropriately measured by the net income lost by rights holders as a consequence of reduced sales of legal copies. As suggested in the text, some of the estimates attempting to measure this loss are problematic because of their methodology. A variety of problems arise:

- The loss to rights holders is not equal to the street price; it is instead the fraction of the wholesale price that represents pretax profits and royalties to the manufacturers, producers, and talent whose incomes depend on the number of authorized sales. Other aspects of manufacturing and distribution costs are just that—costs that the industry avoids if fewer copies are sold (though these costs may be small compared with the cost of original production and distribution). While piracy creates economic consequences for individuals other than rights holders, such as the loss of profits by retailers and sales tax revenue for government, the net loss due to each of these is a small fraction of the gross sales price of copyrighted products.[1]
- The number of additional authorized copies that would be sold is not equal to the number of illegally duplicated copies. Pirates typically sell their wares at prices substantially discounted from street prices; the substantial price discounts induce some people to purchase the product who would not otherwise do so. In addition, some unauthorized copies are produced for noncommercial reasons (e.g., making a copy for a friend). There is a substantial difference between getting a copy for free from a friend and having to pay the street price; hence some of these copies would not be purchased if the consumer had to pay something approximating the street price.
- Street prices are affected by the extent of illegal commercial copying. The availability of inexpensive, high-quality illegal copies reduces the demand for legal copies to the extent that some users buy illegal copies instead of legal ones. Interestingly, the effect on the street price of legal copies can either be positive or negative. The street price will *rise* if most price-sensitive consumers switch to illegal copies while the most price-insensitive consumers do not. The resulting market for legal copies will have less price-sensitive demand, thereby causing the manufacturer's profit-maximizing price to increase, which partially offsets the reduction in sales attributable to piracy.[2]
- By contrast, the street price will *fall* if consumers do not differ very much in price sensitivity. In this case all consumers are equally likely to buy from a pirate if given a chance, so that the effect of piracy is to make the demand for legal copies more price elastic.[3] If demand is more elastic, the profit-maximizing monopoly price falls and the proper calculation of the loss to rights holders must include

---

[1]Retail profits are approximately 2 percent of retail prices, and studies of tax incidence indicate that about half of the incidence of sales taxes is on producers rather than consumers.
[2]"Elasticity" is the precise term in economics for price sensitivity.
[3]"Consumer" here is taken in the purely economic sense, setting aside legal and ethical questions for the moment.

continued

BOX 5.4 Continued

profits lost on legally sold copies (because piracy forced the price down), as well as profits lost from pirated copies.

• Assuming that the extent of unauthorized copying can be estimated with reasonable accuracy—a nontrivial assumption—one should not assume that all unauthorized copies are illegal and, hence, represent piracy. The proper scope and legal definition of illegal copying is a matter of some disagreement and controversy, so different parties produce different estimates of this number. For example, most authorities agree that it is legal to make a backup copy of software (in case the original is damaged or destroyed). More controversial is whether a consumer can legally copy material for multiple uses, such as making a copy of a videotape they own in order to have a copy for personal use near each of two VCRs in their house.[4] Different opinions on the legality of these actions leads to different statistics on the extent of and hence economic consequences of piracy.

• This preceding analysis provides an appropriate foundation for building an estimate of the loss of profits from illegal copying. One first calculates the profit per unit sale for products in the absence of any illegal duplication (call it P), and then multiplies it by the number of unit sales (S) to derive the total profit for rights holders under no piracy ($T = P \times S$).[5] Then, one adjusts the net profit per unit of sale to account for price changes because of illegal copying (P'), and multiplies this number by the new number of legal copies sold (S') to derive the total profit for rights holders with piracy occurring ($T' = P' \times S'$). The difference between these numbers ($T-T'$) is the basic profit lost to rights holders from illegal duplication.

• Additional profit losses can also accrue. The expectation of illegal copying may cause some products not to be marketed at all, because the manufacturer does not believe that legal sales would be sufficient to recover the costs of production and distribution. In this case the loss to rights holders is the profits and royalties that would have been earned had the product been created and brought to market. Consumers also suffer a cost in this situation, equal to the difference between the value they would have placed on this product less the price they would have paid for it.[6]

---

[4]The 9th Circuit court decision in *Recording Industry Association of America v. Diamond Multimedia Systems* in June 1999 gave recognition in passing to the notion of "space-shifting" of music for personal use (i.e., an individual making a copy of a legally owned musical work in order to use the copy in a different place). No such position is currently on record for videos.

[5]By substituting "royalty" for "profit," one can derive analogous numbers for creators.

[6]There are of course also losses from piracy that do not (directly) concern profits. Counterfeits, for example, result in a loss of reputation for the author whose work has been copied. Counterfeit copies of movies can degrade the reputation of the movie maker in the eyes of viewers who see those badly made copies, while counterfeit software can result in harm to the reputation of the software maker when the unsuspecting purchaser is denied technical support. Here we are concerned solely with lost profits and their appropriate measurement, as such figures are the focus of reports widely circulated by trade organizations.

NOTE: Several committee members who earn their livelihoods in the copyright industries believe strongly that although the text in this box may reflect economic theory, it does not reflect the realities of their industry. For example, no motion picture distributor would reduce the terms of its licenses to theaters because pirated videos were on the street.

are significant losses that, if avoided, might result in increased production. It is also clear that uncontrolled digital dissemination could have very serious repercussions for the copyright industries.

A number of committee members conclude that, despite the extensive statistics available, there is a paucity of reliable information of the quality that might be generated if the subject were investigated by a disinterested third party. They conclude that such information is sorely needed.

However, even given the caveats above concerning methodology, the committee believes that the available information suggests that the volume and cost of illegal copying is substantial.

Although this section is concerned with the economics of piracy, the committee also believes that, regardless of whether the extent of illegal copying is financially significant to all industries that produce copyrighted products, the laws against illegal copying should be strictly enforced. Economic harm, after all, is not the only reason for enforcing copyright protection (or any other law with economic consequences). In a 1983 address and article, "The Harm of the Concept of Harm in Copyright," David Ladd, then the United States Register of Copyrights, expressed the following view: "The notion of economic 'harm' as a prerequisite for copyright protection is mischievous because it disserves the basic constitutional design which embraces both copyright and the First Amendment." Mr. Ladd argued for recognition of the fact that copyright protection is a sine qua non of a civilized society and, accordingly, merits recognition independent of economic impact.

This view is not unanimously endorsed by the committee, as some committee members believe that the constitutional basis for intellectual property protection in authorizing laws was meant to encourage strictly instrumental purposes. Even so, the committee as a whole recognizes that many creators believe that their works, as expressions of their individuality, deserve to be protected and controlled by rights holders, quite independent of the economic consequences. Because people differ in the weight they give to this argument, the committee believes that copyright policy will never be resolved solely by appeal to facts about its economic effects.

Despite the difficulty of finding a universally accepted copyright policy, the committee believes that it is important to conduct research in an attempt to better assess the social and economic impact of both commercial illegal copying for profit and noncommercial personal-use copying.[40] The committee believes that reducing the current state of uncer-

---

[40]The rationale for greater research on noncommercial copying for personal use is discussed in Chapter 4. The issue is raised here because the two spheres interact. For example, social norms and the easy availability of other copying alternatives (e.g., from a friend) affect an individual's likelihood of purchasing an illegal commercial copy.

tainty about the impact of these various phenomena will be important to policy makers and entrepreneurs. Clearly, there are multiple phenomena at work in both the commercial and noncommercial copying spheres, and perhaps there are differing behaviors among different demographic groups, geographic locations, and, perhaps, even cultures.[41] These multiple phenomena may include how much the difficulty of making the illegal copy affects the frequency of copying, the effect on consumer decision making of the price and availability of legitimate copies, the personal sense of the moral or ethical dimensions of the copying involved, the degree of law enforcement or legal scrutiny directed at the behavior, peer group or social opprobrium or encouragement, and so on. Society needs to understand better what these multiple phenomena are and how they operate in the real world, so that appropriate responses can be formulated.

THE IMPACT OF GRANTING PATENTS FOR INFORMATION INNOVATIONS

Historically, information innovations have been excluded from the purview of patent law, based on a notion that Congress had meant for only industrial processes to be patented. Documents were deemed unpatentable, as were improved ways for calculating, organizing information, and managing organizations. However, a great deal has changed in recent years, and it seems that nearly all information innovations may now be patented, as long as they meet the patent law's requirements for novelty, nonobviousness, and utility and can be precisely defined in claims.

Computer software was the first digital information product to challenge the traditional interpretation of patent concepts because of its dual nature as both a literary work (the textual source code) and a machine (i.e., a useful device). Programs have a dual nature because they are textual works created specifically to bring about some set of behaviors. They have been characterized as "machines whose medium of construction happens to be text" (Samuelson et al., 1994).[42]

The "printed matter" and "mental process" rules were initially invoked to deny patent protection to computer software, as on occasion was the "business method" rule. In its 1972 *Gottschalk v. Benson* decision, the U.S. Supreme Court ruled that an innovative method for transforming binary coded decimals into pure binary form could not be patented, even

---

[41]For example, see Chapter 1, Box 1.6, "A Copyright Tradition in China?"

[42]See *Intellectual Property Issues in Software* (CSTB, 1991b) for a greater articulation of the issues and implications concerning the use of the patent regime for software.

though the patent applicant intended to carry out the method by computer and one of the two claims before the Court was limited to computer implementations.[43]   Drawing on the "mental process" line of cases, the Supreme Court announced that mathematical algorithms could not be patented. One factor that clearly disturbed the Court about the prospect of patenting the Benson algorithm was that a patent would preempt all uses of it, including apparently the teaching of it. In 1978, the Supreme Court in *Parker v. Flook* denied patent protection for an algorithm useful for calculating "alarm limits" (i.e., dangerous conditions) for a catalytic converter plant.[44]   The Court did not think that this algorithm, any more than the Pythagorean theorem or any other purely mathematical method, could become patentable merely because it might be applied to a particular useful end.

The turning point in the long struggle over patents for information inventions came with the Supreme Court's 1981 decision in *Diamond v. Diehr*, which upheld the patentability of an improved rubber curing process, one step of which required a computer program.[45]   Because *Diehr* involved a traditional technological process and had so deeply divided the Court, patent administrators and the courts continued to struggle over how broadly to construe the *Diehr* decision.

In the late 1980s, the tide turned in favor of patents for computer-program-related inventions because of their technological character. Source code listings might still be regarded as unpatentable under the printed matter rule, but as soon as a program has been put in machine-readable form, recent precedents would seem to regard it as patentable subject matter.

Most recent program-related patents are, however, for more abstract design elements of programs. In the late 1980s and through the 1990s, it became increasingly common for courts to uphold patents for data structures, applied algorithms, information retrieval, and business methods carried out by computer programs. The denouement of the legal controversy over software-related patents in the courts and in the U.S. Patent and Trademark Office (PTO) may be the U.S. Supreme Court's decision in early 1999 not to review the *State Street Bank* decision. The Court upheld a patent attacked on grounds that the claims covered an algorithm and a business method. However, patents continue to be controversial in the information technology industry (Box 5.5).

In *State Street Bank and Trust Co. v. Signature Financial Group*, the U. S. Court of Appeals for the Federal Circuit issued an opinion that has been

---

[43]*Gottschalk v. Benson*, 409 U.S. 63, 34 L. Ed. 2d 273, 93 S. Ct. 253 (1972).

[44]*Parker v. Flook*, 437 U.S. 584, 57 L. Ed. 2d 451, 98 S. Ct. 2522 (1978).

[45]*Diamond v. Diehr*, 450 U.S. 175, 67 L. Ed. 2d 155, 101 S. Ct. 1048 (1981).

BOX 5.5
SightSound.com

There is an interesting intersection between the controversy surrounding the practice of patenting Internet business models and the uploading and downloading of musical recordings in digital formats, including MP3. An Internet multimedia distributor, SightSound.com, has claimed that two patents it holds (U.S. Patent 5,191,573 filed in 1990 and granted in 1993, and Patent 5,675,734 filed in 1996 and granted in 1997) cover the digital distribution of audio and video recordings. SightSound has claimed that its ownership of the patents for the sale and distribution of the music and video content over the Internet gives it the right to prevent any third party from exploiting a business model involving the selling, via download, of digital content sound files. SightSound has sent legal demand notices claiming that its patents "control, among other things, the sale of audio video recordings in download fashion over the Internet," and demanding that digital music sites, such as MP3.com, Platinum Entertainment, Amplified.com, and GoodNoise Corp. (now Emusic, Inc.), enter into patent licenses with SightSound that would give SightSound a royalty of 1 percent of the price per transaction, as charged to the customer, for all such Internet sales. AT&T's a2b Music has reportedly already entered into such a patent license with SightSound. The chief technology officer of AT&T's a2b Music has stated, "We licensed our technology to them, and as part of that deal we protected ourselves against patent claims. This whole area of patenting Internet business models is becoming scrutinized. I have trouble seeing how an auction on the Internet could get a patent." Currently, SightSound has sued music site NK2, Inc., for alleged patent infringement. The Recording Industry Association of America, through its spokesperson Lydia Pelliccia, has stated, "The validity of the patents is certain to be challenged."

In an interesting intersection between patent law and the concerns of copyright proprietors about the protection of content in cyberspace, SightSound has suggested that the enforcement of its patents could aid copyright owners in other protection efforts such as the Secure Digital Music Initiative. In the patent infringement claim letters that SightSound has recently sent out, it has demanded that "if [MP3.com] does not become an authorized licensee, it must immediately cease and desist from selling music, or other audio recordings over the Internet in download fashion," thus using patent infringement claims to "enforce" the potential claims of music copyright owners for contributory or vicarious copyright infringement (Lemos, 1999; *Business Wire*, 1999a,b).

widely regarded as vastly increasing the scope of patent protection available for software,[46] and which led the PTO to begin issuing a number of quite broad patents covering methods for conducting business. In the *State Street* case, the patent in question covered a "hub and spoke" com-

---

[46]*State Street Bank and Trust Co. v. Signature Financial Group*, 149 F.3d 1368, 47 U.S.P.Q.2d 1596 (Fed. Cir. 1998).

puterized business system, which allowed mutual funds (spokes) to pool their assets into an investment portfolio (the hub). The court held that the calculation by a machine of a mathematical formula, calculation, or algorithm is patentable if it produces a "useful, concrete and tangible result." In reaching this conclusion, the court rejected what it termed "the ill conceived [business method] exception" to patentability.

Patents that were issued in the wake of *State Street* that attracted widespread public attention include Patent No. 5,794,210, which was issued to an online Internet marketing company, Cybergold, and covers the practice of paying consumers to view advertisements on the Internet. Other examples include a patent issued to Priceline.com covering reverse sellers auctions (which allow potential purchasers to specify the various items they wish and terms on which they are willing to purchase and then allows Priceline to find a seller) and a patent issued to cover a method for embedding Web addresses in e-mail and news group postings (see Box 5.6).

The effects of this substantial de facto broadening of patent subject matter to cover information inventions are as yet unclear. Because this expansion has occurred without any oversight from the legislative branch and takes patent law into uncharted territories, it would be worthwhile to study this phenomenon to ensure that the patent expansion is promoting the progress of science and the useful arts, as Congress intended.

There are many reasons to be concerned. There is first the concern that the U.S. Patent and Trademark Office lacks sufficient information about prior art in the fields of information technology, information design, and business methods more generally to be able to make sound decisions about the novelty or nonobviousness of claims in these fields.[47]

---

[47]On August 31, 1993, the PTO issued U. S. Patent Number 5,241,671 for a multimedia search system to Compton's NewMedia. Not long after the patent was issued, Compton's announced at the Comdex trade show in the fall of 1993 that it had acquired the patent and intended to enforce it and to collect royalties and licensing fees from independent and third-party multimedia developers who utilized a multimedia search system claimed in the Compton's patent. On December 17, 1993, the Commissioner of Patents and Trademarks, in the wake of numerous multimedia developers' complaints about the Compton's patent, took the fairly unusual step of requesting reexamination of the patent. Third parties submitted new prior art references to assist the PTO in determining whether there was a substantial question as to the patentability of the 41 claims granted under the patent. The PTO's examination concerned the issue of whether the 41 claims filed in the patent application were novel and not obvious to someone skilled in the relevant art. The PTO concluded that Compton's claimed inventions had been disclosed or taught in prior art references, and, accordingly, on November 9, 1994, the PTO issued a press release announcing it had formally canceled all 41 claims granted in the Compton's patent. Many commentators believe the Compton's case strongly underscores their concerns regarding the qualifications of the PTO to effectively evaluate prior art in the information technology field.

BOX 5.6
Cybergold and Priceline.com

Nate Goldhaber, the founder and CEO of Cybergold, describes the inspiration behind the company by saying, "The fundamental premise of our company is that attention is a valuable commodity. With literally millions of pages on the Web and hundreds of advertisements vying for people's attention, attention is one of the Internet's great scarcities. Cybergold allows marketers to pay consumers directly for their time and active attention." In its most basic terms, Cybergold is a Web site that has registered over 1 million "members," who can earn cash and enter sweepstakes by visiting Web sites, trying free products, signing up for services, playing games, purchasing goods, and more. Cybergold holds a number of U. S. patents, including U.S. Patent 5,794,210, issued August 11, 1998, which covers a system "which provides the immediate payment to computer and other users for paying attention to an advertisement or other 'negatively priced' information distributed over a computer network such as the Internet . . . . This is the business of brokering the buying and selling of the 'attention' of users . . . . Private profiles may be maintained for different users and user information may be released through advertisers and other marketers only based on user permission. Users may be compensated for allowing their information to be released."

Priceline.com is a Stamford, Connecticut-based Web site launched in April 1998, consisting of a buying service through which consumers name the price they are willing to pay for goods and services ranging from airline tickets to automobiles. The company holds U.S. Patent No. 5,794,207, which covers a system that lets buyers name their price for goods and services while sellers decide whether or not to accept the offers. Priceline.com's name-your-price service for leisure airline tickets sold more than 100,000 tickets between its launch in April 1998 and the end

---

A related concern is the insufficient number of adequately trained patent examiners and inadequate patent classification schemata to deal with this new subject matter. The success of the patent system in promoting innovation in a field depends on the integrity of the process for granting patents, which in turn depends on adequate information about the field. Serious questions continue to exist within the information technology field about the PTO's software-related patent decisions. A number of legal commentators have pointed out that allowing these kinds of patents potentially makes concepts, not technology, the protectable property of the patent holder, "allow[ing] virtually anything under the sun to win patent protection."[48]

---

[48]See, for example, Goldman (1999), Scheinfeld and Bagley (1998), Sandburg (1998), and Sullivan (1999).

of 1998.[1] A second name-your-price service for hotel rooms is now available in approximately 200 U.S. cities,[2] as well as name-your-price service for all-cash new car buyers in the New York metropolitan market. The company has announced plans to launch name-your-price home mortgage services and to branch out into vacation packages and other online financial services.

Priceline launched an initial public offering (IPO) in March 1999, at one point reaching a market capitalization in excess of $18.5 billion. Like other highly sought-after IPOs for Internet start-ups, much of the perceived value of the company is based on its patent, since at the time of this writing the company has no earnings. In September 1999, Cybergold conducted an initial public offering (Bauman, 1999).[3]

[1]In September 1999, Priceline proposed the expansion of their business to include groceries (Wingfield, 1999).

[2]In September 1999, Expedia, Microsoft's travel site, began allowing customers to name their price on hotel rooms (Wolverton, 1999). In October 1999, Priceline initiated patent infringement litigation against Microsoft (Bloomberg News, 1999).

[3]Other noteworthy patents that have been issued in recent years, and which arguably illustrate the general expansion of patent protection for software and Internet business methods, include the following patents and patent holders: Amazon.com (U.S. Patent Number 5,727,163) (covering secure method for communicating credit card data when placing an order on a nonsecure network); Juno Online (U.S. Patent Number 5,848,397) (covering method and apparatus for scheduling the presentation of messages to computer users); Egendorf (U.S. Patent Number 5,794,221) (covering Internet billing method); The AG Group (U.S. Patent Number 5,787,253) (covering apparatus and method of analyzing Internet activity); Interactive Media Works (U.S. Patent Number 5,749,075) (covering method for providing prepaid Internet access and/or long-distance calling, including the distribution of specialized calling cards); and Excite (U.S. Patent Number 5,577,241) (covering information retrieval system and method with implementation extensible query architecture).

Second, the tradition of independent creation in the field of computer programming may run counter to assumptions and practices associated with patents as they are applied to its traditional domains. When someone patents a component of a manufactured system, for example, it will generally be possible for the inventor to manufacture that component or license its manufacture to another firm and reap rewards from the invention by sale of that component. Rights to use the invention are cleared by buying the component for installation into a larger device.

But there is little or no market in software components. Programmers routinely design large and complex systems from scratch. They do so largely without reference to the patent literature (partly because they consider it deficient), although they generally respect copyright and trade secrecy constraints on their work. With tens of thousands of programmers writing code that could well infringe on hundreds of patents without their knowing it, there is an increased risk of inadvertent infringement.[49] An added disincentive to searching the patent literature is the danger that learning about an existing patent would increase the risk of being found to be a willful infringer. The patent literature may thus not be providing to the software world one of its traditional purposes—providing information about the evolving state of the art. Much the same could be said about the mismatch between patents and information inventions in general.

Third, although patents seem to have been quite successful in promoting investments in the development of innovative manufacturing and other industrial technologies and processes, it is possible that they will not be as successful in promoting innovation in the information economy. One concern is that the pace of innovation in information industries is so rapid, and the gears of the patent system are so slow, that patents may not promote innovation in information industries as well as they have done in the manufacturing economy. The market cycle for an information product is often quite short—18 months is not unusual; thus, a patent may well not issue until the product has become obsolete. If information inventions continue to fall within the scope of patents, then, at a minimum, the patent cycle-time needs to be improved significantly. Patent classification systems for information innovations may also be more difficult to develop and maintain in a way that will inform and contribute to the success of the fields they serve.

One final reason for concern is that developing and deploying software and systems may cease to be a cottage industry because of the need for access to cross-licensing agreements and the legal protection of large corporations. This in turn may have deleterious effects on the creativity of U.S. software and Internet industries.

---

[49]This is in contrast to the copyright framework, where infringement requires a demonstration that some (conscious or unconscious) plagiarism has occurred. For example, independent creation is a defense. Two people could write original but very similar stories (or programs) independently; both would be copyrightable, and neither would infringe upon the other, because the standard for copyright protection is originality, not novelty. Thus an author is not responsible for knowing the entire corpus of literature still within copyright so as not to infringe on it.

6

Conclusions and Recommendations

Three technological trends—the ubiquity of information in digital form, the widespread use of computer networks, and the rapid proliferation of the World Wide Web—have profound implications for the way intellectual property (IP) is created, distributed, and accessed by virtually every sector of society. The stakes are high in terms of both ideology and economics. Not surprisingly, much discussion of these issues has occurred in the Congress, among stakeholder groups, and in the press. But the effects of the information infrastructure extend beyond these institutions; as never before there are also important and direct effects on individuals in their daily life.

The information infrastructure offers both promise and peril: promise in the form of extraordinary ease of access to a vast array of information, and peril from opportunities both for information to be reproduced inappropriately and for information access to be controlled in new and problematic ways. Providing an appropriate level of access to digital IP is central to realizing the promise of the information infrastructure. Ensuring that this appropriate level of access becomes a reality raises a number of difficult issues that in the aggregate constitute the digital dilemma. This report articulates these difficult issues, provides a framework for thinking about them, and offers ways of moving toward resolving the dilemma.

One salient theme in the committee's conclusions and recommendations is an acknowledgment of the multiplicity of stakeholders and forces that must be considered. Intellectual property has a pervasive impact in society, resulting in a corresponding diversity of interests, motivations, and values. Some stakeholders see the issues in economic terms; some in philosophical terms; others in technological terms; and still others in legal, ethical, or social policy terms. There are also a variety of important forces at work—regulations, markets, social norms, and technology—all of which must be considered and all of which may also be used in dealing with the issues. Knowing about the full range of forces may open up additional routes for dealing with issues; not every problem need be legislated (or priced) into submission. Individuals exploring these issues are well advised to be cognizant of all the forces at work, to avoid being blind-sided by any of them; to avail themselves of the opportunity to use any of the forces when appropriate; to be aware of the process by which each of them comes about; and to consider the degree of public scrutiny of the values embedded in each.

The committee believes that the issue of intellectual property in the information infrastructure cannot be viewed as solely a legal issue (as it was, for example, in the white paper *Intellectual Property and the National Information Infrastructure,* IITF, 1995)[1] or through any other single lens. Such an approach will necessarily yield incomplete, and often incorrect, answers. One of the committee's key contributions is to urge an appropriately broad framework for use by policy makers, one that acknowledges the full spectrum of stakeholders and forces.

The first two sections of this chapter focus on the implications for society and individuals that arise from the everyday use of the information infrastructure, with an emphasis on intellectual property that has been published in the traditional sense.[2] The next two sections address research and data collection that are needed and near-term actions that can be initiated to help in getting beyond the digital dilemma. The last section offers guidance on and principles for the formulation of law and public policy.

A significant portion of the committee's deliberations can be characterized as spirited and energetic discussions expressing a range of perspectives on controversial issues. For some of those issues, a summary of alternative perspectives is provided, with the intent of exposing the core issues to aid future discussion. That this committee, a diverse and balanced group of experts, had difficulty in achieving consensus in many areas, despite extensive briefings, background reading, and deliberations,

---

[1]When the IITF's white paper was written, the Web was only beginning to be widely used by the general public; hence some aspects of the digital dilemma touched on here (e.g., business models) had yet to develop.

[2]The committee was unable to address some important subjects (e.g., the cable television industry) thoroughly because of the limited time and resources available.

should serve as a caution to policy makers to contemplate changes to law or policy with the utmost care.

THE DIGITAL DILEMMA: IMPLICATIONS FOR PUBLIC ACCESS

Public access to published works is an important goal of copyright law. The traditional model of publication—the distribution of physical copies of a work—has been effective as the fundamental enabler of public access.[3] Enough copies of a work are usually purchased (e.g., through libraries and other institutions and by private individuals) that it becomes part of the social, cultural, and intellectual record and is thus accessible to sufficiently motivated members of the public. There is also a long-standing (if not always explicitly articulated) understanding that this social and cultural record will continue to accumulate, be preserved, and be available for consultation. At least since the modern era of public libraries, broad access to a college education, and mass media, such information has become increasingly available. Yet there are aspects of the information infrastructure that, although vastly increasing access in some ways, also have the potential to diminish that access, which is a valuable component of our social structure.

The Value of Public Access

Public access, and the social benefits that arise from it, may be an undervalued aspect of our current social processes and mechanisms. As one example, while the first-sale rule enables access that may result in loss of revenue for publishers (because some people or organizations who are able to borrow a book would have purchased it instead),[4] the larger social benefits—an informed citizenry and the democratization of information and knowledge—can be substantial.[5] Those benefits also have a significant and longer-term impact in encouraging the creation of new knowl-

---

[3]The traditional model of publication has been more applicable to some forms of information (e.g., books and magazines) than others (e.g., first-run movies and television broadcasts).

[4]The first-sale rule stipulates that the initial sale of a copy of a work exhausts the copyright owner's right to control further distribution of that copy. An individual, library, or other entity is free to give away, lend, rent, or sell its copies of books and many other materials (17 U.S.C. sec. 109).

[5]Some materials (e.g., academic journals) are rarely bought by individuals and hence would not represent any substantial lost revenue. There may also be some countervailing effect, because some people who get access to a book through borrowing are motivated to buy it; lending is in effect a form of advertising. The point here is that even if there is some degree of loss, the benefits must also be considered.

edge and new works. Being well informed and educated has value that increases with the population of others similarly informed and educated, and ultimately contributes to a larger potential market for authors and publishers. Hence, the public access to material that is made possible in the hard-copy world by the first-sale rule can be worth more to society than the modest revenue lost to publishers. Beyond the economic issues, an informed citizenry and informed discourse are vital to the health of a free and democratic society.

Public access may suffer, however, as the evolution of the information infrastructure compels a reexamination of the first-sale rule and other mechanisms for achieving access. As one example of the difficulties digital information brings, a single online copy of a work available from a digital library could diminish the market for the work much more than the distribution of hard copies to traditional libraries. One reasonable response of publishers might be to avoid making some works available to libraries in digital form, resulting in a net decrease in the accessibility of information. Other challenges to ensuring access arise from the changing nature of publication, the growing use of licenses rather than sale of works, and the use of technical protection mechanisms. As a consequence, historically simple provisions such as the first-sale rule become much more complex in the digital environment, involving difficult questions with respect to technology and business practices.

> *Conclusion:* **The tradition of providing for a limited degree of access to published materials that was established in the world of physical artifacts must be continued in the digital context. But the mechanisms for achieving this access and the definition of "limited degree" will need to evolve in response to the attributes of digital intellectual property and the information infrastructure.**

Consequences of the Changing Nature of Publication and the Use of Licensing and Technical Protection Services

In liberating content from its medium of presentation, digital information challenges many long-held assumptions about copyrighted works, most notably those regarding the nature and character of publication. In the physical world publication is public, irrevocable, and provides a fixed copy of the work; in the digital world none of these may be true.

Publication has traditionally been public in the sense noted above (i.e., that works are widely distributed and become part of the cultural record). Publication is irrevocable in the sense that works may go out of print, but once published can never subsequently be effectively withdrawn from circulation and become "unpublished." Publication also implies a stability of the work. Copies distributed provide a stable snapshot of the work at a particular moment; subsequent editions only add to this record. In the digital world, however, documents published by being posted on the public Internet can be removed from scrutiny at the pleasure of the rights holder. Access can be controlled to allow many levels of dissemination between publication and private distribution, and older versions of a document can be (and are routinely) replaced by newer ones, obliterating any historical record.

The widespread use of licensing and technical protection services (TPSs) also has important implications. Licensing is a familiar mechanism for providing access to some types of digital information (e.g., software) but is relatively new for other types (e.g., research journals). Even where the practice is familiar, it has often stirred controversy, as in the still-developing notion of shrink-wrap licenses. Where licensing is unfamiliar, publishers and their customers are still learning how to establish reasonable licensing relationships. By offering a distribution model different from that represented by copyright and sale, licensing has the potential to open new markets. Some material that has been made available through licensing would not have been published at all in the traditional manner; the restricted distribution of information is thus an important option for the publisher and public.

But the use of licensing also raises significant concerns about the consequences for public access and the maintenance of a healthy corpus of materials in the public domain, particularly where license restrictions differ from legal rules that would otherwise apply. The libraries' role as a permanent repository of material that constitutes a cultural heritage is threatened by a change in the model of distribution from sale to licensing. Libraries could instead become transient, temporary points of access to collections of information that may be available today and gone tomorrow, when licenses expire. Additional concerns arise from the fact that material distributed by license may not become a part of the long-term public record.

Some technical protection services have been developed (and others are being developed) to confront the key problem that digital information seemingly cannot be distributed without the risk of large-scale copying and redistribution.[6] TPSs offer rights holders some assurance that distributing a single copy of a digital work need not result in subsequent unlimited and uncontrollable dissemination. By enabling network distri-

---

[6]Technical protection services are discussed further below under "Moving Beyond the Digital Dilemma: Additional Mechanisms for Making Progress" and in greater detail in Chapter 5.

bution of information products that otherwise would never have been distributed digitally, TPSs could open new markets and substantially increase dissemination of and access to works. Conversely, without such an ability, some rights holders may decide to avoid digital distribution entirely for some works (e.g., investment newsletters), thereby reducing the offerings available to the general public.

But technical protection services may also permit limitations on the distribution of content such that most consumers can only view it—distribution without the ability to save and/or print is now contemplated for several mass market content businesses. Consider the consequences of this model of publication: Information might be distributed but never easily shared, substantially defeating the original intent of publication as an act that leads, eventually, to a contribution to the shared, permanent social and cultural heritage. Time- and audience-limited distribution could increase.[7]

Although limited distribution is a possibility, it may prove to be thoroughly unpopular with consumers if it significantly constrains access to, enjoyment, or use of a product. The marketplace might thus facilitate public access. Nevertheless, policy makers should monitor the situation and be prepared to address the issue in the event that limited distribution models begin to have a significant impact on public access to information.

> *Conclusion:* **The confluence of three developments—the changing nature of publication in the digital world, the increasing use of licensing rather than sale, and the use of technical protection services—creates unprecedented opportunities for individuals to access information in improved and novel ways, but also could have a negative impact on public access to information. Developments over time should be monitored closely.**

Some members of the committee voiced the concern that highly constrained models of distribution undermine the fundamental pact between society and authors that is embodied in copyright, a pact that encourages the creation and dissemination of information for society's ultimate benefit. These individuals are concerned that a limited-distribution model of publication may undermine a constitutional intent, namely that rights be granted to authors for a limited time in exchange for assurance that materials will pass eventually into the public domain and the public record.

---

[7]Time- and audience-limited access has been commonplace for some kinds of IP for many years (e.g., movies exhibited in a theater), but this is a new phenomenon for traditionally published IP.

Recommendation: **Representatives from government, rights holders, publishers, libraries and other cultural heritage institutions, the public, and technology providers should convene to begin a discussion of models for public access to information that are mutually workable in the context of the widespread use of licensing and technical protection services.**

Publication and Private Distribution

In a digital world offering options for distribution other than printing and selling copies, it is not always easy to tell when information has been published and when it has not. The distinction between publication and private distribution is blurred by options such as distribution on electronic mailing lists, posting on password-protected Web sites, or posting on preprint servers available to members of professional societies. Further blurring results from the multiple, finely controlled layers of conditional access that computer systems can provide, offering many degrees of access between public and private. The issue is further complicated by the impermanent nature of digital information, which facilitates the distribution of works in varying states of completion (e.g., posting numerous versions of an article as it evolves). The question of what constitutes publication has significant consequences with respect to public access to the information—facts and ideas in published works are freely available to the public—but deciding whether a work in digital format has been published may be difficult. Although the distinction between public and private may never have been crystal clear in the copyright regime, it has become far murkier in the digital environment.

Conclusion: **The information infrastructure blurs the distinction between publication and private distribution.**

Recommendation: **The concept of publication should be reevaluated and clarified (or reconceptualized) by the various stakeholder groups in response to the fundamental changes caused by the information infrastructure. The public policy implications of a new concept of publication should also be determined.**

Mass Market Licenses

Non-negotiated licenses for mass market items also raise important public access questions. The issue is whether the terms of mass market licenses offered on a take-it-or-leave-it basis would override fair use or other limiting policies of copyright law. The question is controversial and as yet unresolved in the law. The public policies associated with intellectual property law may sometimes be seen as sufficiently important that mass market license terms should not be permitted to override them. For example, public policy favoring competition and innovation may call into question the enforceability of a term in a mass market license for computer software that forbids reverse engineering the software. Similarly, concerns related to free speech may arise if mass market licenses seek to limit criticism of a digital information product or disclosure of its flaws. Part of the intent of fair use is to encourage critical analysis; however, if works are licensed, there is currently no automatic fair use provision and hence no established foundation for criticism.[8]

Some committee members favor subjecting mass market licenses to fair use limitations, viewing fair use and other limiting doctrines of copyright as having an affirmative character—i.e., as providing a right for users under copyright law, rather than solely a defense to infringement. According to this view, rescinding that right in a license should not be possible (even though other rights may, with few exceptions, be waived by agreement). Those who do not favor subjecting mass market licenses to fair use conditions generally perceive copyright as providing default rules that should be overridable by a contract in free market transactions.

> *Conclusion:* **The committee as a whole points out an important underlying legal and philosophical issue—the question of whether fair use is an affirmative right or a defense—and emphasizes the consequences for access that follow from taking one position or the other on this issue.[9]**

Archiving and Preservation of Digital Information

Digital Archives

Archiving our cultural heritage and ensuring a record of intellectual discourse are critical tasks for society. The importance of archiving is discussed in Chapter 3, along with many of the associated problems.

---

[8]Consider the hypothetical case of an electronic commerce software package and an authorized user who discovers a security problem with the software. The vendor may wish to issue licenses that prohibit the authorized user from disclosing such information to third parties.

[9]Similarly, the committee is unable to take a definitive position with respect to the Uniform Computer Information Transactions Act (UCITA). See Chapter 3 for a discussion of the issues concerning UCITA.

Despite several years of intense effort, little practical progress is being made in establishing digital archives. The lack of progress is attributable to several factors:

• Lack of funding for large-scale digital archiving overall and lack of any agreement about how responsibility for providing funding will be divided among government and cultural heritage research institutions. There is also an absence of mechanisms to effectively pool the contributions of the many organizations with some interest in and responsibility for funding archiving.

• Reluctance among major research libraries and archives to make digital archiving a part of their missions or, if there is acceptance of responsibility, the inability to proceed without certainty of funding. Justification of funding is complicated by the difficulty of offering any real access to materials prior to the expiration of copyright, with the result that digital archives may not produce tangible benefits for a century or more, making this investment in the preservation of culture and scholarship a hard sell.

• Insufficient expertise within the most likely archiving institutions; the technical and intellectual problems involved are difficult and experienced individuals correspondingly difficult to find.

• Worry about potential liability for contributing to copyright infringements—for example, fear that any unauthorized use of archived material by a member of the public could result in the archive being held liable for contributing to infringement. Archives are also concerned about liability for copyright infringement, both in the actual processes of capture and management of digital content and in any actions taken to make archived digital materials available to the public prior to the expiration of the term of copyright.

• The daunting scale of the task and the need to develop processes for selecting what will be preserved. Hard intellectual and technological problems exist, some of which require the development of social and scholarly consensus.

• The uncertain relationship between archiving and licensing. The rights to archiving can be negotiated, and indeed many research libraries are starting to do so. These negotiations seem to have been reasonably successful thus far when carried out with scholarly publishers that share an interest with libraries, authors, and readers in ensuring that electronic publications will be archived. The likelihood of success is less clear with mass market publishers and content providers outside the print tradition (e.g., in the music industry). Licensing is simply a contract between a publisher and a client, so the publisher is under no obligation to include provisions for archiving. This situation is unlike that in the print world, where publishing and selling a book (for instance) automatically make archival preservation possible both logistically and legally.

There are significant economic and legal issues to be resolved if archives and libraries are to act as digital archives during the term of copyright. Acquiring works after copyrights expire is clearly ineffective, as it is unlikely that most works will be available for acquisition: Few digital (or traditional) works remain economically viable for 100 years, and thus available in the marketplace.

The Congress, the Administration, and the combined managements of the top research libraries and archives should lead these efforts. Preservation of works on paper and in other physical media has evolved as a responsibility shared by many autonomous institutions. The committee advocates a similar approach to digital archives and encourages practical steps toward distributed digital archives for which existing research libraries and archives share responsibility. The committee believes that starting now is urgent, so that digital archives of significant extent will be established within a decade. The first step is to initiate a process that engages all relevant stakeholders, develop a plan for moving forward, and begin assembling the political constituency that will ultimately be needed to implement the necessary actions.

Conclusion: **Significant economic, technical, and legal issues need to be resolved if libraries and archiving institutions are to be as successful with digital information as they have been with hard-copy information.**

Recommendation: **A task force on electronic deposit should be chartered to determine the desirability, feasibility, shape, and funding requirements of a system for the deposit of digital files in multiple depositories. The task force membership should broadly represent the relevant stakeholders and should be organized by an unbiased entity with a national reputation, such as the Library of Congress or some other governmental organization that has a pertinent charter and relevant expertise. The task force should be assigned for a limited term (2 years maximum) and should be charged with the following responsibilities:**

• Determining the desirability, feasibility and general design of a system for the deposit of digital files in multiple depositories;
• Considering incentives for rights holders that encourage

their participation in such a system and encourage them to move material into the public domain earlier than the term of copyright protection if that material is not being exploited;
- Proposing both the legal and procedural framework for the deposit and subsequent use of digital files;
- Addressing the intellectual property and liability concerns of libraries and rights holders;
- Reporting on and recommending to Congress the long-term funding requirements for making the system work. A recommendation on funding is particularly important, because without funding nothing will happen when the task force's study is completed. This committee does not have sufficient information to indicate what the congressional or total funding level should be, but wants to make clear its belief that the total funding needed is substantial; and
- Evaluating other nations' strategies for the deposit and preservation of information and considering how a U.S. system could build on and relate to these other national efforts, recognizing that the creation and dissemination of digital information are global activities, and that preservation of content is thus a global problem.[10]

Preservation

Preservation within the context of the information infrastructure introduces new challenges. Digital information is often stored on media with relatively short life spans: because the medium itself degrades (e.g., magnetic tape) or the relentless advance of formats, hardware, and soft-

---

[10]To illustrate possible outcomes of the task force, one scenario might call for voluntary (or mandatory) deposit of digital works that are protected by copyright in the United States and that are either offered for sale under license or, if distributed free of charge, are protected by a TPS. Such deposited copies would not be made available to the public by the depository as long as they are still offered to the public by the rights holder, except for viewing in the library itself (as is the case with hard-copy works). All deposited copies would be "in the clear" (i.e., with no encryption or other access-limiting mechanism). The depositors would have no technical responsibility for migrating the copies over time. Libraries and other archives would not be held liable for unauthorized access to these files but would be required to take reasonable steps to prevent and stop such violations.

The intent here is to extend into the digital world the traditional balancing act of IP—providing enough control over a work to offer an incentive for creation, yet ensuring that in the long term all work becomes a part of the public intellectual record to the benefit of society as a whole. Providing for deposit of materials "in the clear" may aid in dealing with problems of access that arise from technical protection mechanisms, as well as issues raised by archiving.

ware leaves behind orphan technologies (try finding a way to read old word processor files on an 8-inch floppy disk). As a result, preserving digital information inevitably involves copying it to a new medium and/ or format. If digital information is to be preserved, such copying will need to be permitted under the law. In some instances, literal copying may not be practical or desirable, and the information may need to be adapted for a new format. This migrating process goes beyond simple copying and might be construed as the creation of a derivative work, which will also need to be allowed under the law to enable preservation.

Technical protection services may also pose difficulties for migrating or accessing archived digital information. As far as the committee was able to ascertain, no TPS incorporates any kind of "self-destruct" feature that is triggered when a work legally enters the public domain; rather, a TPS will continue to try to control access forever. In addition, a TPS may try to block attempts to copy or reformat a work, even when such a step is legal as part of the archival management of content. A self-destruct feature may be technically feasible and could be incorporated in future products and services. Access problems could also arise from TPSs that require online authorization or other interaction with network-based servers before they will permit access: What happens if the publisher ceases support of the authorization system, or ceases to exist at all?[11]

Maintaining digital documents in the face of changing hardware and software over even a decade can be a challenging prospect; yet, under certain circumstances, works can remain protected by copyright for more than 100 years after their date of creation. What incentive is there for the publisher to maintain access to a century-old work (which may mean reformatting it, updating the TPS that surrounds it, and so on), especially if it is providing no appreciable income?

The digital preservation problem is becoming a consumer issue as well. Consumers own content in media and formats that are becoming obsolete—for example, vinyl long-playing (LP) record collections today and perhaps audio CDs in the future. Should consumers legally be able to migrate content that they have purchased to new technologies for their own use?

Recommendation: **Congress should enact legislation to permit copying of digital information for archival purposes, whether the copy is in the same format or migrated to a new format.**

---

[11]A recent example is the discontinuance of Divx digital video service.

Access to Federal Government Information[12]

Advances in the capabilities and use of the information infrastructure, most notably the widespread use of the Web, have provided the means for greatly expanded access to federal government information. Agencies are increasingly using the Web to make information available, with some notable successes, including the Government Printing Office's GPO Access system, the Library of Congress's THOMAS system, and the Security and Exchange Commission's EDGAR system.

However, in some parts of the government, the evolution of the information infrastructure has instead been associated with a trend toward the commercialization of government information, increasingly limiting the amounts of information that can be accessed inexpensively by the public. Broad access to and use of publicly funded information are inhibited when distribution agreements curtail the availability of information. In some cases, federal agencies themselves must pay high prices for data or products that were created by another federal agency.

In addition, some agencies that perform research and development may contract with private companies for commercial ventures involving the results of their research. Such relationships can produce incentives to be less open in sharing research results. Other agencies, such as those involved with managing the records of government, may similarly put a low priority on public access and have, in some cases, allowed private parties to copy their internal records for the purpose of packaging and reselling the resulting data products.

Conclusion: **When commercial enterprises add value to basic data, the resulting products deserve copyright protection insofar as these products otherwise satisfy the legal requirements for copyright.**

Recommendation: **As a general principle, the basic data created or collected by the federal government should be available at a modest cost, usually not to exceed the direct costs associated with distribution of the data. When agencies contract with a commercial enterprise to make federally supported primary data available, and provide no other mechanism for access to the data, such agreements should provide for public access at a**

---

[12]The committee recognizes that state and local governments, as well as the governments of other nations, produce and distribute valuable data and information and face many of the same issues as the federal government regarding the dissemination of information. Because of the limited time and resources available, the committee focused its deliberations on data and information produced and distributed by the federal government.

cost that does not exceed the direct costs associated with distribution.[13]

THE DIGITAL DILEMMA:
IMPLICATIONS FOR INDIVIDUAL BEHAVIOR

Perceptions and Behavior of Individuals

Little is known about how frequently individuals duplicate copyrighted materials and whether they pause to question whether this activity may be illegal. It is likely that a large number of people assume that they have the right to duplicate such material and that their view of appropriate conduct is not shaped by any substantive knowledge of intellectual property law. For the past decade or more, most individuals who photocopied books and journals in a library have encountered signs warning them about potential copyright infringement, but little is known about whether these signs have been seen or understood, or have resulted in a change in user behavior. Individuals attempting to copy videotapes are confronted with similar but more threatening on-screen warnings at the beginning of a tape, but little is known about how these warnings have affected behavior. Previous studies, some more than a decade old, have typically examined behaviors and attitudes of only narrow groups of people prior to the widespread use of the Internet.[14]

Anecdotal evidence suggests that most people are not generally informed about copyright in the context of the information infrastructure; instead, myths and misunderstandings abound regarding what is legal and what is not. The committee believes that such misunderstandings extend to contracting arrangements as well. Few people read and understand shrink-wrap or point-and-click licenses, and whether people think they need to take them seriously is unclear.

Conclusion: **A better understanding is needed of the public's perception and behavior concerning digital intellectual property. When popular attitudes and practices are out of synch with laws, the enforcement of laws becomes more difficult, which may instill in people a lack of confidence in and respect**

---

[13]The committee did not address the status of the data and research created by federally supported researchers based at academic or other institutions outside the federal government.

[14]The most significant study thus far, done by the former congressional Office of Technology Assessment (1989), focused primarily on the home recording of audiotape and briefly examined home video recording as well. Research that is more relevant and current is needed.

for the legal system. There are also political dangers associated with criminalizing generally accepted behavior, given the possibilities for discriminatory and selective enforcement.

Recommendation: Research and data collection should be pursued to develop a better understanding of what types of digital copying people think are permissible, what they regard as infringements, and what falls into murky ill-defined areas. Such research should address how these views differ from one community to another, how they differ according to type of material (e.g., software, recorded music, online documents), how user behavior follows user beliefs, and to what extent further knowledge about copyright law is likely to change user behavior.

Fair Use and Private Use Copying

As discussed in Chapter 4, fair use is an established doctrine of U.S. copyright law that has become especially controversial in recent years. One controversy concerns the extent to which private use copying of copyrighted materials can be justified as fair use. Although this issue is not unique to digital intellectual property, the ease with which digital copies can be made and distributed, especially in networked environments, makes private use copying far more extensive in the digital environment and a more significant problem for content owners. A second controversy concerns the viability of fair use and other limitations on copyright in the digital environment.

The end points on the spectrum of perspectives on this issue are that private use copying is almost always fair use as a matter of copyright law and that private use copying is never fair use. Neither position is correct, although the committee concluded that many members of the general public consider the former to be true. Private use copying is sometimes fair use, and it is sometimes illegal; as a separate matter, it is sometimes ethical and sometimes not. Although it may be difficult to accomplish in practice, whether a private use is a fair use should in principle be determined by considering the fair use provision, section 107, of the copyright law (see Chapter 4).

The committee identified several considerations that complicate articulating a specific position on the copying of information for private use. One difficulty is the lack of clarity about what "private" means in the context of copying (e.g., copying in one's home may be "private," but what about copying in a library or a school?). Another is the multifaceted nature of fair use determinations, which makes it difficult to articulate simple, straightforward rules that could guide the conduct of the average citizen. In addition, the private nature of such copying poses serious enforceability problems, and, for some, privacy and other social values may make stricter enforcement socially unacceptable or undesirable. Nevertheless, although enforcement of regulations concerning private use is clearly not easy, the difficulties of enforcing the law do not transform private uses into fair uses when other considerations suggest that they are not.

Conclusion: **A widespread (and incorrect) belief prevails in society that private use copying is always or almost always lawful. This viewpoint is difficult to support on either legal or ethical grounds. It is important to find ways to convince the public to consider thoughtfully the legality, ethics, and economic implications of their acts of private copying.**

What can or should be done concerning this widespread and incorrect belief? The committee identified three distinct views and corresponding actions:

• One view is that copyright law should not concern itself at all with private use copying—that, as a pragmatic position, such copying should be considered lawful. Because private use copying cannot effectively be controlled at reasonable cost and without impeding other important social values, branding it unlawful might encourage a more general flouting of intellectual property law.
• A second view is that much private use copying is fair use and that illegal copying could be discouraged by appropriate copyright education.
• A third view sees most private use copying as illegal under copyright law and suggests that more significant efforts should be undertaken to enforce rules against illegal private use copying. This group would also advocate an extensive copyright education campaign specifically directed against private use copying. Technical protection services may also make private use copying more controllable than it currently is.

As in the print world, in the digital world basic differences in values make it difficult to determine the proper boundaries of fair use. Different stakeholders weigh the values and interests reflected in copyright law and policy differently. Those who find private use copying illegal generally do so on the grounds that the copyright law's exclusive right to reproduce is indeed truly exclusive and does not restrict only copies made for public distribution. They believe that privately made copies are no less an infringement for being private. Those who would hold private use copying to be fair use generally refer to the traditional standards for fair use, notably the claimed lack of impact on the right's holders market for a protected work.

Beyond the issue of private use copying, the committee considered the broader viability of fair use in the digital environment. One rationale for fair use (historically and in the digital environment) has been that transaction costs can make licensing of some copyrighted works uneconomical, leading to failure to use a work even when both the rights holder and potential licensee would agree to such use. The development of the infrastructure for electronic commerce may lower transaction costs for some classes of works. But the committee concluded that the potential for market failure is not the only rationale for fair use or for other exceptions and limitations to copyright law. Other rationales derive from public policy concerns about fundamental human rights, such as freedom of the press, and certain articulated public interests such as preservation of cultural heritage.

Conclusion: **Fair use and other exceptions to copyright law derive from the fundamental purpose of copyright law and the concomitant balancing of competing interests among stakeholder groups. Although the evolving information infrastructure changes the processes by which fair use and other exceptions to copyright are achieved, it does not challenge the underlying public policy motivations. Thus, fair use and other exceptions to copyright law should continue to play a role in the digital environment.**

Conclusion: **Providing additional statutory limitations on copyright and/or additional statutory protections may be necessary over time to adapt copyright appropriately to the digital environment. The fair use doctrine may also prove useful as a flexible mechanism for adapting copyright to the digital environment.**

Recommendation: **Legal, economic, and public policy research should be undertaken to help determine the extent to which fair use and other exceptions and limitations to copyright should apply in the digital environment. As public policy research, legal developments, and the marketplace shape the scope of fair use and other limitations on copyright, and/or demonstrate a need for additional protections, any additional actions that may be needed to adapt the law, educate the public about it, or enforce the law may become clearer.**

Copyright Education

With the expansion of the information infrastructure into everyday life and the widespread acknowledgment of the information revolution's power to transform society, the role of information looms ever larger. Yet the committee believes that the public is not well informed about intellectual property law in general and also that it labors under misconceptions concerning copyright in particular. Because ignorance regarding copyright law, the fundamental philosophy it embodies, and its intent may be a significant factor contributing to misuse of protected material, the committee believes that a copyright education program may prove quite useful.

Promoting respect for copyright in the United States would lay an important foundation by educating society about some of the ground rules on which an information-based society is built and help ensure that inadvertent violations of law are not commonplace events. This process would help maintain the health of the information industries (e.g., software, entertainment) and their contribution to the economy, as well as help content creators understand their rights. Respect for and enforcement of copyright law in the United States would also provide a foundation for U.S. efforts aimed at enforcement of international agreements on IP.

The committee believes that, to be effective, a program of copyright education must clearly communicate that the law is, in its intent and spirit, attempting a fundamentally fair and equitable balancing of interests. The program should emphasize the core goal of IP law, namely, the improvement of society through the advancement of knowledge; should describe the difficult balance between control and dissemination; and should make clear that, in the long term, all intellectual property becomes a part of the shared intellectual heritage, available to everyone. Such a program would describe both the rights granted exclusively to creators and the limits on those rights. The program should include an introduction to fair use and other limitations on copyright law, and their role in accomplishing the larger purpose of the law.

Several factors make carrying out successful copyright education difficult, including the complexity of copyright law, disagreements within the population on some significant copyright issues (e.g., the proper scope of private use copying), varying cultural views on the different classes of information,[15] and the perceived urgency of other matters on the educational agenda (e.g., can children read?). Appropriate educational materi-

---

[15]For example, some people believe that it is more acceptable to copy popular music files in MP3 format for their friends than to copy a word processing software package, even though both acts may be illegal.

als and trained personnel are also lacking, as is empirical research on who infringes and why. All these factors impede development of effective messages for different parts of the population.[16]

Conclusion: **A better understanding of the basic principles of copyright law would lead to greater respect for this law and greater willingness to abide by it, as well as produce a more informed public better able to engage in discussions about intellectual property and public policy.**

Recommendation: **An educational program should be undertaken that emphasizes the benefits that copyright law provides to all parties. Such a copyright education program needs to be planned and executed with care. Appendix F discusses the rationale for and the desirable characteristics of copyright education.**

The committee could not decide how extensive copyright education should be, who should conduct this education, or who should pay for it. However, the committee agreed that copyright education should focus on the basic fairness of the copyright law, should not be oversimplified, and should not be mandated by the federal government.

MOVING BEYOND THE DIGITAL DILEMMA: ADDITIONAL MECHANISMS FOR MAKING PROGRESS

The difficulties posed by the digital dilemma are formidable. However, progress can be made, and, indeed, several avenues for moving forward have already been discussed in this chapter—establishing a Task Force on Electronic Deposit, recommending changes to the law to facilitate the maintenance of digital archives, encouraging stakeholders to work together to develop mutually agreeable public access models for licensing and technical protection services, and increasing the use of copyright education. Additional means for progress are discussed in this section; the following section discusses the need for research and improved data.

Technical Protection Services

Although technical protection services cannot resolve legal, social, or economic issues underlying intellectual property, they can help to enforce

---

[16]This is one reason the committee recommends research to obtain information on public understanding of copyright law, especially for digital intellectual property.

agreed-upon rights, rules, constraints, and responsibilities. Technical protection for intellectual property can play a variety of roles, from helping rights holders collect revenue, to helping safeguard user privacy, to helping ensure information authenticity.

Like any security system, a TPS cannot protect perfectly. Even state-of-the-art systems can be compromised by a sufficiently knowledgeable and determined adversary (who may simply avoid picking the lock by finding ways around it). Hence, with the exception of situations in which security is the overriding concern, TPS design always involves a trade-off between capability and cost, including the cost of the effort of the content distributor (who must use and maintain the system) and the effort of users, who typically experience inconvenience in dealing with the system.[17] While this trade-off often results in a distributor's choosing a TPS of only moderate strength, such a solution is frequently entirely adequate and appropriate.

Conclusion: **Technical protection services need not be perfect to be useful. Most people are not technically knowledgeable enough to defeat even moderately sophisticated systems and, in any case, are law-abiding citizens rather than determined adversaries. TPSs with what might be called "curb-high deterrence"—systems that can be circumvented by a knowledgeable person—are sufficient in many instances. They can deter the average user from engaging in illegal behavior and may deter those who may be ignorant about some aspects of the law by causing them to think carefully about the appropriateness of their copying. Simply put, TPSs can help to keep honest people honest.**

Conclusion: **Technical protection technologies are currently deployed to varying degrees. Some, such as encryption and password protection, are widely deployed. Others, such as Web monitoring, watermarking, time stamping, and rights-management languages, are well developed but not yet widely deployed. Copy prevention techniques are deployed to a limited degree. The copy prevention mechanism used in digital video disks provides a notable example of mature development and consumer market penetration.[18]**

---

[17]For a detailed discussion of these costs see *Trust in Cyberspace* (CSTB, 1999c).

[18]As noted in Chapter 5, the content scrambling system used in DVDs was cracked in November 1999, illustrating that no system is perfect and that determined adversaries can find ways around state-of-the-art systems.

Concern about computer security for electronic commerce is becoming an increasingly significant influence on the design of computing and communications infrastructure. This could lead to the development and widespread adoption of hardware-based, technologically comprehensive, end-to-end systems that facilitate creation and control of digital IP—so-called trusted systems.[19] While the installed base of general-purpose PCs represents a large obstacle for trusted systems, efforts to develop standards for secure PC hardware and operating systems for electronic commerce could provide the needed foundation.[20] Trusted systems for intellectual property would be a by-product of the efforts to create security for electronic commerce. As of 1999, however, there are no widespread deployments of trusted systems.[21]

Although TPSs can be based on software alone, protecting valuable content against highly skilled and determined adversaries requires both hardware and software components. Part of the reason is the extent to which software-only circumvention techniques can easily be shared: They are easily distributed worldwide via the Internet, and even unsophisticated users can apply well-designed software (illicit though its purpose may be). Circumvention techniques that require special hardware or hardware-handling expertise are far less easily shared. TPSs with a hardware component are also more effective at making content usable on only one machine, preventing circumvention through redistribution.

But marketplace and infrastructure challenges make developing and deploying specialized hardware difficult except in a few niche markets. Thus, significant attention has been paid to solutions that rely on software only, including some implementations of secure containers and cryptographic envelopes. Several commercial efforts are under way to build and deploy software-based, end-to-end IP delivery systems.

Conclusion: **Robust, integrated technical protection services based on the vision of trusted systems using specialized hard-**

---

[19]The committee uses the phrase "trusted systems" as it is generally understood within the consumer technology market. "Trusted systems" also has a specific and different meaning within the U.S. defense community.

[20]One such relevant effort is the Trusted Computing Platform Alliance, a collaborative effort founded by Compaq, HP, IBM, Intel and Microsoft. This effort is apparently aimed at just such a goal, trying to provide security at the level of the hardware, BIOS, and operating system. While intended initially as a platform for business-to-business electronic commerce, the technology has clear relevance to protection of intellectual property. For background information, see <http://www.trustedpc.org>.

[21]The large installed base of personal computers with a fundamentally open design has also created a large market in complementary products (e.g., disks, sound boards, video boards). Because these products would also have to be redesigned to ensure security, they add to the obstacle faced by trusted systems.

ware are not likely to be realized anytime soon for the consumer market. Such systems, if indeed they evolve, will more likely find a home first in the business-to-business electronic commerce sector. TPSs based on software only are being deployed widely, although not without costs: Systems that have been commercialized to date require a substantial infrastructure to manage secure identification of users or authorization of actions.

Conclusion: As cryptography is frequently a crucial enabling technology for technical protection services, continued advances in technical protection services require a productive and leading-edge community of cryptography and security researchers and developers.

Although technical protection services are not yet widely used, some industries—most notably entertainment—have made concerted efforts and plan to deploy TPSs to protect their IP (e.g., the Secure Digital Music Initiative discussed in Chapter 2). Industries associated with images (e.g., publishing, news, sports, and entertainment) are adopting watermarking technologies to identify and protect their work. TPSs aimed at text material have begun to appear, but have not yet been widely adopted.

Recommendation: Rights holders might consider using technical protection services to help manage digital intellectual property but should also bear in mind the potential for diminished public access and the costs involved, some of which are imposed on customers and society.[22]

Not every information product need be distributed by digital networks, given the availability of alternative mechanisms offering most of the advantages and fewer risks. High-value, long-lived products (e.g., classic movies like *The Wizard of Oz*) might not be available legally on digital networks such as the Internet while protected by copyright, because the consequences of someone capturing the bits are simply too great, and the technical, legal, and social enforcement costs of ensuring that this does not happen are prohibitive. Simply put, the information infrastructure need not be made safe for mass marketing of every form of content.

The pressure to do so is reduced by the possibility of developing

---

[22]The enforcement associated with TPSs involves costs that are both private (e.g., enforcement efforts through industry associations) and public (e.g., a pro-rata share of operating the court system).

special-purpose delivery devices (such as DVDs) that combine both software encryption and specialized hardware in a manner that makes the decrypted digital content very difficult to capture. While the specific encryption system used in DVDs was cracked late in 1999, it is still the case that making the content accessible only with specialized hardware can offer substantially more security than is possible with the software-only solutions used when content is delivered to general-purpose PCs. Delivering digital content in a physical medium (like a DVD) represents a combination of the advantages of digital content (e.g., compactness, low manufacturing cost) and the advantages of previous distribution media (e.g., books) in which the content was "bound to" the physical object and hence less easily reproduced. As a result, there is less need to risk the consequences of networked distribution for every work.

Making the content available only on some variety of physical substrate may also have useful consequences for public understanding of the law: If classic movies (or other varieties of high-value content) are known to be available only on disks, then any copy found on the Web is clearly an infringement. This variety of bright-line distinction may be of use in making consumers aware of and more respectful of IP rights.

Conclusion: **Some digital information may be distributed more securely using physical substrates rather than by computer networks.**

The Digital Millennium Copyright Act of 1998

The Digital Millennium Copyright Act of 1998 (DMCA) amends the Copyright Act, title 17 U.S.C., to legislate new rights in copyrighted works, and limitations on those rights, when copyrighted works are used on the Internet or in other digital, electronic environments.

The anticircumvention regulations adopted by Congress as part of the DMCA need to be clarified to be more technologically sound and more sharply targeted to the problems the regulations were designed to address. The detailed rationale for undertaking such clarification is complex and is presented in Appendix G; a summary appears below.

Certain key terms of the anticircumvention regulations should be defined more precisely. A notable example is the imprecise concept of "an effective technological protection measure." Is a measure that can be circumvented by anyone who has successfully completed a freshman-level college course in computer science an "effective technological protection measure"? What is the threshold for "effective"? Both content owners and potential circumventors need to be able to determine with reasonable effort whether a particular technological mechanism is covered by the statute and whether a particular act of circumvention is legal. Insofar as these issues are unclear, the statute should be clarified in consultation with appropriate technical experts.[23]

The encryption research and computer security testing provisions of the DMCA are well intentioned and generally reflect awareness of the importance of research and practice in these fields. But the provisions are also technically unsound in some respects and need to be refined to align better with standard practices in these fields and to allow decisions that are more pragmatic in the context of encryption research and computer security testing.[24]

> *Conclusion:* **More legitimate reasons to circumvent access control systems exist than are currently recognized in the Digital Millennium Copyright Act. For example, a copyright owner might need to circumvent an access control system to investigate whether someone else is hiding infringement by encrypting a copy of that owner's works, or a firm might need to circumvent an access control system to determine whether a computer virus was about to infect its computer system.[25]**

> *Point of Discussion:* **Many members of the committee believe in the need to add to the DMCA an exception that would permit circumvention of access controls for "other legitimate purposes." This change would enable judicial discretion in interpreting exceptions to anticircumvention provisions, and would provide needed flexibility in the statute for dealing with legitimate circumvention activities not anticipated by Congress.**

The committee's response to and deliberations regarding regulation of what the DMCA calls circumvention technologies reflected the conflicting views evident across the range of stakeholders. Many committee members felt strongly that developing a tool to accomplish any lawful act of circumvention should be lawful, even if the DMCA does not explicitly authorize the development of such a tool. They also believed that a right to develop such tools is implied in the right to engage in lawful circumvention. Others felt strongly that the existing exceptions to anticircumvention rules in the DMCA adequately protect legitimate user interests and regarded as unnecessary any permitting of tool development to pre-

---

[23]Other examples of vague and/or technically unsound language in the DMCA appear in Appendix G and in Callas et al. (1999).

[24]See Appendix G for specifics.

[25]For additional examples, see Samuelson (1999).

serve fair use. This group feared that allowing development of circumvention tools would inevitably lead to increased infringement.

Some members of the committee felt that people not capable of developing the tools necessary to engage in lawful acts of circumvention should legally be able to acquire the tools from another person, arguing that, in the absence of such an understanding, the explicit right in the DCMA to circumvent would be available only to those with substantial technical skills, thus producing a curious piece of public policy.

Some members of the committee were highly critical of the DCMA's antidevice provisions and suggested that they be repealed; others thought that the provisions should be narrowed in scope. The provisions perceived to be most in need of amendment were sections 1201(a)(2)(ii) and 1201(b)(1)(ii), which outlaw technologies having "only limited commercially significant purposes or uses other than to circumvent" technical protection measures. However, other committee members felt strongly that section 1201 provides adequately for the use of legitimate devices by responsible persons and that, if a larger class of circumvention devices were legitimized, they would inevitably be distributed widely and used for copyright infringement.

Many members of the committee thought that technologies with noncommercially significant purposes should have the same protection under the DMCA as those with commercially significant purposes. Freeware and shareware, for instance, are examples of technologies that do not generally have "commercially significant purposes" yet may have legitimate, socially desirable uses. These committee members thought that the antidevice provisions would be better phrased in terms of technologies having "no apparently legitimate purpose."

Recommendation: **In addition to the currently required Librarian of Congress study of some of the impacts of the Digital Millennium Copyright Act's anticircumvention provisions, broader assessments should be conducted of the impacts of the anticircumvention provisions of the DMCA as a whole. This broader review of the regulations is justified because of their unprecedented character; their breadth; and widespread concerns about their potential for negative impacts on public access to information, on the ability of legitimate users to make noninfringing uses of copyrighted works, on research and development in security technology, and on competition and innovation in the high-technology sector. This review should occur periodically and should include a study of impacts of the antidevice provisions of the DMCA.**

Business Models

As Shapiro and Varian (1998) have pointed out, the appropriate strategy for those in the information business should be to maximize the value of intellectual property, not its protection. Thinking in these terms expands the options available: In addition to the traditional business model of selling digital IP as a product, there are also models that de-emphasize or forgo attempting to control digital information and focus instead on other products or services for which the digital IP is complementary. Additional business models can be developed by asking what forms of value can be derived from the IP that are not so easily reproduced. At the time of this writing, the rapidly evolving Internet has created an effective environment in which to experiment with various business models.

Conclusion: **Both technology and business models can serve as effective means for deriving value from digital intellectual property. Technical protection mechanisms can reduce the rate of unauthorized use of IP, but impose their own costs (in production, service, and sometimes customer effort). An appropriate business model can sometimes sharply reduce the need for technical protection, yet provide a way to derive substantial value from IP. Models that can accomplish this objective range from a traditional sales model (low-priced, mass market distribution with convenient purchasing, where the low price and ease of purchase make buying more attractive than copying), to the more radical step of giving away IP and selling a complementary product or service.**

Recommendation: **Rights holders should give careful consideration to the power that business models offer for dealing with distribution of digital information. The judicious selection of a business model may significantly reduce the need for technical protection or legal protection, thereby lowering development and enforcement costs. But the model must be carefully matched to the product: While the appropriate business model can for some products obviate the need for technical protection, for others (e.g., first-run movies) substantial protection may be necessary (and even the strongest protection mechanisms likely to be available soon may be inadequate).**

The Interaction of Technical Protection Services, Business Models, Law, and Public Policy

The community of authors and publishers is characterized by substantial diversity, ranging from those who make their living from the sale of their intellectual property, to those who make their living by keeping it proprietary and using it themselves (i.e., holders of trade secrets), to those who make their careers by giving it away (e.g., most academic researchers), finding reward in the recognition and the indirect benefits that accrue. The differences across all these groups are substantial. There are differences in motivation, in reward structure, and in the consequences of changes in level of IP protection, differences large enough that an attempt to identify *the* IP solution would be both stifling and counterproductive. Such an approach would also likely focus on the high end of the market, as these products often present the most immediate, compelling, and easily quantified examples of the consequences of IP theft or misuse. But making law or policy by focusing on those examples would be as inappropriate as creating the policy based on the segment of the market that gives away IP and sells auxiliary products or services.

Conclusion: There is great diversity in the kinds of digital intellectual property, business models, legal mechanisms, and technical protection services possible, making a one-size-fits-all solution too rigid. Currently a wide variety of new models and mechanisms are being created, tried out, and in some cases discarded, at a furious pace. This process should be supported and encouraged, to allow all parties to find models and mechanisms well suited to their needs.

Recommendation: Legislators should not contemplate an overhaul of intellectual property laws and public policy at this time, to permit the evolutionary process described above the time to play out.

MOVING BEYOND THE DILEMMA: A CALL FOR RESEARCH AND IMPROVED DATA

As the information infrastructure creates many new opportunities and challenges, it is not surprising that research and data collection are needed to support informed decision making. Several such recommendations are made above, including the need to reevaluate the concept of publication; initiate legal, economic, and policy research concerning fair use; and develop an understanding of and data resources about the perceptions and behavior of the general public regarding digital intellectual property. This section identifies additional areas where research and data collection are needed. The committee urges the funders and managers of research programs to place a high priority on the areas of inquiry articulated in this report.

Illegal Commercial Copying

The U.S. copyright industry associations collect a great deal of data and report extensively on illegal commercial copying. Notwithstanding its volume, there are some reasons to question the accuracy of this information. One difficulty derives from the observation that illegal sales and distribution are frequently private acts; consequently, any data necessarily need to be based on extrapolation from a limited sample. There are also methodological issues concerning how the information is generated and reported. Some studies are based on assumptions that lead to high-end estimates, such as the assumption that all illegal copies displace sales at market prices.

A number of committee members concluded that there is a paucity of *accurate* information and that information reporting needs to be done by a disinterested third party.

> *Conclusion:* **The methodology employed by some trade associations in the analysis of data concerning illegal commercial copying produces high-end estimates of losses in gross revenue by these industries. Trade associations would make a more useful contribution to the debate if they revised their methodology so that their estimates better reflect the losses attributable to illegal commercial copying.[26] Notwithstanding the methodological deficiencies in the reported information, the volume and cost of illegal commercial copying are substantial.**

Multiple, interacting phenomena are at work here. Surely differences exist in the commercial and noncommercial copying spheres, and differences may well exist in behaviors among different demographic groups, different geographic locations, and perhaps even different cultures.[27] The phenomena may include such things as how the difficulty of making an illegal copy affects the frequency of copying; the effect on consumer deci-

---

[26]See the section "Illegal Commercial Copying" in Chapter 5 for specifics.

[27]For example, the 14-year-old hackers who download content from the Web without paying should not be lumped with the counterfeiters who contract with a mass-production factory and distribute CDs throughout the world.

sion making of the price and the availability of legitimate copies; the personal sense of the moral or ethical dimensions of the copying involved; the degree of law enforcement or legal scrutiny directed to the behavior; and peer group or other social opprobrium or encouragement. An improved understanding is needed of what these phenomena are and how they operate in the real world, so that they can be targeted for educational efforts and policy-making actions. Reducing the current state of uncertainty about the impact of these various phenomena will be important to future policy makers and entrepreneurs.

Recommendation: **Research should be initiated to better assess the social and economic impacts of illegal commercial copying and how they interact with private noncommercial copying for personal use.**

The information infrastructure carries both promise and peril for intellectual property; the peril arises from the ability of the technology to make reproduction and transmission of information vastly easier, cheaper, and faster. This, in turn, substantially increases the difficulty of enforcing copyright law. Yet such enforcement is important because of the economic consequences of piracy, the social consequences of laws unenforced, and the belief held by many creators that their works, as expressions of their individuality, ought to be protectable, quite independent of the economic consequences of infringement.

Research on the Economics of Copyright, Use of Patents, and Cyber Law

As a previous CSTB report has observed, "[n]umerous studies have looked at the economic impact of patents, but far fewer such studies have been done on copyright, even though there is currently much legal and policy activity in this area" (CSTB, 1998, p. 49). The committee concurs with this view.

Recommendation: **Research should be conducted to characterize the economic impacts of copyright. Such research might consider, among other things, the impact of network effects in information industries and how digital networks are changing transaction costs.[28]**

---

[28]Network effects (or, alternatively, positive network externalities) arise when a good is more valuable to a user as more users adopt the same good or compatible ones (Tirole, 1988).

The past decade has seen a substantial de facto broadening of items for which patents can be obtained, including information inventions such as computer programming, information design, and business methods. The long-term effects of this trend are as yet unclear, although the near-term consequences are worrisome.

Conclusion: **Because the expansion of patent law to cover information inventions has occurred without any oversight from the legislative branch and takes patent law into uncharted territory, this phenomenon needs to be studied on a systematic basis, empirically and theoretically.**

Recommendation: **Research should be conducted to ensure that expansion of patent protection for information inventions is aligned with the constitutional intent of promoting the progress of science and the useful arts.**

Digital information leads to new kinds of information products and services, which in turn may require legal protection that is difficult to provide through traditional intellectual property law. Digital repositories pose difficult questions about authorship, ownership, and the boundaries among copyright-protected works (e.g., does the information stored on a firm's computer network qualify for protection in the aggregate, or does it consist of many works, some of which qualify while others do not?). Difficulties are also arising with respect to the concept of derivative work, given the mutability of works in digital form and the variety of ways in which digital information can be presented and accessed. Two areas that are particularly likely to generate important legal and policy questions are the status of temporary reproductions and derivative work rights.

• *Status of temporary reproductions.* Should temporary reproductions of copyrighted works in random access memory be controllable by copyright owners? Much of the debate on this question has occurred as though only two possible positions existed on the issue: that copyright owners are and should be able to control all temporary copies of their works, or that, under present law, they have no right to control any of them. Technologists who think about caching, replication, and the like as ways to build an efficient system view the by-product copies that these activities produce as irrelevant, mechanistic artifacts. Meanwhile, in response to lobbying by particular industry groups about specific temporary copying issues (e.g., copies made in the course of a transmission by a telecommunications provider), Congress has legislated some specific privileges. Rather than developing the law on temporary copies on a case-by-case, or lobby-by-lobby approach, it would be desirable to develop a taxonomy of temporary copies made in computer systems, and then to assess the economic significance of each category. From this information might emerge some principles about when temporary copies should or should not be regulated by copyright owners, which could then be adapted into a more general-purpose and flexible rule. For example, there might be a workable distinction between "ephemeral" and other temporary copies.[29] There is also the question of whether "copies," whether temporary or not, are still the most appropriate basis for copyright; see a discussion of this topic below.

- *Derivative work rights.* The dynamic and interactive character of digital information raises a host of questions about how the derivative work right of copyright law should be applied. At least one court has thus far taken a fairly narrow view of the derivative work right in the digital environment.[30] But filtering, framing, "morphing," real-time language translating, and visualization by other than the rights holder are among the many uses of digital works that raise derivative work rights issues that have yet to be settled.

---

[29]A useful principle for distinguishing ephemeral reproduction from ordinary reproductions is that the user can determine the time and circumstances under which an ordinary reproduction is rendered, while this is not the case for an ephemeral reproduction.

[30]In *Lewis Galoob v. Nintendo*, the Ninth Circuit Court of Appeals decided that the maker of a "Game Genie" program did not infringe Nintendo's derivative work right by selling a tool with which users could alter certain aspects of the play of Nintendo games. The court held that the Game Genie was not a derivative work because it did not incorporate a protected work or any part thereof in a concrete or permanent form. This ruling suggests that add-on programs will generally not infringe the derivative work right, but many questions remain about how far derivative work rights should extend in the digital environment.

In the view of one writer (Patry, 1994), the Ninth Circuit Court erred because the right to prepare derivative works is not limited to reproduction in copies. Accordingly, ". . . an unauthorized, unfixed, derivative work will infringe as long as it incorporates a substantial portion of the fixed original work."

In a later case, *Micro Star v. Formgen*, the Ninth Circuit Court ruled that there was infringement because the derivative work at issue was recorded in permanent form. The court also ruled that there was infringement because Micro Star infringed Formgen's story by creating sequels to that story. The court also considered and rejected Micro Star's argument that it was protected by the fair use defense.

Recommendation: Legal research should be undertaken on the status of temporary reproductions and derivative work rights to inform the process of adapting copyright law to the digital environment, and to assist policy makers and judges in their deliberations.[31]

As in the case of patents, digital IP and the information infrastructure create new conflicts and raise questions with respect to other information laws and policies.[32]

Recommendation: Research should be undertaken in the areas that are most likely to intersect with intellectual property law, namely, contract law, communications policy, privacy policy, and First Amendment policy (see Box 6.1 for specifics). The interaction of intellectual property law and contract law is likely to be of particular significance in the relatively near future, as licensing becomes a more common means of information distribution, leading to potential conflicts with the goals of IP law.

The international nature of digital networks contributes to difficulties of applying national laws. Although some work has been done on developing a framework for dealing with conflicts of law, choice of law, and other jurisdictional issues, no international consensus as yet exists on these issues. Thus, legal research should be initiated to help clarify the issues, build consensus, and further the harmonization of intellectual property rules, to promote global information commerce.

Is "Copy" Still the Appropriate Foundational Concept?

The committee suggests above that the notion of copy may not be an appropriate foundation for copyright law in the digital age. Where digital information is concerned, legitimate copies are made so routinely that the act of copying has lost much of its predictive power: So many non-infringing copies are made in using a computer that noting that a copy has been made tells us little about the legitimacy of the behavior. In the digital world, copying is also an essential action, so bound up with the

---

[31]Only a few years ago, proxy caching by online service providers and linking from one Web site to another on the World Wide Web were the subjects of considerable debate. Both are now generally thought to be lawful as a matter of U.S. copyright law, a position enabled in part by legal research that has explored the implications of alternative resolutions.

[32]A new field of law, cyberlaw, is being established, with its own courses and journals.

BOX 6.1
Areas of Legal and Policy Research at the Intersection of
Intellectual Property and Other Information Policies and Laws

Contract law. With the proliferation of licensing as a mode of commercial distribution for digital information, courts will need to decide when and how licensing rules should relate to certain public policies, such as those favoring competition, innovation, and freedom of expression. It will take some time before courts reach any consensus on working out this relationship, determining which policies are fundamental and when public policy can or should override license terms.

Communications policy. Communications and copyright policies inevitably intersect because digital networks are communications systems. One example is an effort by some copyright industry groups to persuade the Federal Communications Commission to require broadcasters to include copyright management information in digital broadcasts. That information might encourage rights holders to distribute work (because its origin was labeled), thereby increasing access, but could also include, among other things, digital signals that tell a consumer's video recorder not to allow recording of programs.

Privacy policy. Information technologies generally make it easier for copyright owners to monitor an individual's use of a work. Copyright management systems in particular are designed for this purpose (e.g., to enable billing). Clearly, legitimate uses of this information exist, but the information can also be misused. The Digital Millennium Copyright Act contemplates this situation and provides some limited protections for privacy concerns raised by digital technologies.

First Amendment policy. With an expansion of IP rights in the digital environment and a possible contraction of fair use and other exceptions, the First Amendment to the U.S. Constitution may take on new importance in mediating between the rights of rights holders and the public. This new role for the First Amendment may apply not only to copyright but to patents and trade secrets as well. With patents, the issues include whether a patent can be infringed by teaching the subject matter of the patent or writing about it. For trade secrets the issues include determining when information is public enough that there are First Amendment concerns associated with its dissemination. Making such a determination may be difficult, given the unclear line between publication and nonpublication in the digital environment. Database protection legislation may also raise First Amendment issues.

way computers work that control of copying provides unexpectedly broad powers, considerably beyond those intended by the copyright law.

> *Recommendation:* **The committee suggests exploring whether or not the notion of copy is an appropriate foundation for copyright law, and whether a new foundation can be constructed for copyright, based on the goal set forth in the Constitution ("promote the progress of science and the useful arts") and a tactic by which it is achieved, namely, providing incentive to authors and publishers. In this framework, the question would not be whether a copy had been made, but whether a use of a work was consistent with the goal and tactic (i.e., did it contribute to the desired "progress" and was it destructive, when taken alone or aggregated with other similar copies, of an author's incentive?). This concept is similar to fair use but broader in scope, as it requires considering the range of factors by which to measure the impact of the activity on authors, publishers, and others.**

The committee recognizes that this undertaking will be both difficult and controversial but suggests, nevertheless, that such an investigation is likely to prove both theoretically revealing and pragmatically useful.

Content Creators and the Digital Environment

The evolution in the information infrastructure presents both potential dangers and opportunities for individual authors of all varieties. A media economy in transition may well produce a realignment of interests, some of which may not be favorable for content creators or publishers. A public policy course of action is needed that maintains a balance between the interests of creators and the interests of those who commercialize intellectual property, with attention given to ensuring that creators will continue to pursue their work.

Because digital works are more malleable than works in other media, new concerns arise about authenticity and integrity. Visual artists have the right to be attributed as creators of their works and have a limited right to protect the integrity of their works, but U.S. copyright law does not expressly protect moral rights of content creators beyond these stipulations.[33] Discussion is warranted about what protection might be available to creators of digital works, including the possible role of moral rights.

---

[33]The attribution and integrity rights are enumerated in sec. 106A of the Copyright Law.

Conclusion: The digital environment will continue to bring significant and unpredictable consequences for content creators. Further analysis is needed to determine the impact of these consequences and whether any steps should be taken to intervene in the marketplace.

Point of Discussion: Many members of the committee believe that a task force on the status of the author should be established. The goal of such a task force would be to preserve the spirit of the constitutional protection and incentives for authors and inventors. Its mission statement might be as follows: "The task force shall examine how technological change has affected and is likely to affect the individual creator, recognizing the importance of preserving the economic well-being of creators, balanced with the principle that a democratic society requires broad access to public information." Such a task force would evaluate the viability of mechanisms that facilitate both distribution and control of work (e.g., rights clearance mechanisms) and examine whether issues should be addressed with government action or kept within the framework of private-sector bargaining. The task force would have significant stature with an appropriate level of charter and a limited lifetime (but not less than 2 years); include a cross-section of content creators, rights holders, and other stakeholders; and be financed by public and private funds.

THE PROCESS OF FORMULATING LAW AND PUBLIC POLICY

The committee has tried, wherever possible, to recommend specific legal and policy actions that will assist in dealing with the digital dilemma. But society is still in the early stages of the ferment brought about by the information infrastructure and still has much to learn about the multiplicity of forces that affect intellectual property. Hence no one can specify with any precision all of the legal or policy actions that will be needed. Where the committee cannot recommend specifics, it has tried to articulate a set of guiding principles that it believes will assist legislators and policy makers in effective formulation and revision of law and policy. This section describes those principles.

Intellectual property and IP protection are primarily conceived as legal constructs, but problems arising in the interaction of IP and the information infrastructure need to be considered in the broader context of other forces as well—markets, social norms, and technology (hardware and software). As discussed above, not every problem requires a legislative solution. Technology, business models, and education can all provide effective mechanisms and means for dealing with problems.

The multiplicity of forces and the new models of content distribution being explored both contribute to the possibility of substitutions. One such substitution is contract law for copyright law: With information products increasingly distributed under license rather than being sold, contract law may begin to substitute for copyright law as the dominant force shaping our information environment. A second substitution has been pointed out by Reidenberg (1998) and Lessig (1999a,b), who note the potential for software to substitute for law. Software can be a form of private regulation, constraining some behaviors just as effectively as legal statutes. The software written to control access to a Web site, for example, can make certain behaviors easy and others nearly impossible.

These sorts of substitutions matter because of what may be gained and what lost in any particular substitution. As the report makes clear earlier, for example, there are both pros and cons in using either contract law or copyright law as a foundation for the information environment. Changing from one to the other should be undertaken with careful consideration of the consequences the shift may bring.

> **Conclusion**: Law and public policy must be crafted to consider all the relevant forces in the digital environment. Initiatives that consider or rely on only one or a subset of the relevant forces are not likely to serve the nation well.

The rapid pace of technological change in the computer industry is the stuff of legend, and it shows no signs of slacking off. This pace is, if anything, increasing. References are made to "Internet time," reflecting the breakneck speed with which business, technology, and social practices change. No appropriately deliberative process has a chance of keeping up. This rapid evolution, particularly in technology, will be an ongoing source of uncertainty and, likely, frustration for policy makers who conceive of and attempt to deal with issues in terms of the extant technology. Such policies are built on shifting sand and run the risk of rapid irrelevance. Even those in the field cannot always cope easily with the pace of change.

> **Conclusion**: Policy makers must conceive of and analyze issues in a manner that is as technology-independent as possible, drafting policies and legislation in a similar fashion. The question to focus on is not so much exactly what device is causing the problem today, as what the underlying issue is. Nor should policy makers base their decisions on the specifics of any particular business model.

Because the information infrastructure makes infringement of IP rights vastly easier, it also makes detection, prevention, and enforcement of laws against private infringement by individuals in their homes far more difficult. As a consequence, individual standards of moral and ethical conduct and individual perceptions of right and wrong become considerably more important in encouraging appropriate behavior. A risk also exists that if IP law is perceived or presented as being so absolute in its prohibitions as to preclude behavior most individuals feel is morally appropriate, then even the more reasonable restrictions in the law may be painted with the same brush and viewed as illegitimate.

Conclusion: **Public compliance with intellectual property law requires a high degree of simplicity, clarity, straightforwardness, and comprehensibility for all aspects of copyright law that deal with individual behavior. New or revised intellectual property laws should be drafted accordingly.**

Conclusion: **The movement toward clarity and specificity in the law must also preserve a sufficient flexibility and adaptability so that the law can accommodate technologies and behaviors that may evolve in the future.**

Principles for the Formulation of Law and Public Policy

In addition to the specific guidelines offered above, the committee developed a broader set of principles for policy makers to use in their decision making. The principles are intended to be general and enduring in nature, reflect areas of general agreement, and incorporate the specific guidelines above. Among other things, the principles may serve as a checklist of important issues to consider during the policy decision-making process.

Recommendation: **Policy makers should use the principles outlined in Box 6.2 in the formulation of intellectual property law and public policy.**

BOX 6.2
Principles for the Formulation of Law and Public Policy

1. There is abiding wisdom in Article I, sec. 8, cl. 8 of the U.S. Constitution, which empowers Congress "to promote the progress of science and [the] useful arts by securing to authors and inventors for limited times exclusive rights in their respective writings and discoveries," that should guide policy making on intellectual property now and into the future.

2. The wisdom of the constitutional clause lies, in part, in enabling Congress to confer exclusive rights on creators as a way to motivate them to invest their resources and efforts to develop socially beneficial works. Its wisdom also lies in its limitation on congressional power to grant rights of perpetual duration or rights so extensive that they would undermine achieving progress in science (by which the founders meant knowledge) and the useful arts (by which the founders meant technological innovation).

3. Intellectual property regimes should be tailored to provide adequate incentives to invest in developing and disseminating innovative works.

4. Intellectual property regimes should also be tailored to balance fairly the interests of creators and the public. As the U.S. Supreme Court noted in its decision in *Sony Corp. of America v. Universal City Studios*, "[t]he monopoly privileges that Congress may authorize are neither unlimited nor primarily designed to promote a special private benefit. Rather, the limited grant is a means by which an important public purpose may be achieved,"[1] namely, providing public access to information and innovative products and services after the period of exclusive control has expired, so as to advance the greater societal good.

5. Although the interests of creators and the interests of those who commercialize intellectual property are aligned in most respects, they may diverge in some ways. A balance of interests should be maintained. Enough deference should be given both to the interests of creators to ensure that creators will not be deterred from pursuing their work, and to the interests of publishers so that they are not deterred from their central role in disseminating works.

6. Appropriately crafted exceptions to and limitations on the exclusive rights conferred on innovators by intellectual property laws are a well-accepted means of accomplishing balance in intellectual property law. The fair use doctrine is an example of this.

7. Although achieving balance in intellectual property law is important for many reasons, it is especially important because knowledge creation and innovation are dynamic and cumulative in character. Providing extensive protection to a first-generation innovator may stifle follow-on innovation that, if developed, would be in

---

[1] *Sony Corp. of America v. Universal City Studios, Inc.*, 464 U.S. 417 (1984).

the public interest. Providing too little protection to a first-generation innovator can also be problematic because he or she may decide that there is too little prospect of reward to make the effort worthwhile.

8. Maintaining a vital public domain of ideas, information, and works in which intellectual property rights have expired is important not only because it enables and promotes ongoing innovation but also because it promotes other societal values, such as education and democratic discourse.

9. Policies that promote low-cost public access to government data and to data developed with governmental funding, especially those of scientific importance, should be preserved in the digital context because they, too, promote constitutional purposes of promoting science and technological innovation as, for example, by enabling the development of value-added information products and services.

10. If Congress decides to create new forms of intellectual property protection, the new regimes should conform to these principles.

11. When revising existing intellectual property regimes, Congress should ensure that the principles enunciated here continue to be respected.

12. Policy makers and judges should respond to challenges that information technologies pose for intellectual property law in a manner that conforms to these principles. It should be reassuring to these actors to know that intellectual property laws have adapted to challenges posed by new technologies in the past and that digital technologies provide some opportunities for greater protection of IP.

13. Intellectual property law, contract law, business models, and technical protection services are generally complementary ways to provide appropriate protection to rights holders in competitive markets. However, each of these protections can be exercised in an abusive manner, for example, by unduly interfering with competition, innovation, and free speech interests. When such abuse occurs, there are and should continue to be legal processes to deal with them.

14. Intellectual property policy is an important component of the information policy of a society, but it is not the only important information policy. Policy makers concerned with forming appropriate IP rules should not ignore other information policy dimensions of their decisions, such as those that concern privacy or civil liberties.

15. Intellectual property law should not be used to address issues other than intellectual property issues. To do so would mislead the public and unnaturally constrain technology developers.[2]

---

[2]Consider, for example, the issue of online personal privacy, which has at times been cast as an IP issue, but should not be. Users of all sorts of information services (e.g., shopping

continued

BOX 6.2 Continued

16. Intellectual property rules should be as technology neutral as possible to maximize the chances that the rules will be flexible enough to enable the law to adapt well to new situations in an era of rapid technological change.

17. Policy makers should strive to make intellectual property rules as simple and as easily comprehensible to the general public as is feasible. Such simplicity and clarity will not only enable compliance with legal norms but also assist potential participants in the marketplace in knowing when and how they must clear rights to engage in certain activities.

18. There will, however, always need to be some flexibility in intellectual property rules. Experience with the fair use limitation on copyright, for example, has provided a workable framework for applying general norms in a manner that responds to particular situations where rights of copyright owners and other legitimate interests may be in tension.

19. Laws and policies should be formulated based on a realistic assessment of their success. Attempts to develop perfect laws and policies are unlikely to be practical or even possible.

20. Creating rules of law imposes costs on society. The costs of enhanced intellectual property rules, including the costs of enforcement, should be carefully weighed in comparison with the benefits of such rules

---

services, government services, and even e-mail) regularly supply personal information in order to use those services (such as addresses, phone numbers, and social security numbers). These users are entitled to know how the service provider intends to use this information and are entitled to refuse to use the service if they don't agree with the provider's practices.

It has been suggested by some that if each piece of personal information were regarded as the intellectual property of the person being described, users would have a legal framework in which to object to service-providers' information practices, on the grounds that "information about me is my intellectual property and you cannot use it without my permission." Therein arises the temptation to use IP law to "solve" the online privacy problem.

Copyright, trademark, and patent law were designed to "promote the useful arts," not protect privacy. The two goals do not align: Legal, commercial, and artistic considerations that must be taken into account when deciding whether something should be protected under IP law often do not apply to users' personal information. Although originality is a requirement for copyright protection, it hardly seems to make sense to require an individual to display originality in order to protect personal information such as address, phone number, and social security number and to demand that service providers use that information appropriately.

Put somewhat differently, personal information should not have to be copyrightable, trademarkable, or patentable to be deserving of protection. In the Internet Age, technologists and policy makers are challenged to develop a legal framework in which privacy is addressed as such on its own terms.

CONCLUDING REMARKS

Intellectual property will surely survive the digital age. It is clear, however, that major adaptations will have to take place to ensure sufficient protection for content creators and rights holders, thereby helping to ensure that an extensive and diverse supply of IP is available to the public. Major adaptations will also be needed to ensure that the important public purposes embodied in copyright law, such as public access, are fulfilled in the digital context. Considering the vitality of the participants, the committee is optimistic that workable solutions will be forthcoming in time.

The committee has been cautious about major legislative initiatives because it is early in the evolution of digital intellectual property and much remains unknown—both because of the yet-to-come evolution in the information industries, user communities, and technologies and because of the need for research and data collection to improve knowledge and understanding of the issues. Under such circumstances, major changes in legal regimes and public policy are ill-advised.

Bibliography

Abad Piero, J.L., N. Asokan, M. Steiner, and M. Waidner. 1998. "Designing a Generic Payment Service," *IBM Systems Journal*, 37(1):72-88.

Abadi, Martin, Michael Burrows, Butler Lampson, and Gordon Plotkin. 1993. "A Calculus for Access Control in Distributed Systems," *ACM Transactions on Programming Languages and Systems*, 15(4):706-734.

Abate, Tom. 1998. "Record Labels Fear Move to Digital Format Will Encourage Piracy," *San Francisco Chronicle*, September 10, p. B1. Available online at <http://www.sfgate.com/cgibin/article.cgi?file=/chronicle/archive/1998/09/10/BU3722.DTL>.

Acken, John M. 1998. "How Watermarking Adds Value to Digital Content," *Communications of the ACM*, 41(7):75-77.

Alford, William P. 1995. *To Steal a Book Is an Elegant Offense*. Stanford, CA: Stanford University Press.

Alrashid, Tareq M., James A. Barker, Brian S. Christian, Steven C. Cox, Michael W. Rabne, Elizabeth A. Slotta, and Luella R. Upthegrove. 1998. "Safeguarding Copyrighted Contents: Digital Libraries and Intellectual Property Management," *D-Lib Magazine*, April. Available online at <http://www.dlib.org/dlib/april98/04barker.html>.

American Library Association. 1988. *Less Access to Less Information by and About the U.S. Government: A 1981-1987 Chronology*. Washington, DC: American Library Association.

American Library Association. 1992. *Less Access to Less Information by and About the U.S. Government: A 1988-1991 Chronology*. Washington, DC: American Library Association.

Anderson, Lessley. 1998. "RIAA Setback: Rio Gets a Go," *The Industry Standard*, October 27. Available online at <http://www.thestandard.net/articles/article_print/0,1454,2254,00.html>.

Anderson, Ross. 1993. "Why Cryptosystems Fail," pp. 215-227 in *Proceedings of the 1st ACM Conference on Computer and Communications Security*. New York: Association for Computing Machinery.

AsiaBizTech. 1999. "Artists Set Up Digital Media Copyright Protection Group," February 8. Tokyo: Nikkei BP BizTech, Inc. Available online at <http://www.nikkeibp.asiabiztech.com/Database/1999_Feb/08/New.02.gwif.html>.

Association of American Universities (AAU). 1994. *Report of the AAU Task Force on Intellectual Property Rights in an Electronic Environment.* Available online at <http://www.arl.org/aau/IPTOC.html>.

Atkins, D., P. Buis, C. Hare, R. Kelley, C. Nachenburg, A.B. Nelson, P. Philips, T. Ritchie, and W. Steen. 1996. *Internet Security: Professional Reference.* Indianapolis, IN: New Riders Publishing.

Aucsmith, D. 1996. "Tamper Resistant Software," pp. 317-334 in *Proceedings of the 1st International Information-Hiding Workshop.* Lecture Notes in Computer Science, 1174. Berlin: Springer.

Bauman, Larry. 1999. "New Issues, Including Interactive Intelligence, Cybergold, and eGain, Weather the Tech Sell-Off," *Wall Street Journal,* September 24, p. C7.

Bayeh, E. 1998. "The Web Sphere Application Server Architecture and Programming Model," *IBM Systems Journal,* 37(3):336-348.

Bearman, David, and Jennifer Trant. 1998. "Authenticity of Digital Resources: Towards a Statement of Requirements in the Research Process," *D-Lib Magazine,* June. Available online at <http://www.dlib.org/dlib/june98/06bearman.html>.

Benkler, Y. 1999. "Intellectual Property and the Organization of Information Production." Draft: October. New York: New York University Law School.

Besen, Stanley M., and Leo J. Raskind. 1991. "An Introduction to the Law and Economics of Intellectual Property," *Journal of Economic Perspectives,* 5(1):3-27, Winter.

Blaze, M., J. Feigenbaum, and J. Lacy. 1996. "Decentralized Trust Management," pp. 164-173 in *Proceedings of the 17th Symposium on Security and Privacy.* Los Alamitos, CA: IEEE Computer Society Press. Available online at <http://www.research.att.com/~jf/pubs/oakland96proc.ps>.

Blaze, Matt, J. Feigenbaum, Paul Resnick, and M. Strauss. 1997. "Managing Trust in an Information-Labeling System," *European Transactions on Telecommunications,* 8(5):491-501, September.

Blaze, Matt, J. Feigenbaum, and Martin Strauss. 1998. "Compliance Checking in the PolicyMaker Trust-Management System," pp. 254-274 in *Proceedings of the 2nd Financial Cryptography Conference.* Lecture Notes in Computer Science, 1465. Berlin: Springer.

Bloomberg News. 1999a. "6 Companies to Push New Digital Standard," Bloomberg News, *New York Times,* February 18.

Bloomberg News. 1999b. "Priceline.com Sues Microsoft," Bloomberg News, *New York Times,* October 14. Available online at <http://www.nytimes.com/library/tech/99/10/biztech/articles/14suit.html>.

Brackenbury, I.F., D.F. Ferguson, K.D Gottschalk, and R.A. Storey. 1998. "IBM's Enterprise Server for Java," *IBM Systems Journal,* 37(3):323-335.

Branscomb, Anne W. 1994. *Who Owns Information? From Privacy to Public Access.* New York: Basic Books.

Briggs, B. 1998. "Lotus eSuite," *IBM Systems Journal,* 37(3):372-385.

Buel, Stephen. 1999. "Grateful Dead Lets Fans Swap Concert Recordings on the Web," *San Jose Mercury News,* May 12, p. 1A.

Business Software Alliance / Software and Information Industry Association (BSA/SIIA). 1999. *1999 Global Software Privacy Report.* Washington, DC: BSA/SIIA.

Business Wire. 1999a. "Liquid Audio and 48 Leading Companies Form Coalition to Add Authentication Mark to Digital Music Formats Including MP3," *Business Wire (via Lexis/Nexis),* Redwood City, CA, January 25. Available online at <http://www.businesswire.com>.

Business Wire. 1999b. "SightSound.com Links Patent Licensing Program to SDMI, Targeting Companies for Provisional Agreements," *Business Wire (via Lexis/Nexis),* Beverly Hills, CA, January 28. Available online at <http://www.businesswire.com>.

Callas, J., J. Feigenbaum, D. Goldschlag, and E. Sawyer. 1999. "Fair Use, Intellectual Property, and the Information Economy (Panel Session Summary)," pp. 173-183 in *Proceedings of the 3rd Financial Crypto Conference.* Lecture Notes in Computer Science, 1648. Berlin: Springer.

Caruso, Denise. 1999. "A New Model for the Internet: Fees for Services," *New York Times,* July 19. Available online at <http://www.nytimes.com/library/tech/99/07/biztech/articles/19digi.html>.

Chamberland, L.A., S.F. Lymer, and A.G. Ryman. 1998. "IBM VisualAge for Java," *IBM Systems Journal,* 37(3):386-408.

Chapman, Gary. 1999. "Singapore Wires Its Hopes to Net with Ambitious Schools Program," *Los Angeles Times,* February 22, p. C1.

Cheng, P.C., J.A Garay, A. Herzberg, and H. Krawczyk. 1998. "A Security Architecture for the Internet Protocol," *IBM Systems Journal,* 37(1):42-60.

Clark, Don. 1999. "Digital Books on the Web Move Closer to the Market," *Wall Street Journal,* August 31, p. B1.

Cleary, Sharon. 1999. "Music Sites Dish Out Songs, Stats, and Bios," *Wall Street Journal,* April 29, p. B10.

CNN Newsroom. 1998. 4:30 a.m. ET Broadcast, October 20. Atlanta, GA: Cable News Network.

Cohen, Julie E. 1996. "A Right to Read Anonymously: A Closer Look at 'Copyright Management' in Cyberspace," 28 *Connecticut Law Review,* pp. 981-1039.

Cohen, Julie E. 1998. "Copyright and the Jurisprudence of Self-Help," 13 *Berkeley Technology Law Journal,* pp. 1089-1143.

Colonial Williamsburg Foundation. 1999. "Electronic Field Trips." Available online at <http://www.history.org/>.

Commission on Preservation and Access and the Research Libraries Group (CPA/RLG). 1995. *Preserving Digital Information: Final Report of the Task Force on Archiving of Digital Information.* Washington, DC: CPA/RLG. Available online at <http://www.rlg.org/ArchTF/tfadi.index.htm>.

Computer Science and Telecommunications Board (CSTB), National Research Council. 1991a. *Computers at Risk: Safe Computing in the Information Age.* Washington, DC: National Academy Press.

Computer Science and Telecommunications Board (CSTB), National Research Council. 1991b. *Intellectual Property Issues in Software.* Washington, DC: National Academy Press.

Computer Science and Telecommunications Board (CSTB), National Research Council. 1994. *Information Technology in the Service Society: A Twenty-First Century Lever.* Washington, DC: National Academy Press.

Computer Science and Telecommunications Board (CSTB), National Research Council. 1996. *Cryptography's Role in Securing the Information Society.* Washington, DC: National Academy Press.

Computer Science and Telecommunications Board (CSTB), National Research Council. 1998. *Fostering Research on the Economic and Social Impacts of Information Technology: Report of a Workshop.* Washington, DC: National Academy Press.

Computer Science and Telecommunications Board (CSTB), National Research Council. 1999a. *Being Fluent with Information Technology.* Washington, DC: National Academy Press.

Computer Science and Telecommunications Board (CSTB), National Research Council. 1999b. *Funding a Revolution: Government Support for Computing Research.* Washington, DC: National Academy Press.

Computer Science and Telecommunications Board (CSTB), National Research Council. 1999c. *Trust in Cyberspace*. Washington, DC: National Academy Press.

Cox, B. 1994. "Superdistribution," *Wired*, Vol. 2, September. Available online at <www.wired.com/wired/archive/2.09/superdis.html>.

Cox, Ingemar J., Joe Kilian, Tom Leighton, and Talal Shamoon. 1995. *Secure Spread Spectrum Watermarking for Multimedia: Technical Report 95-10*. Princeton, NJ: NEC Research Institute, Inc., December 4. Available online at <http://www.neci.nj.nec.com/tr/index.html>.

Craver, Scott, Nasir Memon, Boon-Lock Yeo, and Minerva M. Yeung. 1997. "On the Invertibility of Invisible Watermarking Techniques," pp. 540-543 in *Proceedings of the IEEE Signal Processing Society 1997 International Conference on Image Processing (ICIP'97)*. Los Alamitos, CA: IEEE Computer Society Press.

Craver, Scott, Boon-Lock Yeo, and Minerva M. Yeung. 1998. "Technical Trials and Legal Tribulations," *Communications of the ACM*, 41(7):44-45.

Crews, Kenneth, and John O'Donnell. 1998. "Indiana University, Online Copyright Tutorial." Indianapolis, IN: Indiana University Copyright Management Center, August 3.

Dean, Drew. 1997. "The Security of Static Typing with Dynamic Linking," pp. 18-27 in *Proceedings of the 4th ACM Conference on Computer and Communications Security*. New York: ACM Press Books.

Dean, Drew, Ed Felten, and Dan Wallach. 1996. "Java Security: From HotJava to Netscape and Beyond," pp. 190-200 in *Proceedings of the 1996 IEEE Symposium on Security and Privacy*. Los Alamitos, CA: IEEE Computer Society Press.

Diffie, W., and M. Hellman. 1976. "New Directions in Cryptography," *IEEE Transactions on Information Theory*, IT-22(6):644-654, November.

Doyle, Jim. 1994. "Milpitas Couple Indicted in Computer Porn Case," *San Francisco Chronicle*, February 14, p. A12.

Dwork, Cynthia. 1999. "Copyright? Protection?" pp. 31-48 in *Mathematics of Information Coding, Extraction, and Distribution*, G. Cybenko, D. O'Leary, and J. Rissanen, eds. New York: Springer.

Dyson, Esther. 1995. "Intellectual Value," *Wired*, 3(7):136-141 and 181-185.

Einhorn, Bruce. 1999. "Asia Logs On," *Business Week*, International Edition, February 1, p. 34.

Ellison, Carl. 1999. *SPKI Certificate Documentation*. Available online at <http://www.pobox.com/~cme/html/spki.html>.

Feigenbaum, Joan. 1998. "Towards an Infrastructure for Authorization: Position Paper." Invited Talk at the USENIX E-Commerce Conference. Available online at <http://www.research.att.com/~jf/pubs/usenix-ecommerce98.ps>.

Finlayson, Gordon. 1999. "Australia Senate Passes Net Control Bill," *ZD Net*, May 27. Available online at <http://www.zdnet.com/zdnn/stories/news/0,4586,2266715,00.html>.

Foster, Ed. 1998. "What's in a Name? Not Even the Real Pirates Want Us to Call It Software Piracy," *Infoworld*, 20(52):79.

Frank, Robert H., and Phillip J. Cook. 1995. *The Winner-Take-All Society*. New York: Free Press.

Geller, Paul. 1993. *International Copyright Law and Practice*. New York: Matthew Bender.

Gilder, George. 1993. "Metcalfe's Law and Legacy," *Forbes ASAP*, September 13. Available online at <http://www.seas.upenn.edu/~gaj1/metgg.html>.

Gladney, H.M. 1997. "Access Control for Large Collections," *ACM Transactions on Information Systems*, 15(2):154-194, April.

Gladney, Henry M. 1998. "Safeguarding Digital Library Contents and Users: Interim Retrospect and Prospects," *D-Lib Magazine*, July/August. Available online at <http://www/dlib.org/dlib/july98/gladney/07gladney.html>.

Gladney, Henry M., and Jeff Lotspiech. 1998. "Safeguarding Digital Library Contents and Users: Storing, Sending, Showing, and Honoring Usage Terms and Conditions," *D-Lib Magazine*, May. Available online at < http://www.dlib.org/dlib/may98/gladney/05gladney.html>.

Gleick, James. 1998. "The Digital Attic: Are We Now Amnesiacs? Or Packrats?" *New York Times*, April 12, sec. 6, col. 1, p. 20.

Goldman, Joel S. 1999. "The State Street Bank Case—Its Implications to Financial Institutions," *Intellectual Property Today*, January 8, p. 8. Available online at <http://www.lawworksiptoday.com/current/magindx.htm>.

Gomes, Lee. 1999. "Free Tunes for Everyone! MP3 Music Moves into the High-School Mainstream," *Wall Street Journal*, June 15, p. B1.

Gottschalk, K.D. 1998. "Technical Overview of IBM's Java Initiatives," *IBM Systems Journal*, 7(3):308-322.

Gray, Jim, and Andreas Reuter. 1992. *Transaction Processing: Concepts and Techniques*. Morgan Kaufmann Series in Data Management Systems. San Mateo, CA: Academic Press.

Hamblett, Mark. 1999. "Freelancers Win Victory Over Reprints," *New York Law Journal*, September 28, p. 8.

Hardy, I. Trotter. 1998. *Project Looking Forward—Final Report: Sketching the Future of Copyright in a Networked World*. Washington, DC: U.S. Copyright Office. Available online at <http://lcweb.loc.gov/copyright/reports/>.

Hayton, Richard J., Jean M. Bacon, and Ken Moody. 1998. "Access Control in an Open, Distributed Environment," pp. 3-14 in *Proceedings of the IEEE Symposium on Security and Privacy*. Los Alamitos, CA: IEEE Computer Society Press.

Hazan, Victor. 1970. "The Origins of Copyright Law in Ancient Jewish Law: Introduction," *Copyright Society of the USA Bulletin*, pp. 23-28.

Hellweg, Eric. 1999. "Musical Discord," *Business 2.0*, September. Available online at <http://www.business2.com/articles/1999/09/content/break.html>.

Henderson, Carol C. 1998. *Libraries as Creatures of Copyright: Why Librarians Care About Intellectual Property Law and Policy*. Available online at <http://www.ala.org/washoff/copylib.html>.

Herzberg, Amir. 1998. "Safeguarding Digital Library Contents: Charging for Online Content," *D-Lib Magazine*, January. Available online at <http://www.dlib.org/dlib/january98/ibm/01herzberg.html>.

Herzberg, Amir, and Dalit Naor. 1998. "Surf'N'Sign: Client Signatures on Web Documents," *IBM Systems Journal*, 37(1):61-71.

Hinds, Michael deCourcy. 1988. "Personal but Not Confidential: A New Debate Over Privacy," *New York Times*, February 27, sec. 1, col. 1, p. 56.

Holderness, M. 1998. "Moral Rights and Authors' Rights: The Keys to the Information Age," 1 *Journal of Information, Law, and Technology*. University of Warwick: Coventry, UK: Strathclyde University: Coventry, UK. February 27. Available online at <http://elj.warwick.ac.uk/jilt/infosoc/98_1hold/>.

Information Infrastructure Task Force (IITF). 1995. *Intellectual Property and the National Information Infrastructure: The Report of the Working Group on Intellectual Property Rights Information Infrastructure Task Force (IITF)*. Washington, DC: U.S. Department of Commerce.

Jackson, Michael, and Jayne Noble Suhler. 1998. "But My Child Wouldn't Lie or Cheat! Think Again, More Teens Admit to Unethical Behavior," *Dallas Morning News*, October 19, p. 1A. Available online at <http://www.josephsoninstitute.org/98-Survey/98survey.htm>.

Joy, Bill, and Ken Kennedy. 1998. *President's Information Technology Advisory Committee: Interim Report.* August. Washington, DC: President's Information Technology Advisory Committee. Available online at <http://www.ccic.gov/ac/interim/>.

Kamien, Morton I., and Nancy L. Schwartz. 1975. "Market Structure and Innovation: A Survey," *Journal of Economic Literature,* XIII(1):1-37.

Kaminer, Wendy. 1997. "On Work Made for Hire: Declaration of an Independent Writer," Washington, DC: National Public Radio, April 8.

Kaplan, Benjamin. 1967. *Copyright: An Unhurried View.* New York: Columbia University Press.

Kaplan, Carl S. 1999. "In Court's View, MP3 Player Is Just a 'Space Shifter,'" *New York Times on the Web,* July 9. Available online at <http://www.nytimes.com/library/tech/99/07/cyber/cyberlaw/09law.html>.

Kiernan, Vincent. 1999. "Vanderbilt's Television Archive Hopes to Digitize Newscasts for Scholars' Use," *Chronicle of Higher Education,* p. A35, June 4.

King, John Leslie, Rebecca E. Grinter, and Jeanne M. Pickering. 1997. "The Rise and Fall of Netville: The Saga of a Cyberspace Construction Boomtown in the Great Divide," pp. 3-33 in *Culture of the Internet,* Sara Kiesler, ed. Mahwah, NJ: Lawrence Erlbaum Associates.

Kobayashi, M. 1997. *Digital Watermarking: A Historical Survey.* IBM Tokyo Laboratory Research Reports. Yamato, Japan: IBM.

Konen, R. 1998. "Overview of the MPEG-4 Standard." Available online at the MPEG-4 Web page, <http://www.cselt.it/mpeg>.

Koved, Larry, Anthony J. Nadalin, Don Neal, and Tim Lawson. 1998. "The Evolution of Java Security," *IBM Systems Journal,* 37(3):349-364.

Lacy, J., J. Snyder, and D. Maher. 1997. "Music on the Internet and the Intellectual Property Protection Problem," pp. SS77-SS83 in *Proceedings of the International Symposium on Industrial Electronics.* New York: IEEE Computer Society Press.

Lacy, J., N. Rump, and P. Kudumakis. 1998. "MPEG-4 Intellectual Property Management and Protection (IPMP) Overview and Applications." Available online at the MPEG-4 Web page <http://www.cselt.it/mpeg>.

Ladd, David. 1983. "The Harm of the Concept of Harm in Copyright: The 13th Donald C. Brace Memorial Lecture," *Journal of the Copyright Society of the U.S.A,* 30(5):420-432, June.

Lawrence, S.R., and C.L. Giles. 1999. "Accessibility of Information on the Web," *Nature* 400(6740):107-109.

Lehman, Bruce. 1998. *Conference on Fair Use: Final Report to the Commissioner on the Conclusion of the Conference on Fair Use.* Washington, DC: U.S. Patent and Trademark Office.

Lemos, Robert. 1999. "SightSound.com to Music Sites: Pay Up!" January 28. Available online at <http://biz.yahoo.com/bw/990128/ca_sightso_1.html>.

Lessig, Lawrence. 1999a. *Code and Other Laws of Cyberspace.* New York: Basic Books.

Lessig, Lawrence. 1999b. "The Law of the Horse: What Cyberlaw Might Teach." *Harvard Law Review,* Fall issue. Available online at <http://cyber.harvard.edu/lessig.html>.

Lipton, Beth. 1988. "Hollywood Steps Up Copyright Fight," *CNET News.com,* July 29. Available online at <http://www.news.com/News/Item/0,4,24751,00.html>.

Livingston, Brian. 1999. "Digital Encryption of Audio Files Is Cracked: What Will This Mean for the Software Industry?" *Infoworld,* August 30, p. 42.

Lotspiech, Jeffrey, Ulrich Kohl, and Marc Kaplan. 1997. "Safeguarding Digital Library Contents and Users: Protecting Documents Rather Than Channels," *D-Lib Magazine,* September. Available online at <http://www.dlib.org/dlib/september97/ibm/09lotspiech.html>.

Luby, Michael. 1996. *Pseudorandomness and Its Cryptographic Applications.* Princeton, NJ: Princeton University Press.

Lynch, Clifford A. 1997. "When Technology Leads Policy," in *Proceedings of the 131st ARL Meeting. Washington, DC.* Available online at <http://www.arl.org/arl/proceedings/131/lynch.html>.

Madison, James, Alexander Hamilton, and John Jay. 1787. *The Federalist.* New York: The Colonial Press.

Manly, Lorne. 1997. "Off the Record," *New York Observer*, April 7, p. 6.

Markoff, John. 1999. "Bridging Two Worlds to Make On-Line Digital Music Profitable," *New York Times*, September 13, p. 1.

Matheson, L., S. Mitchell, T. Shamoon, R. Tarjan, and F. Zane. 1998. "Robustness and Security of Digital Watermarks," pp. 227-240 in *Proceedings of the 2nd Financial Cryptography Conference*, Lecture Notes in Computer Science, 1465. Berlin: Springer.

McCann-Erickson Agency. 1998. "Total U.S. Advertising Spending by Medium," *Advertising Age*, Neilsen Media Research, April. Available online at <http://adage.com/dataplace/topmarkets/us.html>.

McWilliams, Gary. 1999. "New PCs Say, 'Let Me Entertain You,' and Sales Get Unexpected Lift," *Wall Street Journal*, September 28, p. B1.

Memon, N., and P.W. Wong. 1998. "Protecting Digital Media Content," *Communications of the ACM*, 41(7):34-43.

Menezes, Alfred J., Paul C. van Oorschot, and Scott Vanstone. 1997. *Handbook of Applied Cryptography.* Boca Raton, FL: CRC Press.

Merkle, R.C. 1978. "Secure Communications over Insecure Channels," *Communications of the ACM*, 4(21):294-299.

Mintzer, Fred C., L.E. Boyle, A.N. Cazes, B.S. Christian, S.C. Cox, F.P. Giordano, H.M. Gladney, J.C. Lee, M.L. Kelmanson, A.C. Lirani, K.A. Magerlein, A.M.B. Pavani, and F. Schiattarella. 1996. "Towards On-Line Worldwide Access to Vatican Library Materials," *IBM Journal of Research and Development*, 40(2):139-162.

Mintzer, Fred, Gordon W. Braudaway, and Minerva M. Yeung. 1997. "Effective and Ineffective Digital Watermarks," pp. 9-12 in *Proceedings of IEEE ICIP'97*. Los Alamitos, CA: IEEE Computer Society Press.

Mintzer, Fred, Jeff Lotspiech, and Norishige Morimoto. 1997. "Digital Watermarking," *D-Lib Magazine*, December. Available online at <http://www.dlib.org/dlib/december97/ibm/12lotspiech.html>.

Moore, Geoffrey A. 1995. *Crossing the Chasm: Marketing and Selling High-Tech Products to Mainstream Customers.* New York: Harper Collins.

Murphy, John B. 1998. "Introducing the North American Industry Classification System," *Monthly Labor Review*, 121(7):43-47, July.

National Research Council. 1997. *Bits of Power: Issues in Global Access to Scientific Data.* Washington, DC: National Academy Press.

National Research Council. 1999. *A Question of Balance: Private Rights and the Public Interest in Scientific and Technical Databases.* Washington, DC: National Academy Press.

National Telecommunications and Information Administration (NTIA). 1999. *Falling Through the Net: Defining the Digital Divide.* Washington, DC: U.S. Department of Commerce.

Negroponte, Nicholas. 1995. *Being Digital.* New York: Alfred A. Knopf.

Neumann, Peter G. 1995. *Computer Related Risks.* New York: ACM Press Books.

New York Times. 1998. "Keeping Copyright in Balance" [Editorial], February 21, sec. A, col. 1, p. 10.

Nimmer, Melville B., and Paul Edward Geller, eds. 1988. *International Copyright Law and Practice.* New York: Matthew Bender.

Nimmer, David, Elliot Brown, and Gary N. Frischling. 1999. "The Metamorphosis of Contract into Expand," 87 *California Law Review* 1, pp. 17-77.

Noll, Roger G. 1993. "The Economics of Information: A User's Guide," pp. 25-52 in *The Knowledge Economy: Annual Review of the Institute for Information Studies, 1993-94.* Queenstown, MD: Aspen Institute.

Norman, Donald A. 1998. *The Invisible Computer.* Cambridge, MA: MIT Press.

Office of Technology Assessment of the U.S. Congress. 1989. *Copyright and Home Copying: Technology Challenges the Law,* OTA-CIT-422. Washington, DC: U.S. Government Printing Office, October.

Office of Technology Assessment of the U.S. Congress. 1993. *Accessibility and Integrity of Networked Information Collections,* OTA-BP-TCT-109, by Clifford A. Lynch. Washington, DC: U.S. Congress Office of Technology Assessment.

Okerson, Ann. 1996. "Buy or Lease: Two Models for Scholarly Information at the End (or the Beginning) of an Era," *Daedalus,* 125(4):55-76.

Ove, T. 1998. "Honesty Is Still the Best Policy," *Pittsburgh Post-Gazette,* October 18, p. A1. Available online at <http://www.post-gazette.com/regionstate/19981018honor2.asp>.

Patrizio, Andy. 1999a. "DOJ Cracks Down on MP3 Pirate," *Wired News,* August 23. Available online at <http://www.wired.com/news/politics/0,1283,21391,00.html>.

Patrizio, Andy. 1999b. "Why the DVD Hack Was a Cinch," *Wired News,* November 2. Available online at <http://www.wired.com/news/technology/0,1282,32263,00.html>.

Patrizio, Andy, and Malcolm Maclachlan. 1998. "MP3 Player Could Skirt Legal Challenge," *TechWeb,* October 21. Available online at <http://www.techweb.com/wire/story/TWB19981021S0003>.

Patry, William F. 1994. *Copyright Law and Practice.* Washington, DC: Bureau of National Affairs.

Philips, Chuck. 1999. "Label Gets Top Spot with Online Spin," *Los Angeles Times,* October 7, p. C1.

Phipps, Jennie L. 1998 "Ripping Off Writers or Justified New Profits?" *Editor & Publisher,* May 30, p. 18ff.

Pollack, Andrew. 1999. "Feature Film to Be Produced for Release on Web," *New York Times,* August 24, p. 1 of business section.

Ramanujapuram, Arun, and Prasad Ram. 1998. "Digital Content and Intellectual Property Rights," *Dr. Dobb's Journal,* 23(12):20-27, December.

Ramstad, Evan. 1999. "Circuit City Pulls the Plug on Its Divx Videodisk Venture," *Wall Street Journal,* June 17, p. B10.

Reidenberg, Joel R. 1998. "Lex Informatica: The Formulation of Information Policy Rules Through Technology," 76 *Texas Law Review,* pp. 553-593.

Reuters. 1999. "HK to Launch Campaign on Intellectual Property," *San Jose Mercury News,* July 28, 9:30 a.m. PDT. Available online at <http://www.mercurycenter.com/>.

Rivest, R.L. 1998. "Chaffing and Winnowing: Confidentiality Without Encryption." Available online at <http://theory.lcs.mit.edu/~rivest/chaffing-980701.txt>.

Rivest, R.L., A. Shamir, and L.M. Adelman. 1978. "A Method for Obtaining Digital Signatures and Public-Key Cryptosystems," *Communications of the ACM,* 21(12):120-126.

Robertson, Chiyo. 1999. "Lycos May Face Lawsuit Over MP3," *ZDNet,* March 25. Available online at <http://www.zdnet.com/zdnn/stories/news/0,4586,2231720,00.html>.

Robinson, Sara. 1999. "Recording Industry Escalates Crackdown on Digital Piracy," *New York Times,* October 4, sec. C, col. 4, p. 5. Available online at <http://www.nytimes.com/library/tech/99/10/biztech/articles/04musi.html>.

Rogers, David. 1999. "Senate Panel Proposes Property-Rights Office," *Wall Street Journal,* June 25, p. A6.

Rosen, R.J., R.J. Anderson, L.H. Chant, J.B. Dunlop, J.C. Gambles, D.W. Rogers, and J.H. Yates. 1970. *Computer Control Guidelines*. Toronto: Canadian Institute of Chartered Accountants. Superseded by Information Technology Control Guidelines, 3rd ed. (1998). Available online at <http://www.cica.ca>.

Rosen, R.J., R.J. Anderson, L.H.Chant, J.B. Dunlop, J.C. Gambles, and D.W. Rogers. 1975. *Computer Audit Guidelines*. Toronto: Canadian Institute of Chartered Accountants, Study Group on Computer Control and Audit Guidelines.

Rothenberg, J. 1999. *Avoiding Technological Quicksand*. Washington, DC: Council on Library and Information Resources.

Rubin, B.S., A.R. Christ, and K.A. Bohrer. 1998. "Java and the IBM San Francisco Project," *IBM Systems Journal*, 37(3):365-371.

Samuelson, Pamela. 1999. "Intellectual Property and the Digital Economy: Why the Anti-Circumvention Regulations Need to Be Revised," 14 *Berkeley Technology Law Journal*, pp. 519-566. Available online at <http://sims.berkeley.edu/pam/papers/dmcapaper.pdf>.

Samuelson, Pamela, Randall Davis, Mitchell D. Kapor, and J.H. Reichman. 1994. "A Manifesto Concerning the Legal Protection of Computer Programs," 94 *Columbia Law Review*, pp. 2308-2431.

Sandburg, Brenda. 1998. "Madness in PTO's E-Commerce Method? It Doesn't Take a Genius to Try Out Old Ideas on the Net, But It Can Win You a Patent," *IP Magazine*, August 27.

Scheinfeld, Robert C., and Parker H. Bagley. 1998. "Shakeout on State Street: The Decision Allows Virtually Anything Under the Sun to Win Patent Protection," *IP Magazine*, November. Available online at <http://www.ipmag.com/98-nov/schein.html>.

Shapiro, Carl, and Hal R. Varian. 1998. *Information Rules: A Strategic Guide to the Networked Economy*. Boston: Harvard Business School Press.

Shy, Oz. 1998. "The Economics of Software Copy Protection and Other Media," *Internet Publishing and Beyond: The Economics of Digital Information and Intellectual Property*, Deborah Hurley, Brian Kahin, and Hal Varian, eds. Cambridge, MA: MIT Press.

Siegel, David. 1996. *Creating Killer Web Sites: The Art of Third-Generation Site Design*. Indianapolis, IN: Hayden Books.

Simon, Herbert A. 1971. "Designing Organizations for an Information-Rich World," in *Computers, Communications, and the Public Interest*. M. Greenberger, ed. Baltimore, MD: Johns Hopkins University Press.

Siwek, Stephen E., and Gale Mosteller. 1998. *Copyright Industries in the U.S. Economy: The 1998 Report*. Washington, DC: International Intellectual Property Alliance.

Sloman, M. 1994. "Policy Driven Management for Distributed Systems," *Journal of Network & System Management*, 2(4):333-360.

Smith, Sean W., and Steve H. Weingart. 1997. "Building a High-Performance, Programmable Secure Coprocessor," *IBM Research Report*, RC21045. Yorktown Heights, NY: IBM T.J. Watson Research Center, November.

Smith, S.W., E.R. Palmer, and S.H. Weingart. 1998. "Using a High-Performance, Programmable Secure Coprocessor," *FC98: Proceedings of the 2nd International Conference on Financial Cryptography*. Anguilla, Brit. West Indies: Springer-Verlag LNCS.

Sollins, K., and L. Masinter. 1994. *Functional Requirements for Uniform Resource Names*. Internet Engineering Task Force, RFC 1737, December. Available online at <http://www.ietf.org>.

Stefik, Mark. 1996. *The Digital Property Rights Language: Manual and Tutorial*. Version 1.02, September 18. Palo Alto, CA: Xerox Palo Alto Research Center.

Stefik, Mark. 1997a. "Shifting the Possible: How Digital Property Rights Challenge Us to Rethink Digital Publishing," 12 *Berkeley Technology Law Journal*, pp. 137-159.

Stefik, Mark. 1997b. "Trusted Systems," *Scientific American*, 276(3):78-81.

Steinbach, S.E., I. Karp, and A.C. Hoffman. 1976. *The House Report on the Copyright Act of 1976, Appendix 4.* Mathew Bender (daily edition, Sept. 21).

Stevens, W. Richard. 1996. *TCP/IP Illustrated.* Reading, MA: Addison-Wesley.

Stone, Martha L. 1998. "Copyright Questions Abound on the Web," *Editor & Publisher*, Dec. 12, pp. 44-45.

Strauss, Neil, and Matt Richtel. 1999. "Pact Reached on Downloading of Digital Music," *New York Times*, June 29, p. B1.

Streitfeld, David. 1999. "Who's Reading What? Using Powerful 'Data Mining' Technology, Amazon.com Stirs an Internet Controversy," *Washington Post*, August 27, p. A01.

Sullivan, Jennifer. 1999. "Net Overloads U.S. Patent Agency," *Wired News*, July 13. Available online at <http://www.wired.com/news/print_version/politics/story/19473.html?wnpg=all>.

Sullivan, Jennifer, and John Gatner. 1999. "Cracked: MS's New Music Format," *Wired News*, August 18, 1:45 p.m. Available online at <http://www.wired.com/news/print_version/technology/story/21325.html?wnpg=all>.

Swisher, Kara. 1999. "Internet Firms Set Lobby Group to Push Views on Privacy, Intellectual Property," *Wall Street Journal*, July 12, p. A22.

Thomas, Karen. 1998. "Teen Ethics: More Cheating and Lying," *USA Today*, October 19, p. 1D.

Thomason, Robert, and Rob Pegoraro. 1999. "Web Site Cuts Deal with Songwriters," *Washington Post*, June 17, p. C01.

Tirole, Jean. 1988. *The Theory of Industrial Organization.* Cambridge, MA: MIT Press.

U.S. Copyright Office. 1998. *Circular 21: Reproduction of Copyrighted Works by Educators and Librarians.* Washington, DC: Library of Congress, U.S. Copyright Office. Available online at <http://lcweb.loc.gov/copyright/>.

U.S. Copyright Office. 1998. *Guidelines for Educational Uses of Music.* Washington, DC: Library of Congress, U.S. Copyright Office.

U.S. Copyright Office. 1999. *Report on Copyright and Digital Distance Education.* Washington, DC: Library of Congress, U.S. Copyright Office.

Walker, Adrian. 1998. "The Internet Knowledge Manager and Its Use for Rights and Billing in Digital Libraries PAKeM98," pp. 147 and 103-117 in *Proceedings of the First International Conference on the Practical Application of Knowledge Management.* Blackpool, UK: Practical Application Co., Ltd.

Warner, Julian. 1999. "Information Society or Cash Nexus? A Study of the United States as a Copyright Haven," *Journal of the American Society for Information Science*, 50(5):461-470.

Wingfield, Nick. 1999. "New Battlefield for Priceline Is Diapers, Tuna," *Wall Street Journal*, September 22, p. B1.

Wired News. 1999. "Millions of Eardrums Go Digital," August 10. Available online at <http://www.wired.com/news/print_version/culture/story/21193.html?wnpg=all>.

Wolverton, Troy. 1999. "Expedia to Customers: Name Your Hotel Room Price," CNET News.com, September 7, 9:35 p.m. PT. Available online at <http://news.cnet.com/news/0-1007-202-113636.html>.

Wyckoff, P., S.W. Laughry, T.J. Lehman, and D.A. Ford. 1998. "T Spaces," *IBM Systems Journal*, 37(3):454-474.

Zeitchik, Steven. 1999. "The Great Ether Grab," *Publisher's Weekly.com*, June 14. Available online at <http://www.publishersweekly.com/articles/19990614_71279.asp>.

Zimmerman, P. 1994. *PGP User's Guide.* Cambridge, MA: MIT Press.

Appendixes

APPENDIX A

Study Committee Biographies

RANDALL DAVIS, *Chair,* has been a member of the faculty at the Massachusetts Institute of Technology since 1978, where he is currently professor of electrical engineering and computer science, and professor of management. Dr. Davis is a seminal contributor to the field of expert systems, where his research focuses on model-based systems (programs that work from descriptions of structure and function and that reason from first principles) supporting a wide range of robust problem solving. In 1990, Dr. Davis served as a panelist in a series of CSTB workshops that resulted in the publication *Intellectual Property Issues in Software* and has served as a member of the Advisory Panel to the U.S. Congress Office of Technology Assessment study *Finding a Balance: Computer Software, Intellectual Property, and the Challenge of Technological Change.* He has advised a variety of law firms on cases involving software copyright and patents. In 1990 he served as expert to the court (Eastern District of New York) in *Computer Associates v. Altai,* a software copyright infringement case that was upheld by the Appeals Court for the 2nd Circuit in June 1992, resulting in a significant change in the way software copyright is viewed by the courts. He is on the board of the Massachusetts Software Council and serves as head of its intellectual property subcommittee. In 1990 Dr. Davis was named a founding fellow of the American Association for Artificial Intelligence and in 1995 was elected president of the association. Dr. Davis has been a consultant to several major organizations and assisted in the start-up of three software companies. After completing his undergraduate degree at Dartmouth College, Dr. Davis completed his Ph.D. in artificial intelligence at Stanford University in 1976.

SHELTON ALEXANDER is a professor of geophysics and former head of the Department of Geosciences at the Pennsylvania State University. His interests include seismology, natural hazards, earth structure and dynamics, geophysical signal analysis, remote sensing, planetary science, and geophysical methods applied to exploration for natural resources and environmental problems. Recent work includes the study of near-surface neotectonic deformation, crustal stress conditions, and paleoseismic indicators associated with geologically recent earthquake activity in eastern North America. Dr. Alexander participated on the NRC study committee that produced *Bits of Power*. He has also served on five committees, a commission, and a panel at the National Academies. Professor Alexander holds a B.S. (1956) from the University of North Carolina; Letters of Completion in Geophysics (1957) from the Sorbonne, University of Paris; an M.S. (1959) in geophysics from the California Institute of Technology; and a Ph.D. (1963) in geophysics with a minor in mathematics from the California Institute of Technology. He is a Licensed Professional Geologist in Pennsylvania.

JOEY ANUFF is currently the editor-in-chief of *Suck.com*. Prior to resuming his position with *Suck.com*, he supervised most of *Wired Digital's* advertising creative material and wrote Net Surf (which analyzed issues and conflicts related to the Web as a medium and a concept) three times weekly in *HotWired*. Mr. Anuff has published a number of articles on digital media in *Spin*, *Might*, and *Wired* and is a co-founder of *Suck.com*. He is the co-editor of *Suck: Worst-Case Scenarios in Media, Culture, Advertising, and the Internet*, published by Wired Books. Mr. Anuff received his B.A. (1993) from the University of California at Berkeley.

HOWARD BESSER has been an associate professor at the University of California at Los Angeles Department of Information Studies since the fall of 1999. Prior to that time, Dr. Besser was an adjunct associate professor at the University of California, Berkeley School of Information Management and Systems, and a researcher at the Berkeley Multimedia Research Center. Previously Dr. Besser was in charge of long-range information planning for the Canadian Centre for Architecture in Montreal and headed information technology for the University of California, Berkeley Art Museum. Dr. Besser's interests include the technical, social, and economic processes for the networked distribution of multimedia information, protection of digital information, social and cultural impact of the information highway, design of digital documents, digitization of still and moving images for conservation and preservation, and scholarly communication. Dr. Besser serves on the Steering Committee of the Museum Education Site License Project. He served on the Task Force on Preservation of Digital Information of the Commission on Preservation and Access and Research Libraries Group and is the principal investigator of the Economics of Networked Access to Visual Information, sponsored by the Mellon Foundation. He recently taught a graduate course on the protection of digital information. Dr. Besser received his B.A. (1976), M.L.S. (1977), and Ph.D. (1988) from the University of California at Berkeley.

SCOTT BRADNER is a senior technical consultant at Harvard University's Office of the Provost, where he provides technical advice and guidance on issues relating to the Harvard data networks and new technologies. He also manages the Harvard Network Device Test Lab, is a frequent speaker at technical conferences, writes a weekly column for *Network World*, teaches for Interop, and does some independent consulting on the side. Mr. Bradner has been involved in the design, operation, and use of data networks at Harvard University since the early days of the ARPANET. He was involved in the design of the Harvard High-Speed Data Network, the Longwood Medical Area network (LMAnet), and NEARNET. He was founding chair of the technical committees of LMAnet, NEARNET, and CoREN. Mr. Bradner is the co-director of the Transport Area in the Internet Engineering Task Force (IETF), a member of the IESG, and an elected trustee of the Internet Society where he serves as the vice president for standards. He was also co-director of the IETF Internet protocol next generation effort and is co-editor of *IPng: Internet Protocol Next Generation* from Addison-Wesley.

JOAN FEIGENBAUM received a B.A. in mathematics from Harvard University and a Ph.D. in computer science from Stanford University. She is currently the head of the Algorithms and Distributed Data Department of AT&T Labs-Research in Florham Park, New Jersey. Her research interests are in security and cryptology, computational complexity theory, and algorithmic techniques for massive data sets. Her current and recent professional service activities include editor-in-chief of the *Journal of Cryptology*, editorial board member for the *SIAM Journal on Computing*, co-director of the 1997 DIMACS Research and Educational Institute on Cryptology and Security, and program chair for the 1998 IEEE Conference on Computational Complexity.

HENRY GLADNEY has been an IBM research staff member since 1963, with widely varied technical and managerial assignments. He is currently working on digital library storage architecture and technical aspects of intellectual property rights management. He focuses on the traditional library sector, managing collaborations with approximately a dozen university research groups and research libraries. In 1989 Dr. Gladney designed the storage subsystem of IRM, a distributed digital library put to early use in massive paper-replacement applications. He helped make this work a cornerstone of the IBM ImagePlus VisualInfo product in 1992 and of the IBM Digital Library offering in 1996. Dr. Gladney received a B.A. in physics and chemistry from Trinity College at the University of Toronto and M.A. and Ph.D. degrees in chemical physics from Princeton University. As well as 48 publications in chemistry, physics, and computer science, he has four patents and two patents pending. He has been a member of the Association for Computing Machinery since 1979 and a fellow of the American Physical Society since 1973.

KAREN HUNTER is senior vice president of Elsevier Science Inc. With Elsevier since 1976, she has concentrated for several years on strategic planning and the electronic delivery of journal information. She was responsible for TULIP (The University Licensing Program) and for the start-up of ScienceDirect. Before Elsevier, she worked for Baker & Taylor and for Cornell University Libraries. She has a B.A. in history from the College of Wooster and master's degrees in history, library science, and business administration from Cornell, Syracuse, and Columbia Universities, respectively. Recent professional activities include being a member of the Copyright Committee of the Association of American Publishers, the Board of the International DOI Foundation, the Advisory Board of the University of Michigan School of Information, and the RLG/CPA National Task Force on Digital Archiving.

CLIFFORD LYNCH has been the executive director of the Coalition for Networked Information (CNI) since July 1997. CNI, which is sponsored jointly by the Association for Research Libraries, CAUSE, and EDUCOM, includes about 200 member organizations concerned with the use of information technology and networked information to enhance scholarship and intellectual productivity. Prior to joining CNI, Lynch spent 18 years at the University of California Office of the President, the last 10 as director of Library Automation, where he managed the MELVYL information system and the intercampus Internet system for the university. Dr. Lynch, who holds a Ph.D. in computer science from the University of California at Berkeley, is an adjunct professor at Berkeley's School of Information Management and Systems. He is a past president of the American Society for Information Science and a fellow of the American Association for the Advancement of Science.

CHRISTOPHER MURRAY is chairman of the Entertainment & Media Department of O'Melveny & Myers, a multinational law firm based in Los Angeles that has over 700 lawyers. He is a specialist in the legal and business aspects of the production, financing, and distribution of motion pictures, television programs, videogames, and music. His other areas of specialization are theme parks, publishing, entertainment company acquisition, and all forms of Internet-based activities. He practices in the fields of copyright, trademark, and merchandising, counting among his clients Time-Warner, Turner, HBO, Castle Rock Entertainment, Sony, and MGM. He also represents individual performers, producers, and executives. He has acted as legal counsel on a wide range of Internet-related matters (from e-commerce and domain name disputes to strategic Internet ventures between companies in disparate industries, to content licensing, to technical rights clearance issues) for companies including IBM, Microsoft, Warner Brothers Online, the Los Angeles County Museum of Art, and two leading digital production studios and special effects creators, Rhythm and Hues and Digital Domain. He also represents Asian and European companies in connection with their media-related activities, including TeleImage and M6 Television (France), C. Itoh, Japan Broadcasting, Hakuhodo, Hyundai, Tokyo Dome, Marubeni, NHK, Tokyo Broadcasting, Japan Satellite Broadcasting, Nomura Securities, Samsung, and Tokuma Enterprises. Mr. Murray serves as an arbitrator for the American Film Marketing Association and is a member of Digital Coast Roundtable, as well as the planning committees for both the USC and UCLA annual entertainment law symposia. He taught entertainment law at Stanford Law School from 1986 to 1990.

ROGER NOLL is the Morris M. Doyle Centennial Professor of Public Policy in the Department of Economics at Stanford University and a nonresident senior fellow at the Brookings Institution. At Stanford, he is also the director of the Public Policy Program, the director of the Program in Regulatory Policy in the Stanford Institute for Economic Policy Research, and a professor by courtesy in the Graduate School of Business and the Department of Political Science. He was previously the associate dean for Social Sciences in the School of Humanities and Sciences. Prior to coming to Stanford in 1984, Dr. Noll was the chairman of the Division of Humanities and Social Sciences and Institute Professor of Social Science at Caltech. He has also served on the staff of the President's Council of Economic Advisers. Among his honors and awards are the book award of the National Association of Educational Broadcasters, a Guggenheim Fellowship, and the Rhodes Prize for undergraduate teaching at Stanford. He currently serves as a member of the California Council on Science and Technology and of the board of directors of Economists Inc. In the past he served as a member of the President's Commission for a National Agenda for the Eighties, the National Science Foundation Advisory Board, the National Aeronautics and Space Administration Advisory Board, the Solar Energy Research Institute Advisory Board, the Energy Research Advisory Board, and the Secretary of Energy's Advisory Board and as chair of the Los Angeles School Monitoring Committee. He is the author of 11 books and over 200 articles. Dr. Noll's research interests include government regulation of business, public policies regarding research and development, the business of professional sports, applications of the economic theory of politics to the study of legal rules and institutions, and the economic implications of political decision-making processes. His recent books include *The Technology Pork Barrel*, written in collaboration with Linda R. Cohen, an analysis of government subsidies of large-scale commercial R&D projects; *Constitutional Reform in California*, in collaboration with Bruce E. Cain, a study of the role and consequences of constitutional design for secondary governments in a federal system; *Sports, Jobs, and Taxes*, with Andrew Zimbalist, an assessment of the contribution of teams and stadiums to local economic development; *Challenges to Research Universities*, an investigation into the economics of the leading American universities; and *A Communications Cornucopia*, co-edited with Monroe Price, a compendium on communications policy. In addition, he is currently undertaking research on federal programs to promote research joint ventures, the policy consequences of the admission of the western states, the economics of legal rules and institutions, the role of federalism in regulatory policy, and international comparative studies of regulation and infrastructural industries. Professor Noll received his undergraduate degree in mathematics from the California Institute of Technology and his doctorate in economics from Harvard University.

DAVID REED is vice president of the Strategic Assessment Department for Cable Television Laboratories Inc. (CableLabs). In this position he is responsible for leading research and development projects addressing telecommunications technology assessment and business, economic, strategic, and public policy issues of immediate interest to member companies. Before joining CableLabs, he served at the Federal Communications Commission as a telecommunications policy analyst in the Office of Plans and Policy, where he worked on video dial tone, personal communications services, and spectrum auction policies. He has published widely in telecommunications journals, books, and magazines. Dr. Reed earned his Ph.D. and M.S. from Carnegie Mellon University and his B.S. from Colorado State University.

JAMES N. ROSSE retired as president and CEO of Freedom Communications Inc. on September 30, 1999, after serving in that position since April 1992. Freedom owns and operates a nationwide group of daily and weekly newspapers, broadcast television stations, trade and consumer magazines, and interactive media. Previously Dr. Rosse was a professor of economics at Stanford University, where he was the university's provost from 1984 to 1992. Dr. Rosse serves on the Advisory Boards of the Center for Economic Policy Research and the School of Humanities and Sciences at Stanford University.

PAMELA SAMUELSON is a professor at the University of California at Berkeley with a joint appointment in the School of Information Management and Systems and in the School of Law, where she is co-director of the Berkeley Center for Law & Technology. She has written and spoken extensively about the challenges that new information technologies pose for traditional legal regimes, especially for intellectual property law. In 1997 she was named a fellow of the John D. and Catherine T. MacArthur Foundation. She is also a fellow of the Electronic Frontier Foundation. As a contributing editor of the computing professionals' journal *Communications of the ACM*, she writes a regular "Legally Speaking" column. She serves on the LEXIS-NEXIS Electronic Publishing Advisory Board and on the editorial boards of the *Electronic Information Law and Policy Report* and the *Journal of Internet Law*. A 1976 graduate of Yale Law School, she practiced law as an associate with the New York law firm Willkie Farr & Gallagher before turning to more academic pursuits. From 1981 through June 1996 she was a member of the faculty at the University of Pittsburgh Law School.

STUART SHIEBER is Gordon McKay Professor of Computer Science in the Division of Engineering and Applied Sciences at Harvard University. Professor Shieber studies communication: with humans through natural languages, with computers through programming languages, and with both through graphical languages. He received an A.B. in applied mathematics summa cum laude from Harvard College in 1981 and a Ph.D. in computer science from Stanford University in 1989. Between 1981 and 1989, he was a computer scientist at the Artificial Intelligence Center at SRI International and a research fellow at the Center for the Study of Language and Information at Stanford University. He was awarded a Presidential Young Investigator award in 1991 and was named a Presidential Faculty Fellow in 1993. He has been a member of the executive committee of the Association for Computational Linguistics, has served on the editorial boards for the journals for *Computational Linguistics*, the *Journal of Artificial Intelligence Research*, and the *Journal of Heuristics*. He is the founder and organizer of the Computation and Language E-Print Archive and is co-founder of Cartesian Products, a company specializing in advanced document image compression and viewing tools.

BERNARD SORKIN is senior counsel at Time Warner Inc., which participates in a range of entertainment industry sectors, from film to cable television, to publishing. Mr. Sorkin has extensive business and legal experience at Time-Warner, where he has been since 1964. Previously Mr. Sorkin was an attorney with Columbia Pictures. His professional activities include service on many committees of various bar associations and on the advisory committee on copyright registration and deposit of the Library of Congress. After graduating from the City College of New York, Mr. Sorkin completed degrees from Columbia University and the Brooklyn Law School.

GARY E. STRONG has served as the director of the Queens Borough Public Library since September 1994. His career spans more than 30 years as a librarian and library administrator, giving him a unique perspective on the knowledge explosion. He was state librarian of California from 1980 to 1994, deputy director of the Washington State Library, and director of the Everett (Washington) and Lake Oswego (Oregon) Public Libraries. He serves on the New York State Regents Advisory Council on Libraries. He has served as chair of the Subcommittee on Intellectual Property of the American Library Association and is a member of the Committee on Copyright and Other Legal Matters of the International Federation of Library Associations. His degrees are from the University of Idaho (1966) and the University of Michigan (1967), and he was named a distinguished alumnus of the University of Michigan in 1984. He is an author, editor, and lecturer on library and literacy topics. He serves on numerous library and community boards of directors and is involved in negotiating several international library cooperation agreements.

JONATHAN TASINI has been president of the National Writers Union/ UAW Local 1981 since 1990. He was the lead plaintiff in *Tasini et al. v. The New York Times et al.*, the landmark electronic rights case. Mr. Tasini is a prominent advocate among creator groups to preserve the historic balance between copyright protection for individual authors and fair use by researchers and the public at large. For the past 15 years, Mr. Tasini has written about labor and economics for a variety of newspapers and magazines, including the *New York Times Magazine*, the *Washington Post*, and the *Atlantic Monthly*. Prior to his election as president, Mr. Tasini served the union in other capacities, including vice president for organizing. Mr. Tasini has been a six-time resident in writing at Blue Mountain Center, New York, and is a graduate of UCLA.

APPENDIX B

Briefers to the Committee

William Arms, Cornell University
Eileen Collins, National Science Foundation
Les Gasser, University of Illinois, Urbana-Champaign (formerly at the
    National Science Foundation)
Trotter Hardy, College of William and Mary
Shira Perlmutter, U.S. Copyright Office

APRIL 30-MAY 1, 1998

Jim Banister, Warner Brothers Online
Steven Benson, Paramount Digital Entertainment
Chris Cookson, Warner Brothers Motion Pictures
Peter Harter, Netscape
Eileen Kent, Consultant
Jim Kinsella, MSNBC
Bob Lambert, The Walt Disney Company
Jeff Lotspiech, IBM Almaden Research Center
Dean Marks, Time-Warner Records
David Pearce, Microsoft
Suzanne Scotchmer, University of California at Berkeley
Nathan Shedroff, Vivid Studios
Hal Varian, University of California at Berkeley

JULY 9-10, 1998

Scott Bennett, Yale University
Aubrey Bush, National Science Foundation
Dan Duncan, Information Industry Association
Julie Fenster, Time Inc.
Anne Griffith, Software Publishers Association
Carol Henderson, American Library Association
Tom Kalil, The White House, National Economic Council
Deanna Marcum, Council on Library and Information Resources
Steve Metalitz, International Intellectual Property Alliance
Tony Miles, National Science Foundation
Patricia Schroeder, Association of American Publishers
Jim Snyder, AT&T Research
Tony Stonefield, Global Music Outlet
Jim Taylor, Microsoft

APPENDIX C

Networks: How the Internet Works

THE INTERNET COMPARED TO THE TELEPHONE NETWORK

Although much of it runs over facilities provided by the telephone companies, the Internet operates using very different technical and business concepts than the telephone system. The telephone network operates in a connection-oriented mode in which an end-to-end path is established to support each telephone call. The facilities of the telephone network along the path are reserved for the duration of each specific call. An admissions control process checks to see if there are sufficient resources for an additional call at all points along the path between the call initiator and the recipient whenever a new call is attempted. If there are enough resources, the call is initiated. If there are insufficient resources at any point on the path, the call is refused, and the user gets a busy signal.

By contrast, the Internet is not a connection-oriented network. It is a packet-based network built on point-to-point links between special-purpose computers known as routers. In the Internet, all data, including special types of data such as digitized voice sessions, is broken into small chunks called packets. Each packet is normally no larger than 1,500 bytes, so an individual data transmission can consist of many packets. The data packets in the Internet follow the format defined by the Internet Protocol

NOTE: In this appendix (unlike the rest of this report), IP does *not* mean "intellectual property." For a further discussion of the Internet Protocol, see *TCP/IP Illustrated* (Stevens, 1996).

(IP) specifications, the basic transmission protocol for the Internet.[1] All IP packets include IP addresses for the sender and the receiver of the packet. Packets travel through a series of routers as they progress from sender to receiver in IP networks. The destination IP address in each packet is used by the routers to determine what path each packet should take on its way toward the receiver. Because the forwarding decision is made separately for each packet, the individual packets that make up a single data transmission may travel different paths through the network. For this reason, someone monitoring the Internet at an arbitrary point, even a point located between a sender and receiver, might not be able to collect all of the packets that make up a complete message. As monitoring takes place closer to the end user's computer or the source of the transmission, the probability of collecting all of the packets of a given message increases. Thus, monitoring the Internet to steal content or to see what content is being transferred for rights enforcement purposes can be difficult.

There is no equivalent to the telephone system's admissions control process deployed in the current Internet (i.e., there is no busy signal). If the computers attached to the Internet try to send more traffic than the network can deal with, some of the packets are lost at the network congestion points.[2] Transmission Control Protocol (TCP), which is used to carry most of the Internet's data and rides on top of IP, uses lost packets as a feedback mechanism to help determine the ideal rate at which individual data streams can be carried over a network. TCP slows down whenever a packet in a data stream is lost; it then speeds up again until packets start being lost again. If there is too much traffic, all the data transmissions through the congested parts of the network slow down. In the case of voice or video traffic, this produces lower-quality transmissions. Thus, network congestion causes all applications using the path to degrade roughly evenly.[3]

Another difference between the Internet and the telephone networks is in the way one service provider exchanges traffic with another. In the case of the telephone networks, the individual long distance providers do not connect to each other. All long distance telephone networks must connect to each local telephone office in which they want to do business. Interconnections between providers are far more complex in the Internet,

---

[1]Individual computers are identified on Internet Protocol (IP) networks using addresses that are 32 bits long. In the Internet these addresses, known as IP addresses, must be globally unique and can theoretically identify over four billion separate computers. (The actual limit is far less than 4 billion because of the inefficiencies inherent in the processes used to ensure that the assigned addresses are unique.)

[2]These packets are not lost forever but are retransmitted until they are received successfully.

[3]For further discussion of TCP, see Stevens (1996).

which has a rich set of interconnections between all types of providers (see below). A final difference between the way the current Internet and telephone networks are operated involves provider-to-provider payments. In the telephone system, there are traffic-based monetary settlements whenever a telephone call involves two or more providers. In the Internet, provider interconnections fall into two broad categories, peering and customer. Currently, Internet provider-to-provider peering arrangements are settlement free. Individual Internet service providers (ISPs) decide if it is in their interest to peer with each other. Peering decisions and specific arrangements are bilateral in nature, and no general Internet peering policy exists. If a particular Internet service provider cannot work out a peering agreement with another provider, then it must become a customer of that provider or of another provider that does have a peering arrangement with that other provider in order to be able to exchange traffic with that provider's customers. Currently, all inter-ISP agreements are bilateral and follow no set model.

George Gilder (1993) summarized the differences between a telephone network and an Internet-like network as being the difference between a smart and a dumb network. Telephone networks include many computers that provide application support services to the users of the telephone network. This makes them "smart." These support computers are required because the user's access device, a telephone, is very simple and must rely on the network to provide all but the most basic functionality. A by-product of this design is that new applications must be installed within the network itself. This process can take quite a long time because the telephone service provider must first be convinced that the service is worthwhile and then must integrate the support software with the existing server software. A dumb network like the Internet assumes that the user will access the network through smart devices, for example, desktop computers. In this type of network, applications are loaded onto the user's computers rather than into the network. This means that new applications can be deployed whenever users decide they want to install a new application. The network itself is designed to merely transport data from one user to another, although some centralized support services, such as the domain name system, are required in the network. These support services are quite simple and are generally not application specific, so they can support a wide range of old and new applications without modification.

The Internet's architecture also means that there is no control over what applications can be run over the Internet.[4] For example, there is no

---

[4]Controls can, of course, be implemented on end-user computers. For example, central information technology departments in large organizations can install software to prevent users from performing certain functions.

general technical way to prohibit users from downloading an application and running it themselves, even if there were a good reason to do so. In a technical sense, Internet users can run any software that they and their friends (if the software is interactive) want to, even if that software can be used to violate the rights of copyright holders or perform some other illegal or unauthorized function (e.g., breaking into a computer).

The trend in the telecommunications industry is to treat both voice and video as data with somewhat different transmission characteristics. Thus, telephony and, to a lesser extent, video are starting to migrate to the Internet. New features are being added to the Internet to support the more stringent timing requirements that are needed for these applications. To the user, Internet-based telephone services can be indistinguishable from those offered by a traditional telephone company, including being able to provide a "fast busy" signal if a new request would overload a network resource. However, a fast-busy-signal type of telephony service is only one of the options—another service provider could offer a differently priced service where a call could still be placed in times of network congestion but the call would be of a lower quality. This trend is likely to cause additional services to be added to the network infrastructure but does not necessarily mean that the Internet will suddenly become a "smart network." For example, there are two different approaches to Internet-based telephony. In one model Internet connections are added to the existing large telephone switches. A user placing a call would connect to the switch, which in turn would connect to the target telephone. In the other model, Internet-enabled telephones connect to each other directly without going through a central switch. The second model fits the traditional Internet model and thus has no point of control or monitoring.

THE PHYSICAL TOPOLOGY OF THE INTERNET

The Internet started with the ARPANET, a research network established by the U.S. Department of Defense in the late 1960s and early 1970s. The ARPANET was a single backbone network that interconnected a number of university, federal, and industry research centers. Initially very simple, the topology got somewhat more complex by the early 1980s when a number of other data networks were established including CSNET (Computer Science NETwork), BITNET (Because It's There NETwork) and Usenet (a dial-up UNIX network).

With the establishment of the National Science Foundation's NSFNET in the late 1980s and with the demise of the ARPANET in 1990, the Internet topology seemed to simplify, became the NSFNET was being used in the same way that the ARPANET had been—as a national backbone network—but the topology was actually not so simple. Instead of interconnecting individual campuses as the ARPANET did, the NSFNET interconnected regional data networks, which in turn connected the individual sites. In addition the regional networks had their own private interconnections, and commercial public data networks began to appear (e.g., ALTERNET). The commercial networks interconnected with each other and to the regional networks. The resulting topology is quite complex with multiple interconnected backbones interconnecting individual sites and regional networks.

The current Internet is even more complex. Individuals and corporations obtain Internet connectivity from ISPs. Thousands of ISPs exist in the United States, ranging in size from "mom and pop" providers of dial-up service for a few dozen customers each to providers who offer services in all parts of the world and have hundreds of thousands or millions of dial-up and thousands of directly connected customers. ISPs are not confined to delivering service over telephone wires and now include cable TV companies and satellite operators. The smallest ISPs purchase Internet connectivity from larger ISPs. The somewhat larger ISPs purchase connectivity from still larger ISPs, frequently from more than one to establish redundant connections for reliability. The largest ISPs generally peer with each other. But peering is not limited to the largest ISPs. ISPs of all sizes peer with each other. Peering is done at regional exchanges so that regional traffic can be kept local and done at national or international exchanges to ensure that all Internet users can reach all other Internet users. Peering is done at public exchanges, MAE-East in the Washington, D.C., area for example, or increasingly at private peering points between pairs of ISPs. The result is that the current Internet consists of thousands of ISPs with a complex web of connections between them. These connections are both purposeful and predictable, as well as random in nature. Traffic flows between Internet users through a series of ISPs (there can be six or more ISPs along the way) but does not pass through any specific subset of large or backbone ISPs, where someone could monitor the traffic to check for illegal use of content or for applications that could be used to violate copyright.

THE LOGICAL ARCHITECTURE OF THE INTERNET

The logical architecture of the Internet is, in one way, quite simple—it is peer to peer, meaning that in theory any computer on the Internet can connect directly to any other computer on the Internet. In practice situations exist where Internet traffic does not travel end to end. For example, some e-mail systems are set up with a local e-mail repository at each location. A user interacts with his or her local repository to retrieve or send mail. The mail repository can then act on the user's behalf and exchange messages with another e-mail repository or directly with another user's computer. But this configuration is not required and cannot be depended on for monitoring Internet e-mail traffic. Most Internet applications do operate in a peer-to-peer mode.

PRICING AND QUALITY OF SERVICE

The Internet started out being "free" in the sense that it was subsidized by governments around the world. Governmental support for Internet connectivity still exists in some parts of the world, including the United States where subsidies support public schools and libraries going "online." Now, virtually all the cost of the Internet in the United States is covered by the private sector, through the fees that users pay to Internet service providers or through advertising revenues that are used to support "free" online access.[5]

Many fee structures are in place throughout the Internet with each ISP deciding what types of pricing models they wish to support. Most ISPs charge large corporate Internet users on a traffic-based basis, the more traffic they exchange with the Internet, the higher the bill. Many ISPs offer individual dial-up users simple pricing options. These can range from a flat monthly fee to a fee per hours of usage. Currently few ISPs offer traffic-based services to individual dial-up users because of the costs associated with billing based on the level of use and the resulting increase in complexity for the customer.

Although the Internet Protocol was originally designed to support multiple levels of service quality in the network, these features have never been widely used. Internet traffic is delivered in a mode known as "best-effort" where all traffic, regardless of importance, gets equal treatment. However, this practice is in the process of changing. The Internet Engineering Task Force (see Box C.1) has been working on a new set of standards that will enable ISPs and private networks to provide different service qualities for different applications. An Internet telephony application could request a low-latency service, whereas a student surfing the Web might be willing to accept longer response times in exchange for a lower-cost service.

Many policy-related issues must be resolved before ISPs will be able to start offering the new features. Policies covering such issues as user authentication, service authorization, and preemptive authority must be developed and agreed to. In addition, a cost-effective infrastructure must

---

[5]For example, free e-mail is now available through Web browsers from sites such as <http://www.hotmail.com> and many others.

BOX C.1
Internet Standards

For the last 10 years, the Internet Engineering Task Force (IETF) has been the main organization creating standards for the Internet, but the IETF does not deal with everything running over the Internet. Individuals, ad hoc organizations, industry consortia, and the traditional standards organizations are also creating their own standards. The IETF tends to focus on infrastructure standards such as TCP/IP, the Web transport protocol http, and the new differentiated services functions and protocols used to transport voice and video across the Internet. Most of the other standards organizations focus on specific other areas, but some overlap is beginning to develop as the Internet becomes more important and as the pace of convergence accelerates.[1]

---

[1]Additional information about the IETF is available online at <http://www.ietf.org>, including documentation relating to Internet standards.

---

be available to do the accounting and billing that will be required. In spite of these requirements, however, some ISPs are already experimenting with the new quality of service (QoS) functions and expect to be offering QoS-based services soon.

The deployment of QoS-enabling technology into the Internet may significantly change current pricing models. This change will affect both ISP-to-ISP connections and the customers of these ISPs. The new QoS capabilities will include the ability for a user to identify some of his or her traffic as having a higher priority than other traffic. To prevent customers from marking all their traffic "high priority" a different fee will likely be charged for higher-priority traffic.[6]

---

[6]This objective can be accomplished by using a subscription-based model by which the user contracts for a specific amount of high-priority traffic on a monthly basis or by measuring the levels of traffic at each priority and charging for what was used. In any case ISP-to-ISP pricing models will have to be changed to deal with traffic of different priority levels. This type of "class-based" quality of service technology will permit the Internet to support a wide array of new applications without having to deploy specific technology to support individual applications. This is a significant advantage to the Internet service providers and the applications developers, but it means that, even in the area of quality of service, there will be no easy way to tell what applications are being used by individual Internet users.

THE FUTURE OF THE INTERNET

As noted above, the trend in the telecommunications industry is toward convergence, as voice, video, and data are increasingly using the Internet Protocol and thus the Internet. Many of the next generation of communications devices including cellular telephones and fax machines are likely to be Internet enabled. Two factors are influencing this general trend toward convergence: the development of "always-on" Internet services and the development of "Internet on a chip" technologies.

A number of the more recent types of Internet connectivity to the home do not use dial-up modems, which users must purposefully activate when they want to have Internet connectivity. Instead, they are always connected, allowing access to or from the home at any time. This constant connectivity means that devices such as an electric power meter inside the home can be reached by the power company over the Internet whenever the power company would like to read the meter. It also means that systems can be developed that could instantly reach Internet-based servers when a user asked for information. One example used in a recent demonstration was a microwave oven that could retrieve recipes over the Internet at the touch of a button. An additional feature of these new always-on types of connectivity is that they are very high speed and thus capable of enabling widespread deployment of new download-on-demand applications, such as music players that allow the user to select from an almost unlimited menu of selections. The player would then retrieve a file of the music for playing. The advent of constantly available high-speed connectivity will go a long way toward reducing and ultimately eliminating the technological barriers to the easy downloading of digital music and video files.

This always-on capability will be well matched to the Internet-on-a-chip technology for which a number of companies are starting to put Internet Protocol software in integrated circuits. These chips are for use not only in appliances and utility meters but also in alarm systems and small appliances such as air conditioners.

At this writing it seems inevitable that in the future the Internet will become the common communications sinew that will tie our world more tightly together than it has ever been.[7] Although the rate of growth may slow in the United States, because of the relatively high penetration of households with some kind of Internet access, expansion in the worldwide use of the Internet seems likely to continue at a high rate.

---

[7]A CSTB report from the Committee on the Internet in the Evolving Information Infrastructure, currently in preparation, discusses the future of the Internet in detail.

APPENDIX D

Information Economics: A Primer

The economics of information applies to any idea, expression of an idea, or fact that is known by one person and is potentially of value to another. Examples include research that discovers or produces new knowledge about nature or a new product design; creative expression by artists, musicians, poets, and novelists; expository prose that provides a new explication, explanation, or interpretation of facts and ideas that are themselves not new; or "signals" (such as trademarks and logos) that identify a product, company, or organization to others. For the purposes of economic analysis, the narrower and more precise definitions of information that are used in other disciplines—notably computer science and mathematical statistics—do not provide economically meaningful distinctions and, therefore, are not addressed here.

The usefulness of information to others can be either as an instrument to an end or as an end in itself. For example, research may lead to new facts about nature that can be used to make new or improved products, but it may also be valuable in its own right as a kind of consumer good. Likewise, some people may use newspapers and magazines to inform their business and personal decisions, while others may read newspapers and magazines simply because they enjoy doing so. For present purposes, the source of value to others is not a crucial distinction. All that

---

NOTE: Other accessible background works include Noll (1993), Shapiro and Varian (1998), and Besen and Raskind (1991).

matters is that someone other than the creator of the information will value having access to it.

INFORMATION CREATION, DISTRIBUTION, AND CONSUMPTION

Creation

The single most important feature of information products is that the creation of the original information content is what economists call a pure public good: The cost of generating new information is independent of how many people eventually gain access to it. The information creation stage involves all the activities necessary to develop a new information product to the point that it can be distributed to others. For example, in written expression, the creation stage might include conducting necessary research; writing the various drafts of a book, story, or essay; and preparing the final manuscript for publication (whether as printed or electronic text). The costs of carrying out these activities do not depend on how many people will eventually read the product. In publishing, these costs are called "first-copy costs" and refer to all the costs incurred in preparing to print and distribute copies of the original expression.

Distribution

Distribution costs are the costs of delivering the information product to consumers. Historically, this activity entails two main steps: reproducing the physical embodiment of the expression and then delivering it to the consumer. For some popular entertainment, such as music CDs, novels, and magazines, the duplication stage consists of manufacturing a large number of copies of the physical embodiment of the information for distribution by mail or retail stores. With the widespread use of digital networks, however, information products are increasingly being distributed (or otherwise made accessible) electronically. In the case of live theatrical performances or broadcasts, the product need only be produced once but is simultaneously delivered to everyone in the audience. For theatrical performances, the duplication and delivery costs consist of the rental cost of the theater facilities and the salaries of nonperforming theater personnel. For broadcasts, the duplication and delivery costs are the costs associated with distributing the program to television or radio stations around the country and then transmitting the program in each community to its audiences.

Distribution costs vary widely according to the nature of the medium. In many cases, such as printed material and audio and video recordings, distribution costs are more or less proportional to the number of people who receive the product. In other cases, such as broadcasting (whether from television stations, radio stations, or direct broadcast satellites), distribution costs depend only on the size of the geographic area to which the signal is transmitted rather than on the number of people who receive the signal. In still other cases, distribution, although not exactly free, is nearly costless. A prominent example is distribution of documents, music, and pictures over the Internet, where the actual costs of the transmissions are often well below one cent. These costs are so tiny that they are typically not worth monitoring and billing on a per-transmission basis.

Consumption

A consumer incurs costs to use an information product, in addition to the purchase price. Such costs refer to the materials and equipment that a consumer must buy to make use of an information product. Examples include a television set, a radio, an audio system, a computer with modem, and transportation to the theater. Although these costs can be regarded as part of the distribution system, separating them from the distribution costs is useful because the consumer, not the information producer or distributor, decides whether to incur them.

The important feature of these consumption costs is that they are highly variable across different types of information products. For example, a consumer must pay much more to use the Internet to read the newspaper at home than to read the hard-copy version. (In the latter case, the only consumption cost is either storing or disposing of the used copy.) An important feature of electronic distribution of information products is that it reduces distribution costs (and, where relevant, storage costs) but usually increases other consumption costs.

INFORMATION SUPPLY: COMPENSATING INFORMATION CREATORS

Although some people create new information as an act of pure self-expression, perhaps seeking recognition but expecting and receiving no compensation (some Web sites and almost all personal conversations are of this type), most participants in the business of information creation expect compensation. They will not produce new information products unless the amount of compensation is sufficient to justify spending time and resources in this endeavor.

An appropriate conceptualization of the first-copy cost mentioned above is the amount that the creators of new information products must be paid to induce their creative effort. If compensation is required to induce effort to generate new information products, information creators can be paid by:

- Setting the price of the information product above the cost of duplication and distribution so that a surplus above the costs of those activities can be paid to the creator of the information—for instance, an author or inventor can be paid a royalty based on the sales revenue of the information product;
- Using payments from a granting agency to defray the first-copy costs of creating the product—for instance, a government research grant to a university to undertake basic research or to a defense contractor to develop the knowledge for a new weapon; or
- Combining an information product with some ancillary product (frequently another information product) and selling the two in combination—for instance, the news and entertainment content of media products is combined with advertising which, in turn, may pay for all or part of both first-copy and distribution costs.[1]

Each of these approaches is likely to be economically inefficient in important respects.

Payment of Royalties

The first approach—to include first-copy costs in the final price of the information product to cover a royalty for the creator—is economically inefficient because it causes the price of a product to depart from the marginal social cost of producing and distributing it. For example, a recording or a mystery novel typically has built into its price royalties for artistic creation of a few dollars per copy. Consequently, the price is two or three dollars above the social cost of producing and distributing one more copy of the product. This higher price will cause fewer copies of the product to be sold, because some potential customers will be willing to pay one dollar more than the cost of manufacturing and distributing one more copy, but not two or three dollars more. Excluding these potential customers from enjoying the product is economically inefficient. Economically efficient production requires that everyone be allowed to buy the product who is willing to pay at least the social cost of providing his or her copy.

Thus, a fundamental property of a royalty-based system of paying for the creation of new information products is that these products will be

---

[1]In some cases, advertising itself can be a product. For example, some magazines are purchased in part for their advertisements (such as some bridal magazines).

less widely disseminated in society than economic efficiency dictates, and hence the new information leads to less improvement in economic welfare than the information is capable of producing. On the other hand, an advantage of this approach is that it guides producers of information products toward products that consumers or users want and are willing to pay for.

The presence of royalties creates an incentive for the creator of intellectual property (IP) to promote the product, because the information creator can derive an increase in income by increasing sales. Individuals who engage in the promotion of IP must expect an increase in revenue that exceeds the cost of promotion in order to engage in promotional activities. These promotional activities are themselves a form of information product—they provide information to consumers that, presumably, increases the likelihood that they will buy the product.

Use of Grants

The main source of distortion in a grant process arises in deciding which activities to reward and then calculating the appropriate payment. A grant system transfers risks and financial responsibility away from the producers of new information to the granting agency. The effect of this transfer is to attenuate the incentive for the creator to produce the new information that the market—the users of information products—would value most highly unless, of course, that user is the granting agency.

In addition, the essence of new knowledge is that the form it will take cannot be predicted in advance. Consequently, grants are typically based on the principle of cost reimbursement rather than on the value of the output. Cost reimbursement weakens the incentive of the contractor to manage the project efficiently to minimize cost. As a result, cost-plus contracts are typically accompanied by an expensive system for monitoring and auditing performance in an attempt to make project management more efficient.

Grants to support the creation of information products can lead to a second type of distortion. If the granting agency is the government (which it often is), then tax revenues must be raised to finance the grants. Taxes (such as sales, income, and property) are like royalties in that they drive a wedge between the social value of the taxed item and its market price, thereby discouraging the taxed activity and creating a loss of economic efficiency.

Combining Information and Ancillary Products for Sale

Media products are the most common examples of combining information with ancillary products. The principal source of support for some commercial Web news and magazine sites is advertising. In effect, viewers are being attracted with the information or entertainment product and then their attention is being sold to advertisers. Advertising revenue then provides the means to cover first-copy costs.[2] In media products, subscription charges to viewers or readers, when they are imposed, rarely cover more than a modest part of the cost of creating the information product.

Bundling an information product with other revenue-generating activities is in some ways similar to a royalty system. The revenues from the ancillary product must cover both the costs of the ancillary product and the primary information product, which means that the price of the ancillary product must contain a markup over its average cost that is similar to a royalty. In some cases, the connection between the intensity of consumers' preferences and the attributes of these information products can be considerably weakened. In the case of "free" television, for example, programming is designed to be just good enough to induce viewers to watch, thereby delivering the attention of viewers to advertisers. Programming that viewers might find more satisfying but that would not increase the number who are watching is not of any greater value to either broadcasters or advertisers, and so, if it is more costly, will not be produced.[3]

---

[2]Patents can be thought of in similar terms. Often a return to the patent holder is generated not by selling the patent itself but by producing and selling the patented product.

[3]The discussion above neglects the theoretical possibility of a perfectly discriminating monopolist who succeeds in pricing the product to each user in each use at exactly the value placed on the product by the user. From a formal standpoint, perfect price discrimination solves the efficient pricing problem, although at the cost of generating a redistribution of wealth from users to suppliers. It can be argued that licensing schemes for IP (instead of outright sale of IP) offer more opportunities for a price-discriminating monopolist and, therefore, may offer gains in economic efficiency. Experience with software products seems to confirm this possibility, including, apparently, the effects of income redistribution.

In practice, perfect price discrimination is impossible. One reason is that a seller cannot possibly know how much each person is willing to pay for IP. The best that sellers can do is to categorize buyers into groups that generally are willing to pay more versus groups that are generally willing to pay less. To the extent that people within these categories differ in the value they place on the product, the problems discussed in this section will arise within the groups. A second impediment to implementing extensive price discrimination is that sellers often cannot prevent arbitrage—that is, the circumstance in which the buyer who places a low value on the product simply resells it to the buyer who values it highly.

The Lessons

The first important lesson from this discussion is that no system for compensating those who create new information can provide a perfect solution to three central economic problems:

- Adequately compensating those who create new information products;
- Maximally disseminating and using the new information in the economy; and
- Selecting the most valuable information products that will be produced.

Hence, information policy inevitably involves trade-offs between these three objectives. The royalty approach is effective in selecting the information products that users value, but it fails to distribute the products as widely as needed to satisfy economic efficiency criteria. The grants approach is capable of solving the distribution problem efficiently, but it lacks an incentive to ensure that users obtain the information products that generate the greatest net benefit to them. Finally, the ancillary product solution is better than the royalty approach but worse than the grants approach on grounds of economically efficient distribution, but is better than the grants approach and worse than the royalty approach when it comes to selecting the products that users want for production.[4]

THE ROLE OF INTELLECTUAL PROPERTY

Intellectual property is the area of law and policy that determines the solution to the trade-off between fostering incentives to create new information and diffusing its benefits throughout society. Intellectual property law inevitably is a two-edged sword in that it both grants and, simultaneously, limits the rights of the producer of new information in order to articulate how this trade-off will be made.

Patents, copyrights, trademarks, and trade secrets are all types of IP protection that give those holding them some degree of control over the use of information. By creating these rights, the government is conferring the power on the rights holder to refuse to allow others to make use of the

---

[4]This appendix focuses on information products that could be sold plausibly in the marketplace. The reader should not infer that the marketplace is the only or optimal mechanism for stimulating the production of socially desirable and valuable works. Works whose creators or rights holders are unlikely to be able to recapture their costs (e.g., some forms of basic research) will not be pursued in the marketplace and will require funding by government or other institutions.

information under certain circumstances. By granting this power, the government is relying on financial incentives to induce the rights holder to allow dissemination of the information, usually through a financial arrangement that is some form of a royalty system or by means of combining its sale with that of an ancillary product.

Of course, under such a system the producer of the new information rarely will be compensated in a manner that precisely equals the cost of producing the information. However, if the number of potential creators of new information products is large, suppliers of new information will continue to produce it as long as the average payment (combining the successes and failures) equals the information's cost. The rationale for limiting the control inhering in these rights is that doing so creates more benefits, in terms of greater use of the information, than harms, in terms of weakening of the incentive to produce the information.

Two useful examples of such limits are the prohibition against leveraging and the requirement to permit "fair use." The prohibition against leveraging means that a property right in information can be used to monopolize the products emanating from the direct use or duplication of that information but cannot be used to acquire a monopoly in some other product. For example, the creator of a new method for making steel is entitled to monopolize the steel industry for the duration of the inventor's intellectual property right in the new method. However, he or she is not entitled to monopolize the production of automobiles as well, which could be achieved simply by refusing to sell steel to all auto makers.[5]

The principle underpinning this prohibition is that the rewards to the creator of knowledge should be limited to the direct-use value of that knowledge, even though granting broader rights might well induce much more innovative effort in the quest to monopolize the entire economy. Various antitrust actions against Microsoft are illustrations of an attempt to clarify and enforce the boundary between legitimate and illegitimate uses of the market power inhering in intellectual property (in this case, copyrighted operating system software).[6]

---

[5]This doctrine is by no means unanimously endorsed. Some scholars of antitrust and intellectual property law believe that no limits should be placed on the use of a patent or copyright to create as extensive a monopoly as possible, to induce maximal innovative effort. In essence, these analysts are advocates of a different solution to the trade-off between innovation and dissemination than is presently embodied in law and practice.

[6]For example, in a complaint against Microsoft filed by the Antitrust Division of the U.S. Department of Justice, the government argued that Microsoft had used its monopoly power in operating systems for personal computers to disadvantage competitors in applications software, such as Internet browsers, in part to monopolize these markets as well and in part to prevent other browsers and office applications suites from becoming integrated products that would threaten Microsoft's operating system monopoly. Regardless of the merits of

In the second example, the fair-use doctrine has many facets, but one that is instructive is the right to characterize and quote copyrighted publications in other publications that evaluate or extend the first. Thus, an author is entitled to decide who can publish the entire text and to prevent others from incorporating part of the text into another publication without attribution. However, he or she is not entitled to prevent another author from making references to or quoting from the text in a work that evaluates, extends, or corrects the first. In addition to other motivations, this doctrine also has an economic rationale—that the value of criticism and extension arising from fair use exceeds the value of a greater incentive to produce new text that might arise from preventing others from criticizing or correcting it.

Given that society has elected to limit intellectual property rights, the economics of information can shed additional light on exactly how one should approach making the trade-off between the conflicting objectives of generating new information and disseminating existing information as widely as possible. In particular, a crucial component in deciding how to make this trade-off is to quantify the magnitude of the effect of incentives on the creation of new information.

If increasing the strength of intellectual property and, hence, the rewards that come from it generates a large supply response, then the case for strong intellectual property protection is great; however, if the amount of effort put forth in creating new information is not sensitive to the rewards derived from it, the case for strong IP rights is weaker. Unfortunately, the amount of hard evidence about the quantitative significance of IP rights is extremely limited. The reason is that changes in IP rights are infrequent and, typically, are limited to a specific type of information, leading to questions about the generality of the lessons learned from any specific case (see the example discussed in Box D.1).

THE ENFORCEMENT ISSUE

Socially beneficial systems of intellectual property rights necessarily create costs as well as benefits for the simple reason that such rights are costly to enforce. For instance, the holder of a copyright or patent needs to pay the costs of negotiating an agreement with a licensee to exploit the information that the right protects. Then, to protect the value of the right, either the creator or the licensee must make certain that others do not engage in unauthorized use.

---

this complaint, the issue is clear: to what extent should the holder of a valuable intellectual property right that confers substantial, presumably legal, market power (here, Windows) be able to use that IP to gain market power in related markets (here applications software)?

BOX D.1
Pharmaceutical Research

The passage of legislation in 1984 to ease the licensing requirements for manufacturing generic drugs substantially weakened the value of drug patents. Prior to the passage of the act, generic manufacturers had to satisfy the same procedural requirements as the inventor of a new drug to obtain a license from the Food and Drug Administration (FDA). These procedures usually cost tens of millions of dollars and take years to complete. Hence, before the act was passed, the effective life of most drug patents was far longer than the official patent life because of the time and cost of obtaining FDA approval and marketing a generic copy.

The 1984 legislation extended patent life modestly but also greatly increased the number of generic brands of drugs that could be realistically marketed. The net effect is widely believed to have been a reduction in the value of a new drug patent; nevertheless, it is unclear that research and development in the pharmaceutical industry declined significantly.

At the same time, the technological base of the drug industry changed dramatically with the introduction of modern molecular biology as a means of creating new drugs. Thus, it is difficult to disentangle the effects of the de facto shortening of the life of a patent from the effects of the innovations in technology on research and development expenditures in the industry.[1]

---

[1] In the entertainment sector, another confounding effect is the so-called "superstar" phenomenon, or what economists Robert Frank and Philip Cook (1995) have called the "winner-take-all society." At any given time, the number of "superstars" in a domain of pop culture is limited by the simple fact that the industry is hierarchical—one person is the best singer, basketball player, or mystery novelist. To some degree, a change in the protection of intellectual property rights will affect the rewards to superstars. Because entry into the business to become a pop culture icon is determined by the average reward to all entrants, a higher payoff to the top star will induce more entry—which will largely lead to a larger number of failures and to little, if any, increase in the social value of the collective efforts of all entrants. For example, if Michael Jordan earns $100 million per year in salary and endorsement income, many 14-year-olds may decide to try to become basketball players rather than learn their algebra, thereby reducing the total productivity of society a decade later, because regardless of how many make the attempt, a decade hence there will still be only one world's greatest basketball player who can earn $100 million. To the extent that this argument is a valid characterization of the entertainment industry, the supply response arising from stronger IP protection can actually reduce aggregate production efficiency by shifting people away from other, more beneficial pursuits.

Obviously, the value of many IP rights exceeds the private costs of enforcement or else license agreements would not exist. Of course, if these enforcement costs are borne by the creator or licensee, they must be recovered in the final price of the product that uses or embodies the information. Hence, like royalties, these costs drive still another wedge between the price and the social cost of the product and, like royalties, are another source of potential economic inefficiency in the dissemination of the product.

Like all forms of property, some part of the cost of enforcing IP rights is paid by taxpayers through government enforcement. Examples are the costs of running the court system (to adjudicate disputes over rights) and the police system (to investigate the failure to honor enforceable property rights). And, as is the case for all forms of property, enforcement is not entirely public. Just as companies pay for private security, they also pay part of the cost of protecting their IP.

An important point about enforcement is that the holder of a private property right has an incentive to maximize the extent of government enforcement, regardless of the efficiency of public-versus-private enforcement. A public enforcement mechanism, as long as it works, is superior to private enforcement from the perspective of the rights holder because the cost of the former can be spread among all members of society through the tax system.

One example of cost shifting was an attempt in the 1980s to outlaw videocassette recorders (VCRs) because they could be used to create unauthorized copies of motion pictures and television programs. Outlawing VCRs would have created two costs. First, the government would have assumed responsibility to ferret out and capture VCRs, much as it bears responsibility for finding and confiscating illegal drugs. Second, consumers would have been forced to bear the cost of forgoing legal and legitimate uses of VCRs, and the VCR manufacturing industry (including holders of VCR patents) would have been forced to abandon the income generated by a product that sells for hundreds of dollars, to protect a product that could be sold for much less and, in the case of television broadcasting, is given to consumers for free.[7]

The important points here are these:

- Like any other kind of property rights, intellectual property rights can be costly to enforce;
- Enforcement, in and of itself, adds nothing to the social value of IP (although enforcement may be important to induce others to create and distribute IP);
- The cost of enforcement is a balance of public and private costs, seldom entirely one or the other; and
- The balance of enforcement costs between public and private, as well as the overall level of cost, must be considered in designing the legal and social institutions for managing IP.

---

[7]As a result, popular feature films usually are not released to the videotape market until after their first theatrical run. These releases on videotape have a negligible effect on revenues from theaters and can be profitably priced low enough so that pirates have little incentive to engage in extensive unauthorized commercial copying.

Appendix E

Technologies for
Intellectual Property Protection

Chapter 5 deals in general terms with mechanisms that can be used to protect digital intellectual property (IP), providing an overview that describes what is and is not easily accomplished. This appendix is intended for those interested in understanding more of the technical detail of how the available mechanisms work. The focus here is on technologies that are useful in general-purpose computers; techniques used in specialized hardware (e.g., in consumer electronics) are covered in Chapter 5. Encryption, digital signatures, the infrastructure needed to make public-key encryption widely usable, techniques for marking and monitoring digital information, and mechanisms for secure packaging of information are described in this appendix.

PRELIMINARIES

The sophistication and power of some of the mechanisms discussed below are impressive, in some cases providing what are, effectively, unbreakable protections. This power, however, should not blind the reader to some of the simple security principles that, if not followed, can derail any system, whether computational or physical. Modern encryption mechanisms that are well designed are analogous to door locks that are, for all practical purposes, unpickable. Install one on your front door and you feel secure. But the best door lock is rendered useless if you leave a window open, your valuables outside the door, or your key lying around. Each of these mistakes occurs frequently with information security: Open

282

windows are provided by other routes into the computer, files are left in the clear, and encryption keys are too often easily guessed or left accessible. Even when all these mistakes are avoided, other techniques may be used to gain unauthorized access, such as "social engineering" (i.e., tricking someone into surrendering the information, or the password or key). These examples make clear that advanced encryption alone, although providing important tools, is only a part of the story.

ENCRYPTION

Encryption is an underpinning for many computing and communications security services because it provides the only way to transmit information securely when others can eavesdrop on (or corrupt) communication channels. The goal of encryption is to scramble information so that it is not understandable or usable until unscrambled. The technical terms for scrambling and unscrambling are "encrypting" and "decrypting," respectively. Before an object is encrypted it is called "cleartext." Encryption transforms cleartext into "ciphertext," and decryption transforms ciphertext back into cleartext.[1]

Encryption and other closely related mechanisms can be used to help achieve a wide variety of security objectives, including:[2]

- *Privacy and confidentiality;*
- *Data integrity:* ensuring that information has not been altered;
- *Authentication or identification:* corroborating the identity of a person, computer terminal, a credit card, and so on;
- *Message authentication:* corroborating the source of information;
- *Signature:* binding information to an entity;
- *Authorization:* conveying to another entity official sanction to do or be something;
- *Certification:* endorsing information by a trusted entity;
- *Witnessing:* verifying the creation or existence of information;
- *Receipt:* acknowledging that information has been received;
- *Confirmation:* acknowledging that services have been provided;
- *Ownership:* providing an entity with the legal right to use or transfer a resource to others;
- *Anonymity;*
- *Nonrepudiation:* preventing the denial of previous commitments or actions; and

---

[1]These terms are used even when the medium involved is not text. For example, one may refer to a "cleartext image."

[2]Adapted from Menezes et al. (1997), p. 3.

- *Revocation:* retracting certification or authorization.

The two major categories of encryption systems are "symmetric key" and "public key."

Symmetric-Key (One-Key) Systems

A symmetric-key encryption system consists of three procedures: a key generator, an encryption function, and a decryption function. The user first runs the key generator to obtain a key; in a well-designed system, the key will look random to the user, that is, it will be indistinguishable from a key chosen uniformly at random from the set of all possible keys. The user then runs the encryption function, using as input the cleartext object and the key. The result is the ciphertext object. Subsequently, anyone who possesses both the ciphertext object and the key can feed them as input to the decryption function and obtain the cleartext object as output.

There are a variety of ways to implement the encryption and decryption functions shown in Figure E.1. One of the simplest (and oldest) is the "shift cipher." The cleartext is a text message, the key is a number k between 1 and 25, and encryption is accomplished by shifting each letter of the cleartext k places to the right (wrapping around from the end of the alphabet back to the beginning when necessary). Thus, if the key is 10, the cleartext "OneIfByLandTwoIfBySea" is mapped to the ciphertext "YxoSpLiVkxnDgySpLiCok." To decrypt, each letter is shifted k places back to the left (wrapping around the other way when needed). Cryptologic folklore has it that Julius Caesar used the shift cipher with k equal to 3; this special case is often referred to as the "Caesar cipher." The shift cipher is completely unusable in a modern computing environment because the key space (i.e., the set of all possible keys) is far too small: Anyone who obtained a ciphertext could simply try decrypting it with each of the 25 possible keys and would recognize the right key when the result was a meaningful text.

A far more powerful result can be accomplished by implementing encryption using a "one-time pad." One variety of one-time pad uses as a key a text of the same length as the cleartext; the characters of the key are used one by one to indicate how to transform each character of the cleartext. The characters of the key could, for example, be used to indicate how far to shift each character of the cleartext. For instance, if the cleartext is "OneIfBy" and the key is "CallMeIshmael," the first character of the key ("C") would indicate that the "O" should be shifted right by 3 (to become "R"), the "n" shifted right by 1 (to become "o"), and so on. A key is "one-

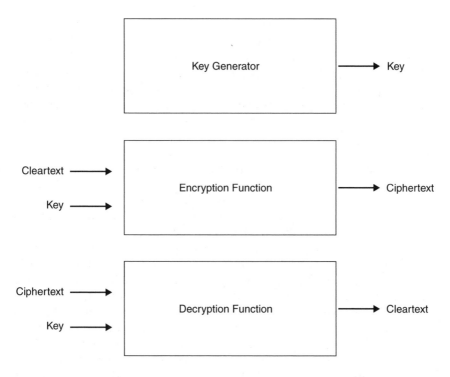

FIGURE E.1   The symmetric-key encryption system.

time" if it is used exactly once; the phrase "one-time pad" comes from the hard-copy pad of such keys used in the intelligence services.[3]

When done by computer, a one-time pad is typically implemented by selecting as a key a random sequence of 0s and 1s that is the same length in bits as the cleartext. To compute the ith bit of the ciphertext, the encryption procedure takes the exclusive-or of the ith bit of the cleartext and the ith bit of the key.[4] Decryption is done in precisely the same way. For example, encrypting the cleartext string "01100010" with the key "11001100" produces the ciphertext "10101110."

---

[3]The most secure keys are truly, uniformly random; hence the example of using readable text as the key is not the best practice.

[4]The exclusive-or is 0 if both bits are 0 or both are 1, and it is 1 if one of the bits is 1 and the other is 0.

The power of a one-time pad arises from both the length of the key and the fact that it is discarded after one use. There are only 25 keys in a Roman alphabet shift cipher, but a one-time pad has 26^N keys for a cleartext of length N, obviously far too many to search in any reasonable amount of time, and if used only once there will be little ciphertext to analyze. Unfortunately, it is usually impractical to use a one-time pad, because its security depends on a new key being generated and used each time a new message is encrypted; this means that the total number of key bits is too large to be practical. The mathematics underlying one-time pads is, however, useful in designing practical cryptosystems; see Luby (1996) and Menezes et al. (1997) for examples.

If a symmetric-key system is well designed, decryption can be done only by having access to the key. That is, it is infeasible for anyone to infer any information about the cleartext from the ciphertext, even if that person has access to the key generation, encryption, and decryption procedures. An adversary can of course always mount an exhaustive search attack to try to find the right key, decrypting using each possible key and testing each output result to see if it is a comprehensible cleartext object (e.g., readable text, viewable image, or sensible sound). If the key space is sufficiently large, an exhaustive search attack will be infeasible. A large key-space comes at the price of longer keys, however, and these make the encryption and decryption processes slower. Thus an encryption system designer must trade off speed of operation against resistance to exhaustive search attacks.

As previously noted, any encryption system must be used very carefully. For a symmetric-key system, in particular, the key generator must be run only in a completely private, reliable computing environment. The person responsible for a cleartext object must also keep the key completely secret, both when using it to encrypt or decrypt and when storing it between uses. The key must be given only to parties with a right to decrypt the ciphertext object, and they must treat the key with the same care. Similarly, everyone who gets the key must keep the cleartext object secure—encryption is useless if the cleartext version is left where it is accessible. Because one user may have many keys, each used in a different application or for a different object in one application, key management can become complex and expensive.[5] The issue of key management is discussed below.

Many commercial IP management strategies plan a central role for symmetric-key encryption systems. Details differ from plan to plan, but the plans all have the following structure in common. Each object is

---

[5]Consider the problem of remembering all your passwords and PINs.

encrypted by the distributor with a key that is used only for it. Ciphertext objects are widely distributed. An object's key is given only to paying customers and other legitimate users; this activity occurs over a different, more secure, but likely less efficient, distribution channel from the one used for the object itself. The product or service that allows paying customers to decrypt and use the object must take responsibility for handling the key and the cleartext carefully.

Examples of symmetric-key encryption systems, including the widely used Data Encryption Standard, can be found in Menezes et al. (1997), which is also a good starting point for literature on the mathematical and engineering foundations of the design and analysis of cryptosystems. Mathematically sophisticated readers can also refer to Luby (1996).

Key Exchange

Anyone using a symmetric-key encryption system must deal with the key exchange problem: If one or more recipients are to be able to decrypt a message, they must get the key, and they must be the only ones to get it. Two distinguishable problems here are evident: authentication and secrecy. The process must ensure that the person legitimately entitled to receive the key is who he or she claims to be (authentication) and that no one else can get the key while it is being transmitted (secrecy). Key exchange is thus a high-overhead operation.

An interesting circularity exists here: If we can ensure authentication and secrecy in transmitting the key, why not use that machinery to send the original message? One answer was suggested above: The key is often far smaller than the thing being encrypted, so the key distribution mechanism can use a more elaborate, more secure, and slower transmission route.

In the context of mass market IP, key exchange can be a large-scale problem: Online distribution of a best-selling novel may mean a few hundred thousand customers, that is, a few hundred thousand key exchanges. Although the encrypted text of the novel may be distributed via high-bandwidth broadcast channels (neither authentication nor secrecy is required), key transmission must be done in a way that ensures authentication and secrecy.

Public-Key (Two-Key) Systems

One way around this problem is the notion of a public-key encryption system, which eliminates the need for key exchange. As in the symmetric-key case, the system consists of three procedures: a key generator, an encryption function, and a decryption function. Here, when a user runs the key generator, he or she produces two keys—a public key and a secret key. The public key should then be distributed and made widely available (perhaps by a directory service), but the secret key must be carefully guarded and kept private (just as in a symmetric-key system).

To send an object to the owner of a public-key/secret-key pair, the sender looks up the user's public key and feeds both the public key and the cleartext object into the encryption function, producing a ciphertext object. When the recipient receives it, he or she feeds the ciphertext object and his or her secret key into the decryption function, which recovers the cleartext object. (See Figure E.2.)

To be successful, a public-key cryptosystem must have the property that each public key corresponds to a unique secret key and vice versa—a ciphertext object produced with a given public key must be decryptable only by the owner of the corresponding secret key. However, determining what the secret key is if all one has is the public key must be infeasible. Logically, a secret key is determined by its corresponding public key, but the time required to compute this uniquely determined quantity should

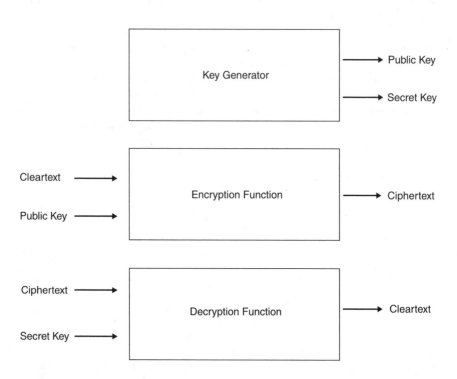

FIGURE E.2   The public-key encryption system.

be longer than an adversary could possibly spend (and is believed to be far longer than a human lifetime for the public-key cryptosystems that are now in use, if an appropriately large key is used).

At first glance, it is not obvious that one could devise a key-generation procedure that is itself efficient but still manages to embed the information needed to determine the secret key into the public key in such a way that it cannot be extracted efficiently. The concept of public-key cryptography was first put forth publicly by Diffie and Hellman (1976) and Merkle (1978) and has given rise to some of the most interesting work in the theory of computation. Interested readers should refer to an introductory book such as the *Handbook of Applied Cryptography* (Menezes et al., 1997) for an overview of cryptographic theory; this handbook also provides examples of public-key cryptosystems that are now in use, including the well-known RSA system (named for its inventors Rivest, Shamir, and Adelman) (Rivest et al., 1978).

Current public-key cryptosystems are considerably slower than current symmetric-key systems, and so they are not used for "bulk encryption," that is, encrypting long documents. In the IP management context, this means that vendors generally do not use public-key systems to encrypt content directly; rather, it is more common to encrypt content using a fast symmetric-key system, then use public-key encryption to solve the key exchange problem. Because the user's key in a symmetric-key encryption system is typically much shorter than the object, the time spent to encrypt and decrypt it using a public-key system is not prohibitive.

One potential obstacle to widespread use of public-key cryptography for IP management (or for any mass market product or service) is the current lack of infrastructure. Public-key cryptosystems were first proposed in the mid-1970s, but only now are developers producing the systems needed for creation, distribution, retrieval, and updating of public keys. Considerable disagreement still exists in the technical community about how to create an effective public-key infrastructure (see below and Feigenbaum, 1998). Other potential obstacles to widespread use of public keys are the same ones that make symmetric-key systems hard to deploy effectively in a mass market service: A user's secret keys must be managed extremely carefully, as must all cleartext objects; otherwise, the property that was protected during transmission can be stolen once it reaches its destination. Furthermore, public-key systems are, like symmetric-key systems, subject to U.S. government export restrictions.

DIGITAL SIGNATURES

Another use of public-key technology, one that is potentially more important for IP management and for electronic commerce in general than public-key cryptosystems, is digital signature. A digital signature scheme involves three procedures: a key generator (with the same structure as the key generator in a public-key cryptosystem), a signing function, and a verification function (Figure E.3). A user who has generated a key pair can feed his secret key and a digital object as input to the signing function, which produces "a signature" (a set of bits) as output. The crucial property of the signature is that it could have been produced only by someone with access to both the digital object and the secret key.

Subsequently, anyone presented with the object and the signature can look up the signer's public key and feed the object, the signature, and the public key into the verification function. The verification function can use this public key to determine whether the signature was produced by the signing function from the object and the secret key that corresponds to the public key.

As before, the success of the scheme depends on the ability to generate public-key/secret-key pairs that cause the signing and verification procedures to work properly and that have the counterintuitive property

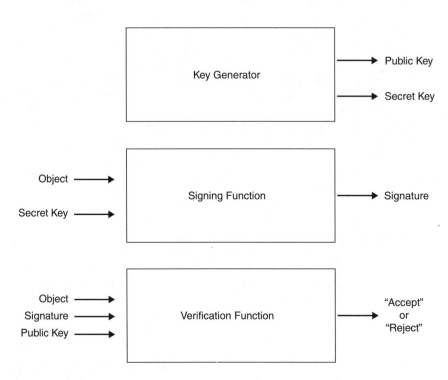

FIGURE E.3   The system for digital signatures.

that the secret key, although uniquely determined by its corresponding public key, cannot feasibly be computed from this public key.

Note that digital signatures can serve as analogues to paper signatures but are different in interesting ways. In the paper world a person's signature depends only on the person doing the signing. In the digital world, the signature is a function of both the person (represented by his secret key) and the document. Each document signed by a given secret key will have a different signature (i.e., a different set of bits). This is necessary, given the digital nature of the documents being signed: If a person's digital signature were the same for each document, as soon as he or she signed one document, the signature could easily be forged. The bits could be removed from the signed document and attached to a different document. (Note that this is not generally feasible with paper documents: Signatures cannot be torn off (or otherwise copied) and pasted elsewhere without disturbing the visual and tactile qualities of the document.) Because digital signatures depend on both the secret key and the document being signed, no one can claim that a signature produced for one document is the signature for another document.

Digital signatures have the potential to play an important role in IP management (as well as electronic commerce more generally), one that may be more important than public-key encryption. As noted earlier, public-key encryption is currently too slow to use to encrypt significant amounts of content, hence its main role in the foreseeable future is likely to be key distribution. Digital signatures, on the other hand, provide assured provenance (only the person in possession of the secret key could have created the signature) and nonrepudiation (the object must have been signed by the possessor of the key, because the signature could not have been created any other way). Knowing the origin of a digital object can be extremely valuable, for example, to ensure that a program you have downloaded comes from a reliable source.

Note that we have phrased this carefully, saying "the person in possession of the secret key" to emphasize that, as with any form of encryption, all the guarantees depend on security of the secret key. A program can determine whether a particular secret key was used to create a signature ("this was signed with Joe's secret key"), but the connection between that secret key and a particular person ("only Joe could have signed this") is a separate issue, one the owner of the key must ensure.

MANAGEMENT OF ENCRYPTION KEYS: PUBLIC-KEY INFRASTRUCTURE

Any product or service that uses public-key cryptography or digital signatures must have a reliable way to determine that it has the right public key each time it decrypts a message or verifies a signature. For example, a digital library may insist that each article it offers be digitally signed by the author so that readers can verify the article's authenticity. This approach requires that the readers be able to trust that the library has correctly associated the authors' names with their public keys. What is the foundation for that trust, given that the library may deal with tens of thousands of authors, most of whom it has only "met" in cyberspace? This and many similar questions fall under the general heading of public-key infrastructure challenges.

The basic building blocks of the simplest approach to public-key infra-structure are "identity certificates" (ID certificates) and "certifying authorities" (CAs), analogous in some ways to the physical identity cards (e.g., passports) issued by authorities such as national governments. In the digital world, an ID certificate is a signed data record containing a public key and the name of its owner (and perhaps some related data items such as an expiration date). An ID certificate is issued by a CA, which signs it with its signature key SK_{CA}. If a user (or a service like a digital library) has a trustworthy copy of the CA's public key PK_{CA}, it can build on its trust in PK_{CA} to develop trust in the ID certificates (and their keys) that the CA has signed.

For example, someone receives a document signed by Alice, together with an ID certificate associating Alice with her public key PK_A. The recipient uses the CA's public key PK_{CA} to verify CA's signature on the ID certificate and, if this succeeds, the recipient can then be confident in using the key in the certificate, PK_A, to verify Alice's signature on the document. The user's trust in the CA consists of trust that his or her copy of PK_{CA} has not been tampered with and trust that the CA has associated the correct public key with the name Alice.

The public-key infrastructure task also requires that there be a means of dealing with keys and certificates that have expired or been revoked. Basic security principles require that both keys and certificates have expi-ration dates (no password should be good forever) and that both are revocable to deal with secret keys that have been compromised.

One standard version of public-key infrastructure calls for official CAs to issue ID certificates for other official CAs, thus forming "hierar-chies of trust." (Having numerous CAs prevents the entire system from being dependent on a single source.) In another approach, anyone can act as a CA (or "introducer") by signing name-key pairs in which he or she has confidence, and "webs of trust" emerge spontaneously as users de-cide whom to go to for introductions (Zimmerman, 1994). More recently, the research community has developed public-key infrastructures in which certificates do more than just bind public keys to names. See, for example, Ellison (1999) for a discussion of "authorization certificates" (in which the public key is bound to a capability or privilege to perform a certain action) and Blaze et al. (1996) or Blaze et al. (1998) for a discussion of fully programmable certificates. Direct authorization of verification keys can be a powerful tool in IP management and in e-business generally, enabling individual customers to do business anonymously (as they often can in the physical world) and enabling businesses to derive much more information from digital signature verification than simply the name of the signer. For an in-depth discussion of this work on expressive public-key infrastructure, see Feigenbaum (1998) and the references cited there.

MANAGEMENT OF ENCRYPTION KEYS: USER AND MIDDLEWARE ISSUES

The security of any encryption scheme depends on keys being difficult to guess (i.e., they should be long and effectively gibberish). The analogous lesson about passwords is not widely heeded: Computer systems today are routinely broken into by "dictionary attacks," routines that simply try as a password all the words in a dictionary, along with common names, birthdays, and so on. The more sensitive the resource to be protected, the more difficult its key must be to guess. The problem is that a good key is basically impossible for any human being to remember.

As a consequence, keys are themselves stored and managed by computing systems and passed across the network; they are in turn encrypted under key-encrypting keys. A protection infrastructure becomes necessary. Because digital systems can be extremely complex, the engineering and management disciplines needed to accomplish such infrastructures have become areas for specialists. A great deal has been written about both the engineering side of system protection (see, for example, Gray and Reuter, 1992) and about practical service operations (see, for example, Atkins et al., 1996).

Because key management can be extremely cumbersome, IP delivery middleware and the end-user applications to which IP is delivered go to great lengths to keep end users unaware of it. When a user clicks on the "play," "view," or "print" commands for an encrypted object, the appropriate decryption keys should be retrieved automatically by the application, which should then decrypt the content and present it as appropriate. The user should not have to do anything explicit to find the right key or apply it and the decryption tool. Similarly, when the user is finished inspecting the content, he or she should not have to do more than select the "close" operation; as a side effect, the system should ensure continued existence of the ciphertext version and (appropriately protected) decryption key(s), but delete all traces of the cleartext.

Unfortunately, the efforts to achieve such ease of use of encryption tend to work against the effectiveness of encryption in safeguarding content, at least in applications that execute in ordinary PCs. No matter how hard the application developers work to find a place on the PC disk safe enough to store cryptographic keys, many PC experts will be sufficiently knowledgeable and skilled to find them. The problem can be solved to some degree by having users store their keys on removable media (e.g., smart cards), but such cards would impose significant burdens on users and distributors. This challenge illustrates that analysis of trade-offs between ease of use and strength of protection is an important and difficult part of IP management systems development.

In addition to posing some technical challenges to the developer of a product or service, the use of encryption also poses legal and political challenges. Encryption systems and products that use them are subject to U.S. government export restrictions, restrictions that are themselves subject to change, making product development and business planning difficult.[6] Although there are currently no restrictions on domestic use of encryption, the U.S. copyright industries clearly would not want to use one set of products and services (with strong encryption capabilities) for the domestic market and another (with weaker, exportable encryption capabilities) for the overseas market. Not only would this arrangement mean higher development and maintenance costs, it is also likely backwards: Serious commercial piracy is more of a problem outside the United States, calling for stronger encryption outside national boundaries. The encryption export issue will thus have to be dealt with if encryption is to play a prominent role in IP management.

Like public-key cryptosystems, digital signature schemes are currently cumbersome to use because of the lack of infrastructure for managing public keys. There is good reason to believe that an appropriate infrastructure will emerge soon, however, because of the enabling role that digital signatures could play in electronic commerce. Unlike encryption, digital signature technology is not encumbered by export restrictions. The freely exportable U.S. government Digital Signature Standard (DSS) uses public-key technology, but it is not a public-key cryptosystem.[7]

---

[6]See *Cryptography's Role in Securing the Information Society* (CSTB, 1996) for a discussion of the market and public policy aspects concerning encryption and U.S. government export restrictions.

[7]At least, there is no obvious way to use it to encrypt things. Technically sophisticated readers should refer to Rivest (1998) for a provocative discussion of the possibility that the claim that a signature scheme "cannot be used for encryption" might be intrinsically difficult to prove—and hence that a crucial distinction made by U.S. export policy may be ill-defined.

DSS and many other signature schemes are covered in basic cryptology textbooks, including Menezes et al. (1997).

MARKING AND MONITORING

A "watermark" is a signal added to digital data (typically audio, video, or still images) that can be detected or extracted later to make an assertion about the data. The watermark signal can serve various purposes, including:[8]

- *Ownership assertion*: To establish ownership over some content (such as an image), Alice can use a private key to generate a watermark and embed it into the original image. She then makes the watermarked image publicly available. Later, when Bob claims he owns an image derived from this public image, Alice can produce the unmarked original and demonstrate the presence of her watermark in Bob's image. Because Alice's original image is unavailable to Bob, he cannot do the same. For such a scheme to work, the watermark has to survive common image-processing operations (e.g., filtering or cropping). It also must be a function of the original image to avoid counterfeiting attacks.
- *Fingerprinting*: To avoid unauthorized duplication and distribution of publicly available content, an author can embed a distinct watermark (or fingerprint) into each copy of the data. If an unauthorized copy is found later, the authorized copy from which it was made can be determined by retrieving the fingerprint. In this application, the watermark should be invisible and invulnerable to attempts at forgery, removal, or invalidation.
- *Authentication and integrity verification*: Although authentication can be done through cryptographic techniques, the advantage of using a verification watermark is that the authenticator is inseparably bound to the content, simplifying the logistical problem of data handling. When the watermarked data is checked, the watermark is extracted using a unique key associated with the source, and the integrity of the data is verified through the integrity of the extracted watermark.
- *Content labeling*: The watermark embedded into the data contains further information about the contents. For example, a photographic image could be annotated to describe the time and place the photograph was taken, as well as identification of and contact information for the photographer.
- *Usage control*: In a closed system in which the multimedia content needs special hardware for copying and viewing, a digital watermark can

---

[8]Adapted from Memon and Wong (1998).

be inserted to indicate the number of copies permitted. Every time a copy is made, the watermark can be modified by the hardware, and at some point the hardware would not create any more copies of the data. An example is the digital video disc.

• *Content protection*: In certain applications, a content owner may want to publicly and freely provide a preview of the multimedia content being sold. To make the preview commercially worthless, the content could be stamped with a visible watermark that is very difficult to remove.

Currently, no universal watermarking technique exists that satisfies all requirements of all applications; instead the specific requirements of each watermarking application depend on the protection objectives, the kind of object and its digital size, and possibly on the kind of distribution channel. Below, some watermarking techniques are described, using images as a motivating example. Keep in mind that many of these techniques are applicable to other forms of content, including video and audio.

Watermarking

Digital watermarks are embedded in digital objects (images) so that owners and perhaps end users can detect illegitimate copying or alteration. Digital watermarks can be made either "perceptible" (by people) or "imperceptible." A "fragile" watermark is damaged by image distortions and thus serves to detect alterations made after the watermark is applied. A "robust" watermark survives distortions such as trimming away most of the image and thus can serve as evidence of provenance.[9] Both kinds can be embedded in most varieties of digital object. Watermarks are currently of most interest for images, audio signals, and video signals.

A watermark is a digital signal, added to or removed from the original object, that does not interfere unduly with the intended use of the altered object and yet carries a small amount of information. "Invisible watermarks" are imperceptible to people but can be detected by appropriate software. A technical protection service (TPS) that uses watermarking can provide a content distributor with a way to mark content before distribution and track what happens to it subsequently.[10] Users of

---

[9]The embedded marks should be short so that they can be repeated many times throughout the work. Long marks are not robust because small changes to the work can cause the mark to be lost or damaged. Embedded marks survive normal file transfer and copy operations, but, as with other IP protections, watermarks can sometimes be removed by a determined, knowledgeable user.

[10]Note that many watermark applications require a network infrastructure and possibly also a rights management system to accomplish the monitoring function.

works may also benefit, because successful watermark detection can demonstrate the source of the content and that it has not been altered subsequently.

Even when designed to be subtle, watermarks do modify the content that carries them and therefore may be unacceptable for applications that need very high fidelity content. Conversely, that same modification can enable some business models: Some companies allow free distribution of lower-resolution works with watermarks, using those as samples of their work, while charging for higher-resolution works without watermarks.

A fragile watermarking scheme has two procedures, one for watermark insertion and one for watermark extraction. The input to the insertion procedure consists of the unmarked object, the watermark, and a key associated with the creator of the object (or another authorized party in the distribution chain); the output is a watermarked object. The input to the extraction procedure consists of the watermarked object and the key used during insertion. If the object has not been altered since it was marked and the correct key is used, the output of the extraction procedure is the watermark; if the object has been altered or the wrong key is used, the extraction procedure outputs an error message. Some fragile watermarking schemes can identify the unauthorized alteration; others detect only that alteration has occurred.

In a robust watermarking scheme, it is assumed that the marked object may be altered in the course of its normal use. For example, robustly watermarked images may undergo compression and decompression, filtering, scaling, and cropping. The inputs to and outputs from a robust watermarking insertion procedure are the same as in the case of fragile watermarking. The inputs to the detection procedure are the watermarked object (which may have been legitimately altered in the course of normal use), the watermark that was inserted into the object, and the key. The detection procedure then indicates whether the object contains a mark that is "close to" the original watermark. The meaning of "close" depends on the type of alterations that a marked object might undergo in the course of normal use.

In a fingerprinting scheme, there is an additional input to the insertion procedure that depends on the recipient of the specific copy. The output is a marked object in which the mark (the fingerprint) identifies the recipient. Two different customers purchasing the same work would receive objects that appeared the same to human perception but contained different watermarks. If unauthorized copies were later found, the fingerprint could be extracted from those copies, indicating whose copy had been replicated.

Detection as described above does not require the original, unmarked object. Watermarking schemes whose detection portions do not require the unmarked, or "reference," object are called "oblivious" or "public." Schemes that do require reference objects for detection (called "private" schemes) are less powerful but may be the best that are attainable for certain types of objects and certain applications. Robust schemes may be further classified in many ways that are beyond the scope of this report; interested readers should refer to Memon and Wong (1998) or, for a scientifically rigorous treatment, Matheson et al. (1998). (See Figure E.4.)

Watermarked works posted on the Internet can be tracked through the use of "spiders" that search the Web. For example, Digimarc Corporation's MarcSpider service scans the Web and provides online reports of where and when marked images (or their copies) are found. [11] To facilitate copyright compliance by purchasers and licensees, a service for users provides access to up-to-date information about the copyright status of a work. This service is useful in situations in which ownership and terms are dynamic; indeed, it is often the difficulty of obtaining up-to-date terms (rather than the expense of license fees) that causes people to violate licenses.

Watermarking cannot force people to refrain from copying or distributing digitally marked works. Rather, TPSs that use marking and tracking attempt to dissuade violations by making them detectable and traceable to the culprits. Typically, it is only rights holder ownership information that is embedded in marked works, but licensing terms or information about users can also be recorded and carried with the object. As noted above, fingerprinting by adding personal information to the object can add force to the disincentive for infringement, but because it can compromise user privacy and can be done without the knowledge of the licensee, it could also alienate potential customers. Cost-benefit analysis of the trade-off between protecting the vendor's ownership rights and protecting the customer's privacy must be done in a case-by-case manner.

Note that watermarking is only a defense against copying on a large enough scale (e.g., on the scale that would occur in financially viable commercial piracy) to cause the illegitimate copies to be discovered by web crawlers. Realistically, a watermarking scheme will not help an owner detect someone who uses his home PC to make a single copy for private use, because that copy will almost certainly never find its way to the watermark detection process. Digital watermarking is in use today by corporations in the photography, publishing, entertainment, sports, and news industries.

---

[11]Information about Digimarc Corporation and its products can be found online at <http://www.digimarc.com>.

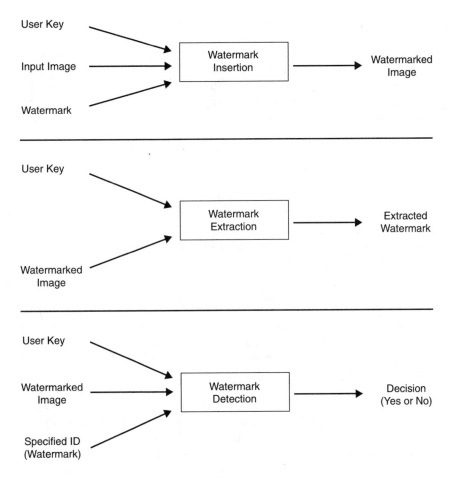

FIGURE E.4 Watermarking procedures. Watermark insertion integrates the input image and a watermark to form the output watermarked image. Watermark extraction uncovers the watermark in watermarked images, a technique usually applicable in verification watermarks. For robust watermarks, the presence of a specified ID (watermark) can be detected using a predefined threshold; a yes or no answer indicates the presence of the ID, depending on whether the output from a signal detection block exceeds the given threshold.

Time Stamps and Labels

The purpose of time stamping in a technical protection system is to fix certain properties of a work (e.g., a description of the content, or the identity of the copyright holder) at a particular point in time. TPSs that use time stamping facilitate copyright protection by affixing an authoritative, cryptographically strong time stamp on digital content that can be used to demonstrate the state of the content at a given time. A third-party time-stamping service may be involved to provide, among other things, a trusted source for the time used in the time stamp. Time-stamping technology is not widely deployed as of 1999.[12]

Digital labels can serve some of the same IP management purposes as digital watermarks. An important difference is that no attempt is made to hide a digital label; it is intended to communicate the terms of use of a digital work and should be noticeable by readers or viewers. In its simplest form, a digital label could take the form of a logo, trademark, or warning label (e.g., "may be reproduced for noncommercial purposes only"). Alternatively, more sophisticated labeling strategies may be used. Like watermarks, labels cannot force compliance with copyright law and licensing terms. Labels may be added to a digital object using digital attachments to files or HTML tags (and they are easily deleted by users who wish to delete them).

Web Monitoring

Web monitoring systematically and comprehensively analyzes thousands of Web pages to find targeted works. Unlike simple Web searching or Web crawling, monitoring tends to be ongoing and in depth and includes analysis and attempts to interpret findings in context.

Effective Web monitoring can detect improper uses of logos or trademarks, as well as piracy and other copyright violations and, more generally, can keep track of how targeted works are being used (including the legal uses). This type of monitoring (and indeed even simpler searching and crawling) is considered by some to be a violation of user privacy; for this reason, some people and organizations take pains to keep monitoring software off their Web sites (e.g., by configuring their firewalls to keep monitors out) or even to feed inaccurate and misleading information to the monitors. Although Web monitoring increases the likelihood of detection of copyright violators, monitoring cannot directly stop violations from happening—its effectiveness lies in the threat of legal action against violators.[13]

---

[12]Although some products do exist, including some by WebArmor at <http://www.webarmor.com> and Surety's Digital Notary Service at <http://www.surety.com>.

[13]As one example, Online Monitoring Services provides these capabilities through its services WebSentry and MarketIntelligence.

CRYPTOGRAPHIC ENVELOPES

A high level of protection of valuable content in the face of determined adversaries must involve special-purpose hardware, because the content of any digital work must eventually be displayed or somehow made accessible to the user for its value to be realized. If all of the processing necessary for display takes place in an ordinary PC, the bits that are displayed can be captured and copied by anyone with sufficient knowledge of PC hardware and software. If, on the other hand, transmission, processing, and display involve special-purpose hardware (e.g., as in traditional cable television systems), capturing the content is much more difficult. Indeed, some visions of the future suggest widespread use of such special-purpose hardware, sometimes termed "information appliances" (see, for example, Norman, 1998).

Special-purpose hardware also reduces the opportunity to modify the device: Security circumventions that require hardware modification are much less likely to be widespread than those that can be accomplished with software. Software modifications can be made easily installable (witness all the programs and upgrades that can be downloaded and installed by relatively unsophisticated users), while tinkering with hardware is typically more difficult and less likely to be widely practiced. Software modifications can, in addition, be widely distributed by the Internet.

Special-purpose hardware also has its costs. It imposes a burden on the content distributor to manufacture and market (or give away) the hardware, and a burden on the user to obtain, learn about, and maintain it.

The desire to avoid this burden has motivated the search for software-only, end-to-end systems to control digital content. A number of schemes are being actively explored and deployed, including efforts by IBM, based in part on its Cryptolopes®; by Xerox, using its Self Protecting Documents®; and by InterTrust, using its Digiboxes®.[14] Several elements are common in these efforts. Each uses some variety of a secured digital container (a cryptographically protected file) holding the content, the vendor's rules for access and use (described in a rights management language), and possibly watermarking or fingerprinting information. Encrypting the information both prevents misuse and ensures authenticity, provenance, and completeness. Not all of the content is necessarily

---

[14]Information on the IBM effort can be found at <http://www.software.ibm.com/security/cryptolope/>, the Xerox effort at <http://www.contentguard.com>, and the InterTrust effort at <http://www.intertrust.com>. As noted in Chapter 2 on music, a number of Internet music-delivery services, including AT&T's a2b system and Liquid Audio, use encryption, rights management, and several of the other techniques discussed here.

encrypted: Encrypted documents may be accompanied by unencrypted abstracts so that the document can be previewed by anyone (a form of advertising). Most of these schemes call for a rights management server that performs tasks such as authorizing requests to use content and tracking use for billing. A typical interaction has the customer purchasing content (e.g., a consulting company report) via the Web, receiving a secured digital container with the content, an indication of the rights they have purchased, and possibly additional information that marks the content as having been distributed to this particular customer.

A customer's use of the information must be cleared by software that checks the rights that have been packaged with the content and checks the identity of the local machine. One approach to checking the rights calls for the local machine to issue a request to the rights management server, which must provide clearance before access can occur and at the same time can record billing information.

One of the challenges of these systems is, as noted above, to provide access to content without losing control of it, a task that is not easily accomplished on a general-purpose PC. One approach to this, used by Xerox, is to encapsulate the content as a runnable (Java) program and have that program perform much (or all) of the rendering (for example, display and printing). Having the specialized software do the rendering rather than the underlying operating system can substantially reduce the opportunity to capture the content, making piracy that much more difficult. Rendering is also done in stages, further reducing the chances for capturing content. Although this is not guaranteed to prevent piracy, it does substantially raise the level of technical skill required by a pirate and may suffice for commerce in a wide variety of digital information.

These secure containers are still a relatively new technology and have yet to be used widely. The level of skill and determination that will go into attacking them will depend on the size and structure of the market for the content they are used to protect; whether they will withstand determined attacks is an open question. For a general discussion of the difficulties of building tamper-resistant software, see Aucsmith (1996).

SUPERDISTRIBUTION

Packaging information in secure containers also enables a concept called "superdistribution," that is, the ability for others to repackage and redistribute content, profiting from their repackaging while respecting the rights of the owners of the original content.[15] As one simple example,

---

[15]The term "superdistribution" has apparently been used historically in a number of ways. The usage here is consistent with current discussions of e-commerce in information

a customer who buys and reads a wide range of consulting company reports on a subject may determine that six of them are particularly useful and that packaging those six together provides added value (e.g., saving someone else the time and effort of finding them). The customer packages those six reports in his own secure digital container, with his own set of rules (e.g., prices) for access. Importantly, those rules are over and above the rules specified by each of the individual reports, which remain "enclosed" in their own (sub)containers. Someone who buys the collection must obtain (i.e., pay for) all necessary rights, including the rights to the collection, and the rights to any of the individual reports. Superdistribution thus enables a chain of value-adding activities, while respecting the rights and restrictions imposed by all the content owners.

products. Earlier uses of the term referred to the distribution of software that metered its use ("meterware"). See, for example, Cox (1994), which describes the approach of Ryoichi Mori of the Japan Electronics Industry Development Association. This work suggested adding a special (tamper-resistant) processor to computers and special instructions in the software to track and bill for use (a system much like the one envisioned for videos with Divx). The interesting suggestion is that where software is currently sold by the copy (and digital copies are difficult to track or control), copies should instead be given away, and only usage should be billed (based in part on the claim that use is easier to track than copying). Mori's work also suggested the possibility of layers of such pay-per-use as different programs call on one another, an idea similar to the notion of superdistribution used above.

APPENDIX F

Copyright Education

WHY IS COPYRIGHT EDUCATION NEEDED?

Information has come to play an increasingly important role in our lives, both at work and at home, yet knowledge of intellectual property law, which provides the basic ground rules for accessing and using information, is neither widespread nor well appreciated. With the emergence of the information infrastructure into everyday life, the opportunity to copy information has increased enormously. The development of the World Wide Web has led to an exponential growth in the volume of digital information available, and the significance of information to the national economy continues to climb, with a frantic pace of exploration in new forms of information businesses. Yet the population of both information consumers and information producers is not particularly well informed about the rules that currently guide the handling of this valuable commodity.

The discussions in Chapters 4 and 5 suggest that there is a substantial amount of infringement of intellectual property rights. This infringement is not only significant economically; it also reflects a mind-set that could, in the long term, ultimately be destructive to the prosperity of an information economy. Some of this infringement no doubt occurs because people do not understand the basic tenets of intellectual property (IP) law. Copyright education could help correct this problem.

Other infringements are carried out by people who know at some level that their activities are unlawful, but have not thought about the long-term consequences of weak respect for intellectual property rights. Education may assist here as well. Others might be persuaded to curb infringing activities if copyright education led to increased social or peer sanctions against infringement. Even modest results would be useful: It would be a step in the right direction if people started thinking about the legality of their actions before making unauthorized copies of protected works. However, the committee is not suggesting that copyright education is likely to influence directly the behavior of commercial pirates, who understand that their behavior is illegal. Finally, although the committee concludes that copyright education would be widely beneficial, this specific recommendation is targeted to the United States.

WHAT SHOULD COPYRIGHT EDUCATION INCLUDE?

Because people tend to obey laws that they understand and think fundamentally fair and sound, copyright education should be based on the fundamental fairness and soundness of intellectual property law. A program of copyright education should describe the core goal of IP law—the improvement of society through advancement of knowledge by encouraging the creation and distribution of a wide array of works. The program should point out that, in the long term, all IP becomes a part of the shared heritage, universally available. In addition, the program should describe the basic means for achieving this goal—time-limited monopolies—and the rationale for providing them (i.e., as a way to provide an incentive to creators, yet ensure that all the fruits of their efforts are eventually disseminated widely). The educational program must communicate these points in a direct, jargon-free manner.

Although intellectual property in general and digital IP in particular are fraught with controversy, several basic principles can be communicated usefully with clear-cut examples. For instance, the basic exclusive rights of copyright, such as reproduction, sale, and public performance, make it clear that reproducing and distributing complete copies of a work (e.g., a computer program) is illegal, even if it is only one copy that is given free to a friend. The program should also note that ease of copying or the risks of detection do not affect the legality of an infringing act, and should perhaps emphasize the role of ethics rather than punishment.

The program should also describe the limits on IP rights by including an introduction to fair use and other limiting principles of copyright and describing their role in accomplishing the larger purpose of the law. The program should acknowledge that fair uses can be made of copyrighted works but that not all private, noncommercial copies are fair uses.

An additional focus may be provided by the common myths and

misconceptions about intellectual property[1] and information generally. For example, people may commonly distinguish between theft of tangible property, which deprives the owner of his or her rightful possession of the property, and appropriation of IP, which enables another person to consume the work even if not depriving the owner of his or her rightful use. The program should stress the concept that harm occurs in both situations, even if the harm from copyright infringement is more intangible. Individual acts of infringement mount up; they matter. Some of them represent lost sales; others represent a kind of unjust enrichment of the consumer. The appeal should be to fairness.

Copyright education should be aimed at more than deterring infringement, by attempting, in addition, to illustrate how intellectual property law provides benefits to anyone seeking to contribute to the information age.

TO WHOM SHOULD COPYRIGHT EDUCATION BE DIRECTED?

Designing an education campaign to appeal to the right audiences is difficult without empirical data about who is infringing and why, and about what people generally know and don't know about copyright. The lack of such data motivates the suggestion of researching public knowledge about and attitudes toward copyright. Although some appropriate target audiences are fairly obvious (e.g., students), more information about these audiences and their views is a crucial foundation for an effective program. Education should go beyond the circle of likely infringers and encompass individuals who have some influence over potential infringers or who are part of the culture that tolerates infringing acts (see Box F.1).

WHO SHOULD FUND OR CONDUCT COPYRIGHT EDUCATION?

Disagreement is likely on the issues of funding copyright education and choosing the appropriate venue for such education. Some constituencies would like to require schools to include a module on copyright in every grade, from kindergarten through postgraduate work (see, for example, the white paper *Intellectual Property and the National Information Infrastructure* (IITF, 1995)). Others have doubts about the appropriateness of such an extensive campaign. One concern is that a federal government requirement for copyright education in schools would raise the issues of whether federal funds should be allocated for such a purpose and whether

---

[1]See Chapter 4 for a discussion that includes numerous examples.

BOX F.1
Would Instructors Benefit from Copyright Education?

One reviewer of this report suggested that educators, too, need copyright education, because at times they encourage students to do things that are questionable regarding copyright:

> It is not uncommon for professors, including law professors, to assign article readings to their students and let them know that the articles may be found in journals in the library or that a copy of each article may be found on reserve in the library. Photocopying machines are made readily available in the library or in the school hallways. For the professor to duplicate course packets without permission of each and every publisher would be deemed in violation of agreements developed among the publishers and the library community. However, for a professor to encourage each student to acquire their own copies seems to skirt the legal issue. The professor can always argue that he or she assumed that students would legally acquire copies, and yet knows that each student will not ask the publisher for permission to make a personal copy. There is always the additional argument that a personal copy could be construed as fair use . . . even though hundreds of personal copies are being made for a single class.

The same reviewer also provided some insight into one rationale for infringement in the scientific community, suggesting that:

> Scientists are often forced to give up exclusive copyright in their works (and sometimes pay page charges) in order to have their articles published in the lead[ing] journals. This is particularly true at the beginning of a scientific career. Since they no longer have copyright, researchers are legally obligated to ask publishers for permission to distribute [photo]copies of their authored articles to their own students and to their close research associates, as well as to distribute the articles of their close associates.

> The pricing structure that seems to maximize profits for scientific publishers is one in which journals are acquired by only a segment of elite academic libraries that can afford them. As a result, many academic scientists and researchers don't have convenient access to many of the journals they publish in (particularly those professors just beginning to build their careers at mid-level or poorer universities) and certainly not to those journals peripheral to their primary areas of interest. Therefore many scientists knowingly breach the letter of the 'law' because of their disagreement with the monopolistic practices being imposed. There is strong feeling in the academic community that publishers are using heavy handed practices and are not playing fair.

continued

BOX F.1 Continued

The committee does not endorse a particular view on this issue or the specific comments of this reviewer. However, the two examples are useful in illustrating the question of whether the issue is ignorance of copyright law, or a deliberate violation of the law that is believed to be justified because of the purpose of the infringement or the perceived unfairness of the marketplace.

---

the federal government should encourage specific content to be included in curricula (which is traditionally determined at the local level). Having the federal government pay for the campaign would raise concerns, because it would likely be seen as a subsidy of the information industries.[2] Why should taxpayers grant such a subsidy? Other government-funded public education campaigns are motivated by issues of public health and safety, which are clearly not at issue here.

There are, however, precedents for industry organizations to undertake mass media campaigns designed to educate people about the consequences of their actions. For example, the "Buy American Goods" campaign some years ago was funded by labor organizations and sought to protect American jobs by making people more aware of the consequences of buying foreign-made goods. Raising consciousness about the overall impact of individual decisions may be a plausible approach, one in keeping with previous industry-supported campaigns.[3]

Although information industry organizations probably should continue their educational campaigns in the mass media, the reach of these campaigns may not be sufficient. One way to extend the reach of such a campaign may be to establish a copyright education program using both private and public funds, at a nonprofit, unbiased institution, which could assist organizations in promoting copyright education at the local level. This approach would have several benefits, including encouraging

---

[2]However, this concern is not shared worldwide. For example, Hong Kong's government announced a major campaign to educate people about IP rights infringement. Hong Kong leader Tung Chee-hwa said during a July 1999 trip to the United States that fears of unchecked piracy remained the biggest obstacle to investment in the territory, notorious for its trade in pirated CDs and copycat products (Reuters, 1999).

[3]The Business Software Alliance has an ongoing campaign that includes spot radio announcements, aimed primarily at software users in institutional environments (in both the public and private sector). The Recording Industry Association of America has a campaign aimed at a younger audience; its "Byte Me" Web site is an effort to stem the distribution of illegal copies of popular music in MP3 format.

grassroots efforts to promote respect for copyright. Because they could be tailored to address the needs and concerns of the communities they serve, grassroots programs might have a greater chance of success than mass media or other national campaigns. Grassroots programs may also be more credible to their constituencies than self-serving industry-initiated efforts.

The committee does not recommend making copyright education mandatory. Some schools may choose voluntarily to integrate copyright education into their curriculum, perhaps as a component of lessons on personal ethics. Students may be able to understand copyright principles better through examples that connect directly to their everyday life, such as considering how they would feel if a story they had written was published by someone else in a school newspaper without crediting them as the author. Because copyright education will be a new subject for schools, appropriate teaching materials may be lacking. In that case, some governmental funding, perhaps even at the federal level, could be used to support development of materials or guidelines for instruction. The National Science Foundation (NSF), for example, has a Science Education program that sometimes convenes panels of experts to provide teaching guidelines to assist in better science education for school curricula. Perhaps a comparable though smaller-scale effort (not necessarily by NSF) could be undertaken to develop appropriate guidelines for teaching about copyright.

Copyright education need not be pursued in isolation. A typical user of digital IP on the information infrastructure should have familiarity not only with copyright issues but also with topics such as "netiquette," spamming, and privacy, as well as an understanding of how to use the technology itself. Copyright education might profitably be conducted as a component of larger efforts in information technology literacy.[4]

SOME CAUTIONARY NOTES

Copyright education must be planned with care; otherwise, it may easily prove ineffective or even backfire. One danger is oversimplifying the message, as was done, for example, in *Intellectual Property and the National Information Infrastructure* (IITF, 1995), which anticipated teaching kindergarten students to think of IP as property just like the toys they play with.

Oversimplified messages about copyright will obscure the genuine and legitimate debate about how far copyright law extends, for example,

---

[4]See *Being Fluent with Information Technology* (CSTB, 1999a) for an extended discussion of these issues.

in the regulation of private-use copying. In the copyright area, a difference may exist between the law as it appears on the books and the law as it is actually carried out, akin to the laws pertaining to speed limits.

In addition, one size does not fit all in copyright education. The unauthorized copying of entertainment works may be judged differently than the unauthorized copying of certain educational materials, particularly when specific purposes and other factors are weighed in the mix. Even entertainment works are not immune from fair uses being made of them as suggested by the Supreme Court's decisions in *Sony Betamax* (off-the-air videotaping as fair use) and *Campbell v. Acuff-Rose* (rap parody of a Roy Orbison song as fair use).[5,6]

Finally, a copyright campaign using heavy-handed, preachy, anti-piracy rhetoric may backfire because it insults the public, rather than appealing to people's better judgment. Some lessons from prohibition in the early 20th century should be remembered: Heavy-handed rhetoric and enforcement practices bred less respect for the law, not more, and left people feeling justified in flouting the law.

---

[5]*Sony Corporation of America v. Universal City Studios, Inc.*, 464 U.S. 417 (1984).
[6]*Campbell v. Acuff-Rose Music*, 114 S.Ct. 1164 (1994).

APPENDIX G

The Digital Millennium Copyright Act of 1998 and Circumvention of Technological Protection Measures

INTRODUCTION

The World Intellectual Property Organization (WIPO) treaty seeks to harmonize different countries' treatment of the ownership and protection of intellectual property, in order to enable the growth of global commerce in information goods and services. The Digital Millennium Copyright Act of 1998 (DMCA)[1] is the implementation of the WIPO treaty by the U.S. Congress.

As articulated in Chapter 6, many members of the committee believe that the DMCA, although well intentioned and well written in many respects, has some significant flaws with respect to its handling of technical protection mechanisms and circumvention. This appendix, endorsed by those committee members, describes those flaws and suggests ways in which the law's approach to circumvention could be improved.

Simply put, the DMCA makes it illegal, except in certain narrowly defined circumstances, to circumvent an "effective technical protection measure" used to protect a work. The DMCA seemingly makes it illegal (again, except in certain narrowly defined circumstances) to distribute software or other tools used in an act of circumvention, even if this particular act of circumvention is covered by one of the exceptions and, hence, is legal.

---

[1]Public Law 105-304. Relevant excerpts are found in the addendum to this appendix; the full text is available online at <http://frwebgate.access.gpo.gov/cgi-in/getdoc.cgi?dbname=105_cong_public_laws&docid=f:publ304.105>.

Given that it is already illegal to infringe copyright, why did the U.S. Congress, in writing the DMCA, feel it necessary to criminalize "circumvention"?

It is a fundamental premise of the DMCA that, for the foreseeable future, the digital-content distribution business will be an important and growing part of the U.S. economy and that technological protection measures will be needed for the success of that business. The DMCA's anticircumvention provisions respond to the (presumed) economic importance of these developments by giving content owners a property right over the technological protection *mechanisms* they deploy, in addition to their existing rights over the *content* that these mechanisms protect. In the physical world, the theft of a tangible object is roughly analogous to copyright infringement; "breaking and entering" the room in which that object is stored is roughly analogous to circumvention. In the words of Callas et al. (1999), it is reasonable to assume that Congress's goal was "[t]o make it a more serious crime to infringe a work that the owner has actively tried to protect than to infringe one that the owner merely stated ownership of." Interpreted as an incentive for copyright owners to protect their own property, rather than to rely solely on the police and the courts, this is a perfectly understandable goal.

Unfortunately, it is far from clear that the DMCA's anticircumvention provisions will have primarily positive effects on content distributors and other interested parties. One problem is that circumvention is a bread-and-butter work practice in the cryptology and security research and development (R&D) community, yet this is precisely the technical community that content distributors are relying on to make effective technological protection measures. If this community is hindered in its ability to develop good products, is it wise to encourage owners to use these products?

It is of course possible that anticircumvention laws will be interpreted by distributors not as incentives to use effective protection measures but, rather, as incentives to do just the opposite—use insufficiently tested, possibly weak protection technology, and increase reliance on the police and the courts to punish people who hack around it. This would result in some cost shifting: Instead of owners and distributors paying for good technology to protect their property, the public at large would likely pay for a greater portion of this protection through the law-enforcement system, although some of the increased costs in enforcement may be borne by the antipiracy efforts of the various information industry associations.

This appendix begins by explaining how the cryptology and security R&D community works and what role circumvention plays in that work. The relevant sections of the DMCA are excerpted and some commentary given on their shortcomings, suggesting ways in which they could be improved. Formal recommendations on this subject can be found in Chapter 6.

HOW THE CRYPTOLOGY AND SECURITY R&D COMMUNITIES WORK

Understanding the interaction of intellectual property and technical protection services requires an understanding of the research and development process in cryptology and security.[2] A distinguishing feature of these disciplines is that they proceed in an adversarial manner: One member of the R&D community proposes a protection mechanism; others attack the proposal and try to find its vulnerabilities. Using this approach, serious vulnerabilities can be discovered and corrected before the mechanism is fielded and relied on to protect valuable material.

Like most scientific and engineering communities, the security R&D community does both theoretical and experimental work. The theory of cryptology and security is substantial and still evolving, touching on some of the deepest and most challenging open questions in the foundations of computation.[3] A goal of this theory is to study concepts such as privacy, security, tamper resistance, integrity, and proof in a manner that is both mathematically rigorous and relevant to the construction of secure products and services.[4]

One purpose that this study serves is rigorous analysis of security mechanisms. When a technique for protecting digital assets is put forth, there are often follow-up papers demonstrating technical flaws that prevent it from living up to its claims. Sometimes, a purely theoretical analysis is sufficient to show that a proposed protection mechanism is flawed. For example, a follow-up theoretical paper may show that a mathematical assumption made in the original proposal is false or that the class of adversaries against which the proposed mechanism was shown to be "secure" is weaker than the classes of adversaries that exist in the real world.

If pencil-and-paper analysis fails to find flaws in a protection system, should the system be considered secure? No. Before a system is deployed and valuable digital assets are entrusted to it, it should be analyzed experimentally as well. There are several basic reasons that a system that

---

[2]In addition to providing the scientific and engineering foundation for IP management, these disciplines are also widely applicable in other domains, ranging from military system command and control to privacy protection for personal correspondence.

[3]Mathematically sophisticated readers should refer to, for example, Luby (1996) for an introduction to this theory.

[4]A survey and analysis of the policy and market aspects of cryptography may be found in *Cryptography's Role in Securing the Information Society* (CSTB, 1996).

has survived all pencil-and-paper attempts to break it could still fail in real use:[5]

• Theoretical analysis of a proposed security mechanism may fail to demonstrate that the mechanism has a flaw but fall short of proving that it is secure in a mathematically rigorous sense. The failure to prove that something doesn't work is not of course equivalent to a proof that it does work.

• Even if a proposal is proven to satisfy a formal security criterion, an implementer may make a mistake in a particular hardware or software implementation of that proposal. Fielded implementations, not abstract specifications, are what real customers will use, and hence implementations must be tested.

• Abstract, provable security criteria may be too costly for product vendors to develop. Developers of secure products make compromises that entail informed guesses about how their products will actually be used, how much money and cleverness will actually be put into attacking them, and with which other products they will interact. Experimentation is needed to test the accuracy of guesses.

A crucial part of experimental security R&D is circumvention (i.e., attack on hardware and software that is claimed to be secure). A research or development team builds a piece of hardware or software, claims that it protects the relevant digital assets, and then challenges the security community to refute its claim (e.g., through vendor challenges). An integral portion of the "security community" comprises nonprofessionals, who can be among the most effective circumventors.

Vigorous, expert attacks should be carried out under the same conditions in which the secure hardware or software will be used or, if those conditions are unknown or infeasible to simulate in the laboratory, under conditions that are as realistic as possible. If such attacks have not been carried out, the allegedly secure system should be regarded as untested and potential users should be as wary as they are of any untested product or service.

In addition to their methodological role in basic research in cryptology and security, experimental attacks on secure hardware and software play an important and growing role in commercial practice. Responsible vendors assemble and fund internal "tiger teams" that try to circumvent a security mechanism before a product relying on the mechanism enters the marketplace. If security is a critical feature of a product or service that a

---

[5]See *Computers at Risk: Safe Computing in the Information Age* (CSTB, 1991a) and *Trust in Cyberspace* (CSTB, 1999c) for additional discussion.

vendor has offered, prudent customers (a small minority of customers), before signing a large contract with that vendor, often demand the right to have their own security experts or third-party security consultants evaluate the product or service. Such an evaluation should include vigorous experimental attempts to circumvent the security mechanism. These evaluations may also be done by potential strategic partners and by industrial standards bodies, as well as by direct customers. Security consulting firms that routinely attempt circumvention to evaluate products include Network Associates, Counterpane Systems, and Cryptography Research, Inc.[6]

The evolution of the Sun Microsystems' Java programming system illustrates the importance of experimental circumvention to progress in the security R&D world. When Sun launched this innovative system, one of the most important claims it made was that server-supplied executable content could be run safely from any Java-enabled Web browser. Java programmers were supposed to be able to develop software that could be run on any hardware and software platform that supports the Java virtual machine (JVM) and the JVM was supposed to be secure enough to prevent any Java program that had been through its byte-code verifier from damaging the host machine on which it was running. Dean et al. (1996) were skeptical of this broad claim, performed some experimental attacks, and indeed managed to circumvent the JVM security mechanism. Sun Microsystems and Netscape shipped some quick fixes soon after those circumvention attempts succeeded and were publicized. Shortly thereafter, Dean (1997) wrote a more comprehensive analysis of the underlying problem, and Sun's subsequent Java Development Kit, version 1.1, adopted Dean's suggestions.[7]

Numerous examples of attacks, both theoretical and experimental, on proposed security mechanisms can be found in, for example, the proceedings of the International Association for Cryptologic Research (IACR) Crypto and Eurocrypt conferences, the Institute for Electrical and Electronics Engineers (IEEE) Symposium on Security and Privacy, the Association for Computing Machinery (ACM) Conference on Computer and Communications Security, the *Journal of Cryptology*, and several comprehensive books, including the one by Menezes et al. (1997). See Anderson (1993) for a thorough and highly readable account of failures in fielded

---

[6]Information is available at <http://www.nai.com>, <http://www.counterpane.com> and <http://www.cryptography.com>, respectively.

[7]This discussion should not be construed to mean that all of the security issues with Java have been resolved; it is included to serve as an example of the role that experimental circumvention plays in improving security.

automated teller machine security systems. Examples of actual attacks may be found in the Risks-Forum Digest.[8]

Although the security and cryptology community regards the right to "attack" technical protection services as a fundamental part of its work, crucial both to research and to commercial practice, it does not assert that those who are successful in breaking protection services have a right to steal intellectual property that those systems were deployed to protect. Although the pursuit of knowledge about the actual security of products and services that are advertised as secure is a respected and valued activity, the exploitation of that knowledge to commit crimes is not.

At this time, R&D security and cryptology community members are not required to be licensed or have any other special legal or administrative status by the government or by a professional society, to perform experimental circumvention. If a company, university, or government laboratory wants to hire a particular person to test the strength of technical protection services, it is free to evaluate that person's qualifications according to its own criteria; if a person wants to pursue these activities as an amateur, he or she is free to do so, as long as he or she does not do anything illegal. The people who do this sort of work, whether for a living or as a hobby, have a broad range of academic and professional backgrounds, and the field thrives on the multidisciplinary and unpredictable nature of the skills needed to be a good circumventor. For this reason, strong opposition exists in the security R&D community to the idea of developing a licensing process for circumvention activity and trying to use the process to strengthen copyright owners' control over the fate of their property. The effect of a licensing process might just be the opposite (i.e., in fact to weaken the protection for owners). The technical community feels strongly that there is no appropriate licensing body (i.e., there is no group of people well qualified to judge who is likely to be a competent and responsible circumventor) and that any licensing process likely to be developed would have the effect of stifling creativity and dissemination of circumvention results, ultimately degrading the state of the art of technical protection.[9]

---

[8]A discussion list of the ACM Committee on Computers and Public Policy, moderated by Peter G. Neumann, is available online at <http://catless.ncl.ac.uk/Risks/>. Also see *Computer Related Risks* (Neumann, 1995).

[9]The legal status of circumvention activity and the software and hardware tools developed by circumventors is an area in which analogies between intellectual property and some sorts of physical property break down. For example, one has to be a licensed locksmith to practice lock-picking or even to possess lock-picking tools. Otherwise, one is guilty of the crime of possession of burglary tools. There are many possible explanations for this difference in the status of tools that could be used to steal things. For example, it may be that there is an appropriate licensing body for locksmiths and that this licensing

Like other security research results, discoveries of technical flaws in IP protection services should be published in scientific journals and conference proceedings. These publication fora enforce quality control and objectivity, and the ethics of publishing circumvention results in these fora is noncontroversial in the security and cryptology R&D community. Publication in journals and conference proceedings is also inherently slow: At least 6 months, and sometimes as much as 2 years, passes between the submission of a paper and its appearance in print. During the interval between submission and publication, the circumventor can inform a vendor about the flaws in its system, and the vendor can take whatever steps he or she thinks are necessary before the flaws are reported in a paper.

In the 1990s, an alternative, more controversial publication strategy has emerged in the security and cryptology world: the popular media. Now that tens of millions of people are using the Internet and the World Wide Web, privacy, authenticity, anonymity, denial of service, and other security issues are of interest to the general public, and mainstream media report on them. Substantial coverage in the mainstream media, most notably in the *New York Times*, often catapults a researcher into stardom, with predictable consequences for job offers and promotion. This is quite unlike the traditional model of career advancement of researchers coming in proportion to one's standing in a meritocracy regulated by objective peer review. Because its career-enhancing potential is so huge, many security and cryptology researchers actively seek mainstream media coverage when they discover flaws in well-known products and services.

This form of publication is highly controversial in the security R&D community, with both benefits and drawbacks. The advantages of media coverage of results are considerable: Well-written popular articles can raise public awareness of the importance of computer security in general and IP protection in particular. Media coverage also forces vendors of flawed products to pay attention to the problem, denying them the option of hoping that customers won't discover that the tool may not be offering the advertised protection.

But the disadvantages are also considerable. Many popular articles are not well written and, through mistakes or exaggeration, give the impression that a product has been completely broken, when, in fact, the technical flaw that has been discovered is difficult to exploit and may not be practically important in the short run (even if it is potentially important in the long run and hence interesting to researchers). Widespread media coverage may also function as encouragement to criminals to exploit a newly discovered flaw. The security and cryptology community is

---

requirement does not have a chilling effect on lock development; if such is the case, then the two fields of endeavor really are not analogous, even if some of their potential effects are.

divided on the question of whether the pluses outnumber the minuses. Many in the community believe that each case must be considered separately, because no general code of ethics governs all of them.

Experimental circumvention often entails the development of hardware or software that breaks technical protection features of intellectual property (IP) management systems. The ethics governing distribution of these tools are similar to those governing their use: The developer may share his tools with other researchers so that his results can be reproduced and improved upon; he or she may not share them with pirates or anyone else whose goal is illegal appropriation of other people's property, rather than advancement of the state of the art of technical protection (or some other legal goal, including, of course, all legal circumventions defined in the DMCA).

Although most researchers may subscribe to the code of ethics described above, it is clear that there are others who do not. And once a particular circumvention technique becomes available on the Internet, its wide distribution occurs in a very short time span.[10]

DISCUSSION AND CONCLUSIONS

The general approach taken by the Digital Millennium Copyright Act (see addendum below) is to make circumvention illegal except under certain conditions. The legislative approach favored by the crypto and security community is to make *circumvention* legal, while making certain *uses* of circumvention illegal (including, of course, the theft of IP). The DMCA is a fairly good compromise for legislation that makes circumvention illegal except under certain conditions. The relevant sections do a reasonable job of carving out exemptions for the circumvention activities that the community now performs in the daily course of its work. However, there are issues that need to be addressed.

The essential and pervasive problem with the DMCA is that it is vague and uses crucial terms in ill-defined or misleading ways. As a consequence, a practicing circumventor, whether a researcher or criminal, is left without a clear definition of what a "technological protection mechanism" is or of what it means for one to be "effective." Although this may seem like an academic quibble, the example given in Callas et al. (1999) shows that, it is, on the contrary, a real-world concern. Some time ago

---

[10]For example, Microsoft launched the Windows Media Audio (WMA) format as an alternative to the popular MP3 technology. WMA files can be encoded to restrict playback to a single PC, time period, or number of plays. Almost instantly, cracking software that removes all playback restrictions began making its way around newsgroups and Internet Relay Chat sessions. See Sullivan and Gartner (1999).

there was a computer file system in which one could indicate that a particular file should not be copyable (i.e., there was a "don't copy" flag that could be set); the system's copy command would refuse to copy files on which this flag was set. Undoubtedly, a large fraction of computer users, when presented with a "cannot copy" error message, would conclude that there was no way for them to copy the file and would give up. Anyone with a rudimentary knowledge of computer programming, however, would know that it is trivial to write a program that opens a file, reads the file's contents, and writes them to another file. So is the "don't copy" flag an "effective technological protection measure" or not? Is the exercise of rudimentary programming knowledge that circumvents the flag always, sometimes, or never illegal under the DMCA?

There are several other examples of vague or inaccurate language in the law:

1. Circumvention activity is done by crypto and security R&D people in the course of research, development (of products and services), and consulting. Most of these activities are covered in 1201(g) ("Encryption Research") and 1201(j) ("Security Testing"). Roughly speaking, 1201(g) covers research, and 1201(j) covers development and consulting. However, this division of the material is artificial. It is inaccurate to associate the word encryption with research and the word security with development and consulting. All technical aspects of cryptology and security have to undergo research, development, and consulting. In particular, section 1201(j) should not concern itself only with "accessing a computer, computer system, or computer network." The discussion of "breaking out of the Java sandbox" above is a prime example of "security testing," but it is not an example of "accessing a computer, computer system, or computer network." The Java system security work was done by Professor Ed Felten and his students as a research project at Princeton, but Sun Microsystems could have justified the same project under the rubric of "security testing" before Java was released (and might regret that it didn't).

2. Section 1201(g)(2)(C) is too vague and will leave many well-intentioned crypto and security people unsure about what to do:

> . . . it is not a violation of that subsection for a person to circumvent a technological measure as applied to a copy, phonorecord, performance, or display of a published work in the course of an act of good faith encryption research if . . .
>
> (C) the person made a good faith effort to obtain authorization before the circumvention . . . .

Important questions that are left unanswered include:

a. From whom is one supposed to obtain authorization? For example, suppose that a software vendor sells a digital library product, the owner of a valuable collection uses that product to control access to the collection, and a computer security expert wants to test the rights-management feature of the digital library product by attempting to get access to the collection without paying for it. Should he or she make a good-faith effort to get authorization from the software vendor, the collection owner, or both?
b. In the same example, suppose that one party grants authorization to circumvent but the other doesn't? Suppose it is the collection owner who has hired the computer security expert to test the product before deploying it; must they make a good-faith effort to get authorization from the vendor to test the product? If the vendor does not authorize the testing, may the collection owner and the security expert still test the product if they purchase it? Must they even seek authorization if the product is available and they buy it in the retail market?
c. Suppose that a request for authorization to circumvent simply goes unanswered. How long must a requester wait for an answer before he is considered to have made a good-faith effort?

3. Section 1201(g)(3)(B) is anathema to the multidisciplinary, extra-institutional culture of the crypto and security community and might inhibit some of that community's best work:

> (3) Factors in determining exemption.

> In determining whether a person qualifies for the exemption under paragraph (2), the factors to be considered shall include . . .

> (B) whether the person is engaged in a legitimate course of study, is employed, or is appropriately trained or experienced, in the field of encryption technology . . . .

Amateurs can be some of the best circumventors. Indeed, bugs in protection services are sometimes found by accident. Users may not even know that they are attempting to circumvent; they may simply do something that should work, see that it doesn't, and thus discover a flaw in the protection system. It would be against the interests of all concerned, including the content owners (who want flawed protection services to be fixed), to chill this type of unpredictable, nonprofessional circumvention activity.

4. It is unclear that the U.S. Copyright Office and the National Telecommunications and Information Administration of the U.S. Commerce Department can fulfill the responsibility conferred on them in Section 1201(g)(5):

> (5) Report to Congress—Not later than 1 year after the date of the enactment of this chapter, the Register of Copyrights and the Assistant Secretary for Communications and Information of the Department of Commerce shall jointly report to the Congress on the effect this subsection has had on—

> (A) encryption research and the development of encryption technology;

> (B) the adequacy and effectiveness of technological measures designed to protect copyrighted works; and

> (C) protection of copyright owners against the unauthorized access to their encrypted copyrighted works.

> The report shall include legislative recommendations, if any.

These bodies have little (if any) expertise in cryptology and few (if any) connections to the cryptologic research community.

Congress's implementation of the WIPO treaty provides a cautionary tale about the pitfalls of legislating in the high-tech arena. The extent that digital content distribution will prove to be important to the U.S. economy will not be known until major investments are made by distributors and major experiments are played out in the marketplace. Similarly, the importance of technological protection to the success of the content distribution business can only be determined in real-world competition. In the meantime, Congress has decided in advance that both are important and that the way to solve the problem raised by these important developments is to criminalize a set of activities that are valuable and standard in the high-tech community. The unintended consequences of criminalizing circumvention might ultimately prove to be more important than the problems that the DMCA set out to solve.

ADDENDUM:
SECTION 103 OF THE DIGITAL MILLENNIUM COPYRIGHT ACT

(a) In General.—Title 17, United States Code, is amended by adding at the end the following new chapter:

CHAPTER 12—COPYRIGHT PROTECTION AND MANAGEMENT SYSTEMS

Sec.
1201. Circumvention of copyright protection systems.
1202. Integrity of copyright management information.
1203. Civil remedies.
1204. Criminal offenses and penalties.
1205. Savings clause.

Sec. 1201. Circumvention of copyright protection systems

(a) Violations Regarding Circumvention of Technological Measures.—
(1)(A) No person shall circumvent a technological measure that effectively controls access to a work protected

[[Page 112 STAT. 2864]]

under this title. <<NOTE: Effective date.>> The prohibition contained in the preceding sentence shall take effect at the end of the 2-year period beginning on the date of the enactment of this chapter.
    (B) The prohibition contained in subparagraph (A) shall not apply to persons who are users of a copyrighted work which is in a particular class of works, if such persons are, or are likely to be in the succeeding 3-year period, adversely affected by virtue of such prohibition in their ability to make noninfringing uses of that particular class of works under this title, as determined under subparagraph (C).
    (C) <<NOTE: Reports. Regulations.>> During the 2-year period described in subparagraph (A), and during each succeeding 3-year period, the Librarian of Congress, upon the recommendation of the Register of Copyrights, who shall consult with the assistant Secretary for Communications and Information of the Department of Commerce and report and comment on his or her views in making such recommendation, shall make the determination in a rulemaking proceeding on the record for

---

NOTE: The material reprinted in this addendum was obtained from the Web site of the U.S. Copyright Office at <http://www.loc.gov/copyright/>. It is intended for use as a general reference, and not for legal research or other work requiring authenticated primary sources.

purposes of subparagraph (B) of whether persons who are users of a copyrighted work are, or are likely to be in the succeeding 3-year period, adversely affected by the prohibition under subparagraph (A) in their ability to make noninfringing uses under this title of a particular class of copyrighted works. In conducting such rulemaking, the Librarian shall examine—

(i) the availability for use of copyrighted works;

(ii) the availability for use of works for nonprofit archival, preservation, and educational purposes;

(iii) the impact that the prohibition on the circumvention of technological measures applied to copyrighted works has on criticism, comment, news reporting, teaching, scholarship, or research;

(iv) the effect of circumvention of technological measures on the market for or value of copyrighted works; and

(v) such other factors as the Librarian considers appropriate.

(D) <<NOTE: Publication.>> The Librarian shall publish any class of copyrighted works for which the Librarian has determined, pursuant to the rulemaking conducted under subparagraph (C), that noninfringing uses by persons who are users of a copyrighted work are, or are likely to be, adversely affected, and the prohibition contained in subparagraph (A) shall not apply to such users with respect to such class of works for the ensuing 3-year period.

(E) Neither the exception under subparagraph (B) from the applicability of the prohibition contained in subparagraph (A), nor any determination made in a rulemaking conducted under subparagraph (C), may be used as a defense in any action to enforce any provision of this title other than this paragraph.

(2) No person shall manufacture, import, offer to the public, provide, or otherwise traffic in any technology, product, service, device, component, or part thereof, that—

(A) is primarily designed or produced for the purpose of circumventing a technological measure that effectively controls access to a work protected under this title;

(B) has only limited commercially significant purpose or use other than to circumvent a technological measure that effectively controls access to a work protected under this title; or

[[Page 112 STAT. 2865]]

(C) is marketed by that person or another acting in concert with that person with that person's knowledge for use in circumventing a technological measure that effectively controls access to a work protected under this title.

(3) As used in this subsection—

(A) to "circumvent a technological measure" means to descramble a scrambled work, to decrypt an encrypted work, or otherwise to avoid, bypass, remove, deactivate, or impair a technological measure, without the authority of the copyright owner; and

(B) a technological measure "effectively controls access to a work" if the measure, in the ordinary course of its operation, requires the application of information, or a process or a treatment, with the authority of the copyright owner, to gain access to the work.

(b) Additional Violations.—(1) No person shall manufacture, import, offer to the public, provide, or otherwise traffic in any technology, product, service, device, component, or part thereof, that—

(A) is primarily designed or produced for the purpose of circumventing protection afforded by a technological measure that effectively protects a right of a copyright owner under this title in a work or a portion thereof;

(B) has only limited commercially significant purpose or use other than to circumvent protection afforded by a technological measure that effectively protects a right of a copyright owner under this title in a work or a portion thereof; or

(C) is marketed by that person or another acting in concert with that person with that person's knowledge for use in circumventing protection afforded by a technological measure that effectively protects a right of a copyright owner under this title in a work or a portion thereof.

(2) As used in this subsection—

(A) to "circumvent protection afforded by a technological measure" means avoiding, bypassing, removing, deactivating, or otherwise impairing a technological measure; and

(B) a technological measure "effectively protects a right of a copyright owner under this title" if the measure, in the ordinary course of its operation, prevents, restricts, or otherwise limits the exercise of a right of a copyright owner under this title.

(c) Other Rights, Etc., Not Affected.—(1) Nothing in this section shall affect rights, remedies, limitations, or defenses to copyright infringement, including fair use, under this title.

(2) Nothing in this section shall enlarge or diminish vicarious or contributory liability for copyright infringement in connection with any technology, product, service, device, component, or part thereof.

(3) Nothing in this section shall require that the design of, or design and selection of parts and components for, a consumer electronics, telecommunications, or computing product provide for a response to any particular technological measure, so long as such part or component, or the product in which such part or component is integrated, does not otherwise fall within the prohibitions of subsection (a)(2) or (b)(1).

[[Page 112 STAT. 2866]]

(4) Nothing in this section shall enlarge or diminish any rights of free speech or the press for activities using consumer electronics, telecommunications, or computing products.

(d) Exemption for Nonprofit Libraries, Archives, and Educational Institutions.—(1) A nonprofit library, archives, or educational institution which gains access to a commercially exploited copyrighted work solely in order to make a good faith determination of whether to acquire a copy of that work for the sole purpose of engaging in conduct permitted under this title shall not be in violation of subsection (a)(1)(A). A copy of a work to which access has been gained under this paragraph—
    (A) may not be retained longer than necessary to make such good faith determination; and
    (B) may not be used for any other purpose.
(2) The exemption made available under paragraph (1) shall only apply with respect to a work when an identical copy of that work is not reasonably available in another form.
(3) A nonprofit library, archives, or educational institution that willfully for the purpose of commercial advantage or financial gain violates paragraph (1)—
    (A) shall, for the first offense, be subject to the civil remedies under section 1203; and
    (B) shall, for repeated or subsequent offenses, in addition to the civil remedies under section 1203, forfeit the exemption provided under paragraph (1).
(4) This subsection may not be used as a defense to a claim under subsection (a)(2) or (b), nor may this subsection permit a nonprofit library, archives, or educational institution to manufacture, import, offer to the public, provide, or otherwise traffic in any technology, product, service, component, or part thereof, which circumvents a technological measure.
(5) In order for a library or archives to qualify for the exemption under this subsection, the collections of that library or archives shall be—
    (A) open to the public; or
    (B) available not only to researchers affiliated with the library or archives or with the institution of which it is a part, but also to other persons doing research in a specialized field.

(e) Law Enforcement, Intelligence, and Other Government Activities.—This section does not prohibit any lawfully authorized investigative, protective, information security, or intelligence activity of an officer, agent, or employee of the United States, a State, or a political subdivision of a State, or a person acting pursuant to a contract with the United States, a State, or a political subdivision of a State. For purposes of this subsection, the term 'information security' means activities carried out in order to identify and address the vulnerabilities of a government computer, computer system, or computer network.

(f) Reverse Engineering.—(1) Notwithstanding the provisions of subsection (a)(1)(A), a person who has lawfully obtained the right to use a copy of a computer program may circumvent a technological measure that effectively controls access to a particular portion of that program for the sole purpose of identifying and analyzing those elements of the program that are necessary to achieve interoperability of an independently created computer program with other programs, and that have not previously been

[[Page 112 STAT. 2867]]

readily available to the person engaging in the circumvention, to the extent any such acts of identification and analysis do not constitute infringement under this title.

(2) Notwithstanding the provisions of subsections (a)(2) and (b), a person may develop and employ technological means to circumvent a technological measure, or to circumvent protection afforded by a technological measure, in order to enable the identification and analysis under paragraph (1), or for the purpose of enabling interoperability of an independently created computer program with other programs, if such means are necessary to achieve such interoperability, to the extent that doing so does not constitute infringement under this title.

(3) The information acquired through the acts permitted under paragraph (1), and the means permitted under paragraph (2), may be made available to others if the person referred to in paragraph (1) or (2), as the case may be, provides such information or means solely for the purpose of enabling interoperability of an independently created computer program with other programs, and to the extent that doing so does not constitute infringement under this title or violate applicable law other than this section.

(4) For purposes of this subsection, the term "interoperability" means the ability of computer programs to exchange information, and of such programs mutually to use the information which has been exchanged.

(g) Encryption Research.—
(1) Definitions.—For purposes of this subsection—
   (A) the term "encryption research" means activities necessary to identify and analyze flaws and vulnerabilities of encryption technologies applied to copyrighted works, if these activities are conducted to advance the state of knowledge in the field of encryption technology or to assist in the development of encryption products; and
   (B) the term "encryption technology" means the scrambling and descrambling of information using mathematical formulas or algorithms.

(2) Permissible acts of encryption research.—Notwithstanding the provisions of subsection (a)(1)(A), it is not a violation of that subsection for a person to circumvent a technological measure as applied to a copy, phonorecord, performance, or display of a published work in the course of an act of good faith encryption research if—

(A) the person lawfully obtained the encrypted copy, phonorecord, performance, or display of the published work;

(B) such act is necessary to conduct such encryption research;

(C) the person made a good faith effort to obtain authorization before the circumvention; and

(D) such act does not constitute infringement under this title or a violation of applicable law other than this section, including section 1030 of title 18 and those provisions of title 18 amended by the Computer Fraud and Abuse Act of 1986.

(3) Factors in determining exemption.—In determining whether a person qualifies for the exemption under paragraph (2), the factors to be considered shall include—

(A) whether the information derived from the encryption research was disseminated, and if so, whether

[[Page 112 STAT. 2868]]

it was disseminated in a manner reasonably calculated to advance the state of knowledge or development of encryption technology, versus whether it was disseminated in a manner that facilitates infringement under this title or a violation of applicable law other than this section, including a violation of privacy or breach of security;

(B) whether the person is engaged in a legitimate course of study, is employed, or is appropriately trained or experienced, in the field of encryption technology; and

(C) whether the person provides the copyright owner of the work to which the technological measure is applied with notice of the findings and documentation of the research, and the time when such notice is provided.

(4) Use of technological means for research activities.—Notwithstanding the provisions of subsection (a)(2), it is not a violation of that subsection for a person to—

(A) develop and employ technological means to circumvent a technological measure for the sole purpose of that person performing the acts of good faith encryption research described in paragraph (2); and

(B) provide the technological means to another person with whom he or she is working collaboratively for the purpose of conducting the acts of good faith encryption research described in paragraph (2) or for the purpose of having that other person verify his or her acts of good faith encryption research described in paragraph (2).

(5) Report <<NOTE: Deadline.>> to Congress—Not later than 1 year after the date of the enactment of this chapter, the Register of Copy-

rights and the Assistant Secretary for Communications and Information of the Department of Commerce shall jointly report to the Congress on the effect this subsection has had on—

(A) encryption research and the development of encryption technology;

(B) the adequacy and effectiveness of technological measures designed to protect copyrighted works; and

(C) protection of copyright owners against the unauthorized access to their encrypted copyrighted works.

The report shall include legislative recommendations, if any.

(h) Exceptions Regarding Minors.—In applying subsection (a) to a component or part, the court may consider the necessity for its intended and actual incorporation in a technology, product, service, or device, which—

(1) does not itself violate the provisions of this title; and

(2) has the sole purpose to prevent the access of minors to material on the Internet.

(i) Protection of Personally Identifying Information.—

(1) Circumvention permitted.—Notwithstanding the provisions of subsection (a)(1)(A), it is not a violation of that subsection for a person to circumvent a technological measure that effectively controls access to a work protected under this title, if—

(A) the technological measure, or the work it protects, contains the capability of collecting or disseminating personally identifying information reflecting the online activities of a natural person who seeks to gain access to the work protected;

[[Page 112 STAT. 2869]]

(B) in the normal course of its operation, the technological measure, or the work it protects, collects or disseminates personally identifying information about the person who seeks to gain access to the work protected, without providing conspicuous notice of such collection or dissemination to such person, and without providing such person with the capability to prevent or restrict such collection or dissemination;

(C) the act of circumvention has the sole effect of identifying and disabling the capability described in subparagraph (A), and has no other effect on the ability of any person to gain access to any work; and

(D) the act of circumvention is carried out solely for the purpose of preventing the collection or dissemination of personally identifying information about a natural person who seeks to gain access to the work protected, and is not in violation of any other law.

(2) Inapplicability to certain technological measures.—This subsection does not apply to a technological measure, or a work it protects, that does not collect or disseminate personally identifying information and that is disclosed to a user as not having or using such capability.

(j) Security Testing.—

(1) Definition.—For purposes of this subsection, the term "security test-ing" means accessing a computer, computer system, or computer net-work, solely for the purpose of good faith testing, investigating, or cor-recting, a security flaw or vulnerability, with the authorization of the owner or operator of such computer, computer system, or computer network.

(2) Permissible acts of security testing.—Notwithstanding the provisions of subsection (a)(1)(A), it is not a violation of that subsection for a person to engage in an act of security testing, if such act does not constitute infringement under this title or a violation of applicable law other than this section, including section 1030 of title 18 and those provisions of title 18 amended by the Computer Fraud and Abuse Act of 1986.

(3) Factors in determining exemption.—In determining whether a per-son qualifies for the exemption under paragraph (2), the factors to be considered shall include—

(A) whether the information derived from the security testing was used solely to promote the security of the owner or operator of such computer, computer system or computer network, or shared directly with the developer of such computer, computer system, or computer network; and

(B) whether the information derived from the security testing was used or maintained in a manner that does not facilitate infringement under this title or a violation of applicable law other than this section, including a violation of privacy or breach of security.

(4) Use of technological means for security testing.—Notwithstanding the provisions of subsection (a)(2), it is not a violation of that subsection for a person to develop, produce, distribute or employ technological means for the sole purpose of performing the acts of security testing described

[[Page 112 STAT. 2870]]

in subsection (2), provided such technological means does not otherwise violate section (a)(2).

Index

A

Access. *See* Public access; Research community access
Access control.
  in bounded communities, 158-159
  enforcing in open communities, 159-164
Adversaries, in defeating technical protection solutions, 13, 313-318
Advertising-based business models, 81-82, 179-181
All-rights language, 36-37, 64
American Society for Composers, Authors and Publishers (ASCAP), 67, 93
American Society of Media Photographers, 68
Anchoring content, to single machine or user, 85, 88, 160-161, 164, 295-302
Anticircumvention regulations, 171-175, 221, 312
  exceptions to, 222, 313-318
Archiving
  large-scale, 119, 207
  the public record, libraries' interest in, 69
Archiving digital information, 9-10, 113-122, 206-209
  fundamental intellectual and technical problems with, 116-119

intellectual property and, 119-121
  lack of progress in, 207-208
  technical protection services, 121-122
ASCAP. *See* American Society for Composers, Authors and Publishers
Attention, as a commodity, 40, 196
Attribution, rights of, *see* Moral Rights
Author-operated models for rights management, 68
Authors. *See* Creators of intellectual property
Auxiliary markets, 82-83, 181-182

B

Balance
  upsetting the existing, 24-25
Berne Copyright Convention, 56, 59
BMI. *See* Broadcast Music, Inc.
Bounded communities, access control in, 158-159
Broad contracts, increasing use of, 64
Broadcast Music, Inc. (BMI), 67
Business models, 14-15, 65-68, 79-83, 176-186, 224, 237
  bringing technical protection services in line with, 176
  dealing with intellectual property, 183-186